CCSP (ISC)2 Certified Cloud Security Professional Exam Guide

Build your knowledge to pass the CCSP exam with expert guidance

Omar A. Turner

Navya Lakshmana

CCSP (ISC)2 Certified Cloud Security Professional Exam Guide

Authors: Omar A. Turner and Navya Lakshmana

Reviewers: Dharam Chhatbar, Eyal Estrin, and Commander Saurabh Prakash Gupta

Publishing Product Manager: Anindya Sil

Editorial Director: Alex Mazonowicz

Development Editor: M Keerthi Nair

Senior Development Editor: Ketan Giri

Presentation Designer: Shantanu Zagade

Editorial Board: Vijin Boricha, Megan Carlisle, Simon Cox, Ketan Giri, Saurabh Kadave, Alex Mazonowicz, Gandhali Raut, and Ankita Thakur

First Published: June 2024

Production Reference: 1210624

Published by Packt Publishing Ltd.
Grosvenor House
11 St Paul's Square
Birmingham
B3 1RB

ISBN: 978-1-83898-766-4

www.packtpub.com

This book is dedicated to my family and colleagues whose unwavering support and encouragement have been the cornerstone of my success. To my family, your patience, love, and understanding during this two-year journey have been my greatest strength. To my colleagues, your collaboration, insights, and steadfast belief in my vision have been invaluable. Together, you have made this achievement possible, and for that, I am eternally grateful.

Omar A. Turner

Contributors

About the Authors

Omar A. Turner is a general manager for cloud security at Microsoft, where he brings over 25 years of experience in supporting, deploying, architecting, and securing solutions for start-ups and globally recognized organizations. He holds numerous certifications, including CISSP, CCSP, CRISC, CISA, and CISM, and he holds B.S. degrees in mathematics and computer science, as well as the Wharton CTO designation. Omar is passionate about cybersecurity enablement and training as well as career mentoring for those looking to start their journey in the amazing and important field of cloud security.

Navya Lakshmana, a cybersecurity professional with a decade of experience in information technology, earned her bachelor's degree in electronics and communication from Visvesvaraya Technological University (VTU) in Bengaluru, Karnataka, India. She is currently employed at Siemens Healthineers, a renowned healthcare service provider that creates advanced medical technology for everyone, everywhere, sustainably. Navya holds distinguished certifications, including CISSP, CCSP, GIAC Cloud Penetration Tester (GCPN), and GIAC Penetration Tester (GPEN).

Beyond her professional endeavors, Navya is dedicated to cybersecurity education. As the founder of CyberPlatter, a YouTube channel, she educates cybersecurity enthusiasts and professionals alike.

About the Reviewers

Dharam Chhatbar is a seasoned information security professional with over 14 years of experience in various verticals of information security, delivering impactful and high-quality risk-reducing work. He has helped secure many banks and retail firms and is currently working at a Fortune number 1 company. He holds a master's degree, is a fervent learner, and has earned several global certifications, including CISSP, GSLC (SANS), CCSP, CSSLP, and CIPM. His key competencies include vulnerability management, security architecture, application security, cloud security, and leading and managing security engineers/vendors. He has also reviewed books on Azure security and CISSP practice questions.

I would like to thank my parents, Bina and Jagdish, for their continued support and encouragement with everything that I do and for motivating me to always achieve my ambitions.

Eyal Estrin is a cloud and information security architect and the author of the book Cloud Security Handbook, with more than 20 years of experience in the IT industry.

He has worked in several different industries (the banking, academia, and healthcare sectors).

He has attained several top security certifications: CISSP, CCSP, CDPSE, CISA, and CCSK.

Commander Saurabh Prakash Gupta, CISSP, CCSP, CISM, GCIH, is a military veteran currently employed as a cybersecurity expert with Bosch Global Software Technologies in Bengaluru, India. Having started his journey as a marine engineer, he then developed expertise in the domains of information technology and information security over more than 20 years. He is currently leading the cybersecurity program for providing consulting and testing services to global customers in automotive, embedded, IoT, OT, cloud, and enterprise IT product domains. Previously, for the Indian Navy, he led the program for software induction and enterprise cybersecurity deployment at the Indian Navy headquarters. He loves traveling and is an avid reader.

Table of Contents

3

Top Threats and Essential Cloud Security Concepts and Controls   41

4

Design Principles for Secure Cloud Computing   55

5

How to Evaluate Your Cloud Service Provider 73

6

Cloud Data Security Concepts and Architectures 87

7

Data Governance Essentials   117

8

Essential Infrastructure and Platform Components for a Secure Data Center   145

9

Analyzing Risks   163

10

11

Planning for the Worst-Case Scenario – Business Continuity and Disaster Recovery   207

12

Application Security   229

13

Secure Software Development Life Cycle   251

14

15

16

17

Cloud Physical and Logical Infrastructure (Operationalization and Maintenance) 327

18

International Operational Controls and Standards   357

19

Digital Forensics   371

20

Managing Communications   385

21

Security Operations Center Management   395

22

Legal Challenges and the Cloud   415

23

24

25

Preface

Cloud security is critically important for enterprises as the adoption of cloud infrastructure and services continues to grow at a rapid pace. As businesses move more and more of their data, services, and applications to the cloud, they need talented and certified professionals to help them secure these cloud environments. Today, cloud computing has moved from being a nice-to-have to being a core competency in the enterprise.

This has led to a high demand for knowledgeable and talented cloud security engineers and architects who can help organizations design, build, and operate secure cloud environments. This, combined with the myriad of security compromises out there, is creating challenges for organizations of all types. Cloud certifications can help organizations identify and develop critical skills for implementing various cloud initiatives. Certifications can also help individuals demonstrate their technical knowledge, skills, and abilities to potential employers to advance their careers.

The goal of this book is to help you pass the Certified Cloud Security Professional (CCSP) certification by ISC2. The CCSP certification is the most sought-after global credential and represents the highest standard for cloud security expertise. It confirms your ability to apply best practices to cloud security architecture, design, operations, and service orchestration.

As you progress through this book, you'll engage with practical and straightforward explanations of cloud security concepts designed to educate you on the challenges security professionals face in cloud environments. The chapters in this book cover the domains of topics relevant to the CCSP exam, including developing a comprehensive cloud security policy, conducting risk assessments for cloud deployments, implementing identity and access management solutions, securing data in cloud storage, and designing disaster recovery plans. Each chapter will guide you through scenarios that test your understanding of the CCSP domains, from architectural considerations to legal and compliance frameworks.

By the conclusion of this study guide, you'll possess a solid understanding of cloud security principles and practices, as well as the confidence needed to apply this knowledge in your current role. You will also be well prepared to take the CCSP exam.

Who This Book Is For

This book is for those who are preparing to take the CCSP exam. It is recommended that you have at least five years of experience in IT, with two of those years being focused on aspects such as cloud security, application security, privacy, or data governance.

What This Book Covers

Chapter 1, Core Cloud Concepts, introduces the most relevant cloud computing characteristics and concepts with regard to cloud service models, cloud deployment models, and the different types of stakeholders in cloud computing.

Chapter 2, Cloud Reference Architecture, covers the cloud reference architecture, cloud service models, cloud deployment models, and cloud capabilities. We will also introduce the shared considerations for cloud deployments and the impact of new and emerging technologies on the evolution of cloud computing.

Chapter 3, Top Threats and Essential Cloud Security Concepts and Controls, describes the common threats to cloud deployments and attack vectors. We will introduce the control frameworks and control types necessary to secure data, network, and virtualization layers for cloud computing.

Chapter 4, Design Principles for Secure Cloud Computing, focuses on the service model security considerations.

Chapter 5, How to Evaluate Your Cloud Service Provider, discusses how to review and understand key cloud service contractual documents from the perspective of cloud service consumers. We will provide the best practices on how to evaluate your CSP based on a set of criteria.

Chapter 6, Cloud Data Security Concepts and Architectures, describes cloud data concepts, cloud data storage architectures, data security, data classification, and cloud data security technologies. We will review the stages of the cloud data life cycle in cloud environments, from creation to safe destruction practices.

Chapter 7, Data Governance Essentials, reviews the most important concepts of governance oversight for data life cycle phases in the cloud environment. We will introduce the concepts of **Information Rights Management (IRM)** and best practices for auditability, traceability, and accountability when it comes to data use in cloud environments.

Chapter 8, Essential Infrastructure and Platform Components for a Secure Data Center, reviews key cloud infrastructure and platform components and the best practices for the secure design of the logical, physical, and environmental components of a modern data center.

Chapter 9, Analyzing Risks, identifies the top risks to the physical, logical, and virtual environments as a cloud consumer and provider. We will discuss how to analyze, assess, and address the risk with safeguards and countermeasures.

Chapter 10, Security Control Implementation, provides an overview of the key concepts of the selection, planning, and implementation of security controls in cloud environments.

Chapter 11, Planning for the Worst-Case Scenario – Business Continuity and Disaster Recovery, discusses how organizations are preparing to withstand disasters and business disruptions to be able to continue the delivery of products and services within acceptable time frames.

Chapter 12, Application Security, reviews development basics, the challenges organizations face, and the common cloud vulnerabilities for web applications.

Chapter 13, Secure Software Development Life Cycle, is dedicated to educating you on the **Secure Software Development Life Cycle (S-SDLC)**, including coverage of topics such as defining requirements, what methodology to use to apply the S-SDLC, threat modeling, and secure coding.

Chapter 14, Assurance, Validation, and Verification in Security, describes key processes as they relate to functional testing, profiling security testing methodologies, QA, and other solutions.

Chapter 15, Application-Centric Cloud Architecture, reviews the important specifics of traditional cloud application architecture, with a focus on essential security components such as WAF, DAM, API gateways, cryptography, sandboxing, and securing virtualized applications.

Chapter 16, IAM Design, focuses on **Identity and Access Management (IAM)** solutions, which are critical elements of securing organizations. This chapter covers identity providers, federated identities, secrets management, and other important IAM solutions.

Chapter 17, Cloud Physical and Logical Infrastructure (Operationalization and Maintenance), reviews the key physical and logical infrastructure configuration requirements for cloud environments. We will also provide an overview of the most common configurations and controls for operational and maintenance activities for physical and logical infrastructures.

Chapter 18, International Operational Controls and Standards, reviews the leading industry standards for **Information Technology Service Management (ITSM)**.

Chapter 19, Digital Forensics, discusses forensic data collection methodologies, evidence management, and other key concepts for the collection, acquisition, and preservation of digital evidence.

Chapter 20, Managing Communications, covers the best practices for the communication channels and procedures that need to be set up if an organization intends to be resilient against impacts of all types. We will review the most common communication channels with vendors, customers, regulators, partners, and other stakeholders.

Chapter 21, Security Operations Center Management, covers the best practices for establishing the primary requirements of a security operations center and how they are informed by the business mission, regulatory and legal requirements, and service offerings. We will review a wide range of tools related to monitoring and logging that are necessary for effective security operations center management.

Chapter 22, Legal Challenges and the Cloud, discusses compliance with legal and contractual requirements. The chapter covers in detail the policies, standards, guidelines, baselines, and procedures that frame decision-making, as well as the roles that delineate authority levels (e.g., shareholders, stakeholders, senior management, service consumers, and service providers).

Chapter 23, Privacy and the Cloud, discusses privacy regulations and country-specific legislation related to PII and PHI. We will review key jurisdictional differences in data privacy.

Chapter 24, Cloud Audit Processes and Methodologies, reviews the most common ways to conduct audits of IT systems, covering the audit process, the methodologies, and the required adaptations for a cloud environment.

How to Get the Most Out of This Book

This book is crafted to equip you with the knowledge and skills necessary to excel in the CCSP exam through memorable explanations of major domain topics. It covers the six core domains critical to cloud security expertise that candidates must be proficient in to pass the exam. For each domain, you'll work through content that reflects real-world cloud security challenges. At certain points in the book, you'll assess your understanding by taking chapter-specific quizzes. This not only prepares you for the CCSP exam but also allows you to dive deeper into a topic as needed based on your results.

Online Practice Resources

With this book, you will unlock unlimited access to our online exam-prep platform (*Figure 0.1*). This is your place to practice everything you learn in the book.

How to Access These Materials

To learn how to access the online resources, refer *to Chapter 25, Accessing the Online Practice Resources* at the end of this book.

Figure 0.1 – Online exam-prep platform on a desktop device

Sharpen your knowledge of CCSP (ISC)2 concepts with multiple sets of mock exams, interactive flashcards, and exam tips accessible from all modern web browsers.

Download the Color Images

We also provide a PDF file that has color images of the screenshots/diagrams used in this book. You can download it here: `https://packt.link/N7BwJ`.

Conventions Used

There are several text conventions used throughout this book.

`Code in text`: Indicates code words in text, database table names, folder names, filenames, file extensions, pathnames, dummy URLs, user input, and X (formerly Twitter) handles. Here is an example: "Since `'1'='1'` is always true, this query will always return all data from the `users` table, giving the malicious user access to all user accounts."

A block of code is set as follows:

```
SELECT * FROM users WHERE username = 'username' AND password =
'password'
```

Bold: Indicates a new term, an important word, or words that you see onscreen. For example, words in menus or dialog boxes appear in the text like this. Here is an example: "**Infrastructure as a Service (IaaS)** offers virtualized computing resources, including **Virtual Machines (VMs)**, storage, and networking. The user controls their infrastructure, while the **Cloud Service Provider (CSP)** oversees the physical hardware."

> **Tips or important notes**
> Appear like this.

Get in Touch

Feedback from our readers is always welcome.

General feedback: If you have any questions about this book, please mention the book title in the subject of your message and email us at customercare@packt.com.

Errata: Although we have taken every care to ensure the accuracy of our content, mistakes do happen. If you have found a mistake in this book, we would be grateful if you could report this to us. Please visit www.packtpub.com/support/errata and complete the form. We ensure that all valid errata are promptly updated in the GitHub repository at https://packt.link/gUxX0.

Piracy: If you come across any illegal copies of our works in any form on the internet, we would be grateful if you could provide us with the location address or website name. Please contact us at copyright@packt.com with a link to the material.

If you are interested in becoming an author: If there is a topic that you have expertise in and you are interested in either writing or contributing to a book, please visit authors.packtpub.com.

Share Your Thoughts

Once you've read *CCSP (ISC)2 Certified Cloud Security Professional Exam Guide*, we'd love to hear your thoughts! Scan the QR code below to go straight to the Amazon review page for this book and share your feedback.

https://packt.link/r/1838987665

Your review is important to us and the tech community and will help us make sure we're delivering excellent quality content.

Download a Free PDF Copy of This Book

Thanks for purchasing this book!

Do you like to read on the go but are unable to carry your print books everywhere?

Is your eBook purchase not compatible with the device of your choice?

Don't worry, now with every Packt book you get a DRM-free PDF version of that book at no cost.

Read anywhere, any place, on any device. Search, copy, and paste code from your favorite technical books directly into your application.

The perks don't stop there, you can get exclusive access to discounts, newsletters, and great free content in your inbox daily.

Follow these simple steps to get the benefits:

1. Scan the QR code or visit the link below:

https://packt.link/free-ebook/9781838987664

2. Submit your proof of purchase.

3. That's it! We'll send your free PDF and other benefits to your email directly.

1
Core Cloud Concepts

In this chapter, you will be introduced to the cloud computing characteristics and concepts of cloud service models, cloud deployment models, and different types of stakeholders in cloud computing. In addition to this, you will learn about the core elements required to provide and use cloud-based solutions.

The chapter will cover the most common cloud computing concepts, such as the customer, the provider, the partner, measurable services, scalability, virtualization, storage, and networking. You'll also learn about the cloud reference architecture that forms the foundation of modern cloud providers. Finally, you'll learn about cloud computing security and design concepts, as well as the cost-benefit analysis of cloud-based systems.

Making the Most Out of This Book – Your Certification and Beyond

This book and its accompanying online resources are designed to be a complete preparation tool for your **CCSP Exam**.

The book is written in a way that you can apply everything you've learned here even after your certification. The online practice resources that come with this book (*Figure 1.1*) are designed to improve your test-taking skills. They are loaded with timed mock exams, interactive flashcards, and exam tips to help you work on your exam readiness from now till your test day.

> **Before You Proceed**
> To learn how to access these resources, head over to *Chapter 25, Accessing the Online Practice Resources*, at the end of the book.

Figure 1.1 – Dashboard interface of the online practice resources

Here are some tips on how to make the most out of this book so that you can clear your certification and retain your knowledge beyond your exam:

1. Read each section thoroughly.

2. **Make ample notes**: You can use your favorite online note-taking tool or use a physical notebook. The free online resources also give you access to an online version of this book. Click the BACK TO THE BOOK link from the Dashboard to access the book in **Packt Reader**. You can highlight specific sections of the book there.

3. **Chapter Review Questions**: At the end of this chapter, you'll find a link to review questions for this chapter. These are designed to test your knowledge of the chapter. Aim to score at least **75%** before moving on to the next chapter. You'll find detailed instructions on how to make the most of these questions at the end of this chapter in the *Exam Readiness Drill - Chapter Review Questions* section. That way, you're improving your exam-taking skills after each chapter, rather than at the end.

4. **Flashcards**: After you've gone through the book and scored **75%** more in each of the chapter review questions, start reviewing the online flashcards. They will help you memorize key concepts.

5. **Mock Exams**: Solve the mock exams that come with the book till your exam day. If you get some answers wrong, go back to the book and revisit the concepts you're weak in.

6. **Exam Tips**: Review these from time to time to improve your exam readiness even further.

By the end of this chapter, you will be able to confidently answer questions on the following topics:

- Cloud computing
- Essential cloud computing characteristics
- Cloud stakeholders
- Key cloud computing technologies and building blocks
- You will now go through each topic above.

What Is Cloud Computing?

Cloud computing significantly altered some of the established IT conventions, even though the majority of the underlying technology and security fundamentals remain the same. Many of the key IT principles addressed in this chapter reaffirm the underlying features that remain constant as cloud computing provisioning and consumption models are embraced. The cloud computing **Software-as-a-Service** (**SaaS**) model uses internet-based computing resources to provide scalable and elastic IT-enabled capabilities to internal or external consumers.

Various cloud service providers, such as **Amazon Web Services** (**AWS**), Microsoft Azure, and Google Cloud Platform, have their own definitions of cloud computing, based on their respective service offerings. The non-regulatory agency of the United States Department of Commerce, the **National Institute of Standards and Technology** (**NIST**), in its **Special Publication** (**SP**) 800-145, provides the most widely used definition for cloud computing, which is cited by IT experts and cloud computing professionals when communicating the basic terminology:

"Cloud computing is a model for enabling ubiquitous, convenient, on-demand network access to a shared pool of configurable computing resources (e.g., networks, servers, storage, applications, and services) that can be rapidly provisioned and released with minimal management effort or service provider interaction. This cloud model is composed of five essential characteristics, three service models, and four deployment models."

> **Note**
>
> You can read about the NIST publication 800-145 cloud computing definition here:
> `https://csrc.nist.gov/publications/detail/sp/800-145/final`.

Now that you are familiar with the definition of cloud computing, it is time to focus on the five essential characteristics of cloud computing.

Essential Cloud Computing Characteristics

Cloud computing, as described by the NIST publication 800-145, is an innovative computing paradigm that delivers computer resources, services, and applications via the internet on demand. It enables users to remotely access, store, and administer data and applications without having to invest in or maintain physical infrastructure or hardware.

As per the NIST publication 800-145, the cloud computing model can be further defined by having five fundamental characteristics, three service models, and four deployment methods:

- The five essential characteristics of cloud computing are as follows:

  - **On-demand self-service**: Cloud services can be deployed and maintained by the user without the service provider's participation

  - **Extensive network access**: Cloud services are accessible over the internet, making them accessible from several devices and places

  - **Resource pooling**: Cloud providers share resources such as storage, computation, memory, and bandwidth to serve several consumers simultaneously

  - **Rapid elasticity**: Cloud resources can be readily scaled up or down to meet variable demands, allowing peak loads to be accommodated without compromising performance

  - **Measured service**: Cloud consumption is monitored, controlled, and reported so that users only pay for the resources they consume

- The three service models are as follows:

  - **SaaS**: The SaaS approach provides internet-based applications that are ready for use. Consumers need not concern themselves with infrastructure, software upgrades, or maintenance.

  - **Platform as a service**: **Platform as a Service** (**PaaS**) provides an environment to create, deploy, and maintain applications. Users can concentrate on application development without thinking about the underlying infrastructure.

  - **Infrastructure as a service**: **Infrastructure as a Service** (**IaaS**) offers virtualized computing resources, including **Virtual Machines** (**VMs**), storage, and networking. The user controls their infrastructure, while the **Cloud Service Provider** (**CSP**) oversees the physical hardware.

- The four deployment models are as follows:

  - **Private cloud**: The cloud infrastructure is devoted to a single enterprise, providing more security and data privacy controls

  - **Community cloud**: This deployment approach supports several enterprises that have common concerns, such as security needs or regulatory compliance

- **Public cloud**: The cloud infrastructure is owned and managed by a service provider, who sells services to the general public or a major industrial group

- **Hybrid cloud**: This model combines two or more of the preceding deployment methods, enabling enterprises to make use of the benefits of each while keeping separate environments

> **Note**
>
> You can find more resources about cloud computing and its characteristics here: `https://nvlpubs.nist.gov/nistpubs/Legacy/SP/nistspecialpublication800-145.pdf`.

As a cloud security expert, it is crucial that you understand these definitions and components in order to create, implement, and maintain security solutions that safeguard sensitive data and guarantee compliance with industry requirements. Cloud security comprises a vast array of techniques and technologies, including identity and access management, encryption, intrusion detection, and secure data transfer that protect cloud-based resources and services. By understanding the specific characteristics of cloud computing, security professionals can better minimize possible risks and vulnerabilities in an environment that is rapidly evolving.

In this section, you learned about the essential cloud computing characteristics. The next section will focus on cloud stakeholders.

Cloud Stakeholders

The **International Information Systems Security Certification Consortium (ISC2) CCSP Common Body of Knowledge (CBK)** identifies multiple cloud computing stakeholders with specific responsibilities, based primarily on the following **International Organization for Standardization (ISO) / International Electrotechnical Commission (IEC)** standards and NIST special publications:

- ISO/IEC 17789 **Cloud Computing Reference Architecture (CCRA)**

- NIST SP 500-292 CCRA

> **Note**
>
> You can read more about the ISO/IEC 17789 CCRA here - `https://www.iso.org/standard/60545.html`, and the NIST SP 500-292 CCRA here - `https://www.nist.gov/publications/nist-cloud-computing-reference-architecture`.

The key differences you need to be aware of concerning the identification of these cloud stakeholders are as follows:

- The ISO/IEC 17789 CCRA defines **three** main roles with multiple sub-roles in each main role

- The NIST CCRA defines **five** key actors

> **Note**
>
> It is important to focus on the cloud service models and cloud delivery models in this chapter. You will learn about the shared responsibility model, the three service models, and the six common deployment models (as mentioned in the NIST definition) in *Chapter 2, Cloud Reference Architecture*.

You will now go through each role and actor of ISO/IEC 17789 CCRA and NIST CCRA respectively.

ISO/IEC 17789 CCRA Roles and Sub-Roles

ISO/IEC 17789 is a standard developed by the ISO and the IEC, providing an extensive framework for CCRA. The purpose of this standard is to establish a common language, concepts, and structure to create, deliver, and manage cloud services across various domains.

ISO/IEC 17789 defines a CCRA that includes numerous roles and sub-roles, representing the major actors within the cloud computing ecosystem. You will learn about the duties and interactions between entities within this environment for effective operation and efficiency.

Cloud Service Customer

A **Cloud Service Customer (CSC)** is an entity that purchases cloud services from a CSP for itself or its users. CSCs can include organizations, departments within organizations, and individuals.

Sub-Roles of the CSC

A **Cloud Service User (CSU)** is an individual or application that utilizes cloud services provided by the CSP on behalf of the CSC.

CSP

A CSP is the entity responsible for supplying, running, and supporting cloud services. CSPs offer various cloud solutions such as SaaS, PaaS, and IaaS that CSCs can access.

Sub-Functions of a CSP

There are three sub-functions of a CSP:

- **Cloud Service Development**: The **Cloud Service Development (CSD)** sub-role is responsible for designing, creating, and deploying cloud services that meet the demands of CSCs.
- **Cloud Service Operation**: The **Cloud Service Operation (CSO)** sub-role is responsible for managing, monitoring, and operating cloud services provided by the CSP. This involves ensuring those services' availability, performance, and security.

- **Cloud Service Support**: The **Cloud Service Support (CSS)** sub-role is responsible for offering technical assistance, troubleshooting, and resolving issues related to cloud services for CSCs.

Cloud Service Partner

A **Cloud Service Partner (CSN)** is an entity that collaborates with the CSP to provide value-added services or support to CSCs. CSNs can be suppliers, resellers, or other organizations working closely with the CSP to improve cloud services as a whole.

Sub-Functions of a CSN

There are two sub-functions of a CSN as listed below:

- **Cloud Broker**: The **Cloud Broker (CB)** serves as an intermediary between the CSC and various CSPs.
- **Cloud Carrier**: The **Cloud Carrier (CC)** facilitates network connectivity between a CSP and the CSCs to guarantee secure, dependable communication.

Cloud Auditor

The **Cloud Auditor (CA)** is an independent body that reviews and validates a CSP and its services' adherence to applicable standards, laws, and best practices.

You will now learn about the key actors as per the NIST CCRA.

NIST Cloud Computing Key Actors

NIST Cloud Computing Reference Architecture (NIST SP 500-292), is a document published by the NIST, with the aim of offering an in-depth framework to comprehend, design, and implement cloud computing services and solutions. This reference architecture is intended to produce a uniform, technology-neutral framework that allows communication, cooperation, and the creation of cloud computing standards among diverse stakeholders, such as CSPs, users, and regulators.

The NIST CCRA is composed of five essential components, often termed as actors. These components describe the fundamental functions and duties inside a cloud computing system, therefore clarifying their interrelationships. The five major elements of the NIST CCRA are as follows.

Cloud Consumer

The cloud consumer is a person, group, or business that utilizes cloud services offered by the cloud provider. The cloud consumer obtains and administers cloud services in accordance with its needs and can access these services through a variety of interfaces and devices.

Cloud Provider

The cloud provider is the entity tasked with making cloud services accessible to the cloud customer. This covers the design, management, and maintenance of the cloud infrastructure, platforms, and applications necessary to offer the services. Cloud providers can provide a variety of service models, including IaaS, PaaS, and SaaS.

Cloud Broker

The cloud broker is an agent that helps cloud customers choose, manage, and integrate cloud services from numerous cloud providers. Cloud brokers can provide value-added services, such as collecting and integrating various offers, negotiating contracts, and maintaining **Service-Level Agreements** (**SLAs**) to guarantee that the demands of cloud consumers are satisfied.

Cloud Auditor (CA)

The CA is an independent, responsible body that assesses and evaluates the cloud services offered by the cloud provider. This involves confirming the cloud services' performance, security, and compliance with industry standards, legislation, and best practices. CAs contribute to the confidence and trust of cloud consumers by verifying that cloud providers achieve the necessary service levels and customer expectations.

Cloud Carrier (CC)

The CC is responsible for delivering the connectivity and transport services required for cloud consumer access to a cloud provider's cloud services. CCs provide the delivery of data and communication between cloud consumers and cloud providers, guaranteeing safe and dependable access to cloud services.

In addition to these core aspects, the NIST CCRA highlights many cross-functional characteristics that are essential to the installation and operation of cloud computing services. They include security, privacy, and compliance, which are vital for ensuring data protection and adherence to applicable laws and regulations.

By providing a structured and thorough reference architecture, NIST SP 500-292 fosters a shared understanding of cloud computing ideas and terminology, enabling stakeholders to make informed decisions and ease the development of interoperable cloud computing solutions. This reference design is a great resource for enterprises intending to adopt cloud computing or to enhance their current cloud-based services.

You will now understand the definitions and specifics of cloud stakeholders as seen from the perspective of two organizations. The ISO/IEC 17789 CCRA, with its focus on the CSC, the sub-role of the CSU, the CSP (with its associated sub-roles), the CSN, and the CA, offers a comprehensive view of the dynamics of each of the aforementioned roles, while the NIST reference architecture looks at the five primary actors of consumer, provider, broker, CA, and CC. Both are equally important, and it is essential to understand the differences between the two for the CCSP exam.

In the next section, you will dive into the key core technologies that allow cloud computing to exist and be used at scale for those requiring the use of the cloud.

Key Cloud Computing Technologies and Building Blocks

Cloud computing technologies enable on-demand, scalable, and adaptable computing resources and services. These hardware, software, and networking components enable enterprises to upgrade their IT infrastructures, reduce costs, and quickly adjust according to changing business demands. The fundamental elements that comprise cloud computing technology are as follows:

- **Compute resources**: Cloud computing relies on compute resources for the execution of applications, services, and workloads. These can be virtualized to provide multiple VMs or containers running on one physical server, providing efficient hardware usage and flexible resource allocation.

- **Storage resources**: Storage resources are essential for storing and managing cloud-based data. They offer various storage solutions, such as block storage, file storage, and object storage, to meet various data types, access patterns, and performance demands. On-demand scalability of cloud storage capacity ensures cost-effective and efficient solutions.

- **Networking resources**: Networking resources provide connectivity between cloud users and services, allowing communication between cloud components. These include virtual networks, routers, load balancers, and firewalls that ensure secure, dependable data transfer inside and across cloud environments.

- **Middleware and runtime**: Middleware and runtime components provide the platform and environment required to deliver, manage, and execute cloud applications and services. This consists of application servers, databases, as well as other platform-level elements that facilitate the creation of applications based on various programming languages and frameworks.

- **Cloud management and orchestration**: Management and orchestration technologies are essential for automating the management and control of cloud resources, services, and applications. They aid in the provisioning, monitoring, scalability, and optimization of these resources to ensure optimal resource allocation and use. Moreover, these solutions offer resource life cycle management – guaranteeing resources are available when needed and relinquished when no longer necessary – thus providing optimal resource life cycle management.

- **Security and privacy**: Securing cloud-based data, applications, and infrastructure requires security and privacy components. To safeguard these resources from potential threats or vulnerabilities, they include encryption, identity and access management, intrusion detection systems, and secure data transmission methods.

- **Service models**: Cloud computing offers three basic service models that define the customer's control scope and level – IaaS, PaaS, and SaaS. Each model isolates different levels of the underlying infrastructure, allowing customers to focus on core business needs while taking advantage of cloud technology benefits.

- **Deployment models**: Deployment models refer to how cloud resources are organized and made accessible to users. The public cloud, private cloud, hybrid cloud, and community cloud are the four primary deployment options. Each offers varying degrees of control, security, and scalability to meet the unique demands and expectations of organizations.

- **Billing and metering**: Billing and metering components enable the tracking and reporting of cloud resource usage, enabling consumption-based pricing so that users only pay for what they use. This pay-as-you-go model offers a flexible yet cost-effective method to access and manage cloud resources.

Although this knowledge may appear basic, it is essential for CCSP candidates to comprehend the fundamental principles of cloud computing. To effectively secure cloud environments, they must possess an in-depth understanding of cloud technologies such as compute resources, storage resources, networking resources, middleware, and runtime, as well as service and deployment patterns. Having this understanding allows them to detect and address potential security risks or vulnerabilities within cloud infrastructures.

Candidates taking the CCSP exam must also be able to evaluate CSPs and suppliers to confirm whether their products meet organizational security and compliance requirements. An understanding of cloud computing building blocks and reference designs such as **NIST SP 500-292** can assist in selecting and managing cloud services effectively.

Summary

In this chapter, you learned the fundamental definitions of cloud computing, the different types of stakeholders involved, the activities, and the technology models and building blocks. These are the core CCSP exam topics.

The next chapter will provide more details regarding the cloud reference architecture, the service models, and the cloud deployment models and capabilities. The chapter will also specify the shared considerations for cloud deployments and the impact of new and emerging technologies on the evolution of cloud computing.

Exam Readiness Drill – Chapter Review Questions

Apart from a solid understanding of key concepts, being able to think quickly under time pressure is a skill that will help you ace your certification exam. That is why working on these skills early on in your learning journey is key.

Chapter review questions are designed to improve your test-taking skills progressively with each chapter you learn and review your understanding of key concepts in the chapter at the same time. You'll find these at the end of each chapter.

> **How to Access These Materials**
>
> To learn how to access these resources, head over to the chapter titled *Chapter 25*, *Accessing the Online Resources*.

To open the Chapter Review Questions for this chapter, perform the following steps:

1. Click the link – `https://packt.link/CCSPE1_CH01`.

   Alternatively, you can scan the following **QR code** (*Figure 1.2*):

Figure 1.2 – QR code that opens Chapter Review Questions for logged-in users

2. Once you log in, you'll see a page similar to the one shown in *Figure 1.3*:

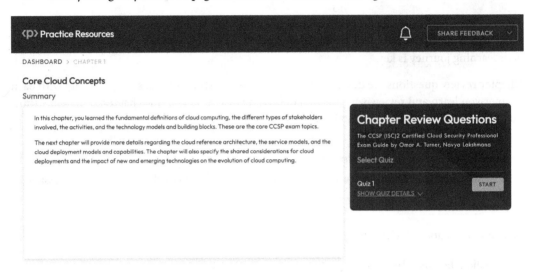

Figure 1.3 – Chapter Review Questions for Chapter 1

3. Once ready, start the following practice drills, re-attempting the quiz multiple times.

Exam Readiness Drill

For the first three attempts, don't worry about the time limit.

ATTEMPT 1

The first time, aim for at least **40%**. Look at the answers you got wrong and read the relevant sections in the chapter again to fix your learning gaps.

ATTEMPT 2

The second time, aim for at least **60%**. Look at the answers you got wrong and read the relevant sections in the chapter again to fix any remaining learning gaps.

ATTEMPT 3

The third time, aim for at least **75%**. Once you score 75% or more, you start working on your timing.

> Tip
>
> You may take more than **three** attempts to reach 75%. That's okay. Just review the relevant sections in the chapter till you get there.

Working On Timing

Target: Your aim is to keep the score the same while trying to answer these questions as quickly as possible. Here's an example of how your next attempts should look like:

Attempt	Score	Time Taken
Attempt 5	77%	21 mins 30 seconds
Attempt 6	78%	18 mins 34 seconds
Attempt 7	76%	14 mins 44 seconds

Table 1.1 – Sample timing practice drills on the online platform

> Note
>
> The time limits shown in the above table are just examples. Set your own time limits with each attempt based on the time limit of the quiz on the website.

With each new attempt, your score should stay above **75%** while your "time taken" to complete should "decrease". Repeat as many attempts as you want till you feel confident dealing with the time pressure.

2
Cloud Reference Architecture

In the previous chapter, you were introduced to the most relevant cloud computing characteristics and concepts with regard to cloud service models, cloud deployment models, and the different types of stakeholders in cloud computing. In this chapter, you will learn about the different types of cloud service capabilities that cloud service providers offer to their clients to meet their business requirements, along with getting an overview of the key service models—Infrastructure as a Service, Platform as a Service, and Software as a Service—and how they map to cloud capabilities.

You will then move on to the five different cloud deployment models and learn how responsibility shifts between the cloud service provider and the cloud service customer depending on the cloud deployment model and cloud services categories in use. The focus will then move to a review of shared considerations for cloud deployments, and an overview of new and emerging technologies that are related to cloud computing.

By the end of this chapter, you will be able to confidently answer questions on the following:

- Cloud service models
- Cloud service models and categories
- Cloud deployment models
- Shared responsibility model
- Shared considerations for cloud deployments
- New and emerging technologies related to cloud computing

You will now go through each topic in detail.

Cloud Service Models

Cloud consumers try to find solutions to their business and technical needs while searching for cloud service provider offerings that meet their functional and non-functional requirements. Cloud customers look at the service models that the cloud service providers offer and their pricing, and based on that information they are able to conduct a cost and benefit analysis for their business case. The **ISO/IEC 17788:2014** (`https://www.iso.org/standard/60544.html`) standard provides an overview of cloud computing along with a set of terms and definitions for cloud computing, which makes it easier to learn about and discuss the many facets of this technology. You will see this later in the chapter.

The standard describes the following three primary types of cloud service models:

- **Software as a Service (SaaS)**

- **Platform as a Service (PaaS)**

- **Infrastructure as a Service (IaaS)**

You will now review each one of these capabilities separately and examine their functionality and benefits.

Software as a Service (SaaS)

In the SaaS model, a cloud service provider makes various software applications available to end users remotely over the internet, typically through a web browser.

According to ISO/IEC 17788:2014, SaaS is distinguished by the following characteristics:

- **Management and control**: In the SaaS model, the cloud service provider is accountable for managing and maintaining all underlying infrastructure, application software, middleware, and data, including **Personal Identifiable Information (PII)**. You typically have very little influence or control over these elements except for certain user-specific application configuration settings.

- **Access**: SaaS applications can be accessed through the internet, enabling users to utilize them from any device that has an internet connection—regardless of their physical location. This provides greater flexibility as well as remote work and collaboration options.

- **Scalability**: This refers to the capacity of SaaS providers to modify their products and services according to customer demands. This helps guarantee that available resources are utilized efficiently and effectively, helping businesses avoid costly purchases, installations, and management of in-house software and hardware—an advantage that may not be immediately evident.

- **Pricing modeled on subscriptions**: SaaS is often sold through a subscription model. This allows customers to pay for the software on a recurring basis (for instance, monthly or annually) rather than making an initial investment in licenses. Organizations thus benefit from costs that are predictable and manageable, plus the freedom to scale up or down depending on requirements.

- **Automatic updates**: The SaaS model places responsibility for maintaining the software's most recent version, including all available features, bug fixes, and security patches, on the cloud service provider. This guarantees users always have access to the most up-to-date version of the application while relieving customers' IT teams of this task.

- **Multi-tenancy**: Multi-tenancy architecture is often employed by SaaS providers, as it permits multiple clients to share an application or infrastructure while still protecting their respective privacy and integrity. This may lead to better resource usage, lower overall costs for customers, and faster release of new features and updates.

Platform as a Service (PaaS)

PaaS is a model in which the cloud service provider provides you with a platform that allows you to construct, run, and maintain their applications without needing to construct, maintain, or manage the underlying infrastructure and middleware. This relieves you of having to build, upgrade, or manage these components themselves.

According to ISO/IEC 17788:2014, PaaS stands out from other cloud computing models for its following attributes:

- **Management and control**: In the PaaS model, the cloud service provider is accountable for managing and maintaining all underlying infrastructure and middleware, such as operating systems, runtime environments, and development tools, including the infrastructure and middleware of applications. You retain ownership of your own applications and data but don't have to concern yourself with overseeing components beneath them on the platform.

- **Tools for application development and deployment**: PaaS offerings typically consist of a collection of tools and services that enable you to design, construct, test, and deploy your own software applications. This could include programming languages, frameworks, libraries, databases, and any other relevant components required for successful development.

- **Scalability**: PaaS providers can scale platform resources to meet changing user demands. This ensures applications can handle increased workloads without customers needing to manage the underlying infrastructure. Scalability is one of the major advantages of PaaS; businesses now have more time and resources for what matters most: developing new applications instead of managing infrastructure.

- **Integration**: PaaS offerings typically feature built-in integration with other cloud services, such as databases, messaging systems, and data storage services. This makes it simpler for you to construct and deploy applications that utilize these resources without needing to manage them independently.

  PaaS solutions come in various configurations. PaaS customers have the flexibility to customize their applications and development environments according to individual needs, while still taking advantage of the managed platform provided by their cloud service provider.

- **Pay-as-you-go pricing**: PaaS services typically follow a pay-as-you-go pricing model, where customers only pay for resources they actually utilize. This pricing structure helps businesses save money while better aligning IT spending to actual usage patterns.

Infrastructure as a Service (IaaS)

IaaS is an internet-based model of cloud computing that delivers virtualized computing resources. You have the freedom to access, configure, and manage infrastructure components such as virtual machines, storage, and networking with this service model without needing to purchase or maintain hardware. Using the IaaS model enables businesses to scale resources according to demand, optimize costs, and focus on core business operations rather than managing IT infrastructure.

IaaS stands out from other cloud computing models by virtue of the following characteristics, as defined by ISO/IEC 17788:2014:

- **Pooling of resources**: IaaS providers utilize multi-tenant architectures to pool their available resources such as compute, storage, and networking in order to better serve their customers. This shared model allows them to efficiently allocate those resources among multiple customers while optimizing both utilization and cost.

- **Rapid elasticity**: IaaS offers customers the power to quickly scale back infrastructure resources in response to changes in demand. This flexibility allows organizations to adjust quickly to evolving requirements and workloads, leading to improved agility and flexibility.

- **Measured service**: IaaS providers typically offer a pay-as-you-go pricing model, in which customers are charged according to how many resources they actually utilize. This type of service is known as **measured service**. With this setup, organizations only pay for what resources are consumed, thus helping optimize costs and promote efficient resource usage.

On-demand customers have access to, configure, and manage their infrastructure resources through self-service portals, **Application Programming Interfaces** (**APIs**), or management tools provided by the IaaS provider. This enables customers to have more control over their own resources with less manual intervention from the service provider.

IaaS services can be accessed over the internet from various devices, such as laptops, smartphones, and tablets. This wide network access enables users to manage and interact with their infrastructure resources from any location, thus improving overall accessibility and making remote work simpler.

This section discussed cloud service models in relation to software, platform, and infrastructure. The next section will be an extension of this discussion—cloud service categories and cloud service models. There you will again compare the ISO/ IEC 17788 standard and NIST cloud computing reference architecture and also see ISO cloud service categories related to cloud service model definitions offered by NIST.

Cloud Service Models and Categories

Cloud computing services are often provided as one of three main service models, also known as service categories. In order to pass the CCSP exam, the (ISC)2 CBK requires you to know the cloud service models and be able to describe their differences.

The NIST Special Publication 800-145 titled **The NIST Definition of Cloud Computing** defines three fundamental cloud computing service models as follows:

- **IaaS**: IaaS is the capability provided to the cloud consumer to provision processing, storage, networks, and other fundamental computing resources where you are able to deploy and run arbitrary software that can include operating systems and applications. The consumer not only manages or controls the underlying cloud infrastructure but also has control over operating systems, storage, and deployed applications; and possibly limited control of select networking components (e.g., host firewalls).

- **PaaS**: PaaS is the capability provided to the cloud consumer to deploy onto the cloud infrastructure consumer-created or acquired applications created using programming languages, libraries, services, and tools supported by the provider. The consumer does not manage or control the underlying cloud infrastructure including network, servers, operating systems, or storage, but has control over the deployed applications and possibly configuration settings for the application-hosting environment.

- **SaaS**: SaaS is the capability provided to the cloud consumer to use the provider's applications running on a cloud infrastructure. The applications are accessible from various client devices through either a thin client interface, such as a web browser (e.g., web-based email), or a program interface. The consumer does not manage or control the underlying cloud infrastructure including network, servers, operating systems, storage, or even individual application capabilities, except for limited user-specific application configuration settings.

> **Note**
>
> You can access these definitions in NIST Special Publication 800-145 at `https://csrc.nist.gov/publications/detail/sp/800-145/final`.

Figure 2.1 shows the cloud service models with their key components and hierarchical structure where SaaS is being built on top of PaaS and IaaS:

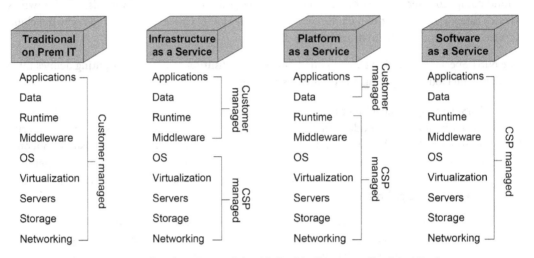

Figure 2.1 – Cloud service models with SaaS built on top of IaaS and PaaS

The three fundamental service models—IaaS, PaaS, SaaS—are also part of an extended list of cloud service categories described in ISO/IEC 17788 Standard. A **cloud service category**, according to the ISO/IEC 17788 standard, is a collection of cloud services that have a certain set of characteristics. A cloud service model can include capabilities from one or more cloud service models as described in the previous section.

The extended list of cloud service models mentioned in the ISO/IEC 17788 standard is as follows:

- **Compute as a Service (CompaaS)** is the computing capability provided to cloud customers to use the provider's processing resources needed to deploy and run customer software.

- **Data Storage as a Service (DSaaS)** is the capability provided to the cloud customer to use the provider's data storage and related capabilities.

- **Network as a Service (NaaS)** is the capability provided to the cloud customer to use the provider's data transport connectivity and related network capabilities.

- **Communications as a Service (CaaS)** is the capability provided to cloud customers to use the provider's network and communications capabilities for real-time interaction and collaboration (e.g., **Instant Messaging (IM)**, voice, and video teleconference solutions).

- **Anything as a Service (XaaS)** covers additional extended capabilities and services provided to cloud customers where the service provider takes responsibility for installing, maintaining, and operating the service and customers pay according to the service usage levels (e.g., Database as a Service, Desktop as a Service, Email as a Service, Identity as a service, and so on).

You have now learned about the fundamental cloud service models and categories. It is very important that as you prepare for the CCSP exam, you have a good knowledge of the three fundamental cloud service models and are able to identify and describe additional cloud service models/categories, based on the ISO/IEC 17788 standard.

The next section will give you an overview of the well-established four cloud deployment models and the additional fifth cloud deployment model recently introduced to the CCSP CBK.

Cloud Deployment Models

Another perspective for looking at cloud computing has to do with the ownership level of cloud infrastructure components rather than looking at the service types being provisioned and utilized. For the CCSP exam, you will be expected to know and learn the following cloud deployment models:

- Public cloud
- Private cloud
- Hybrid cloud
- Community cloud
- Multi-cloud

At a high level, you have the first two opposing models, with clear differences in the approach to the ownership level of cloud computing infrastructure components: the public cloud versus the private cloud. Then there are the **hybrid** and **community clouds**, with varying levels of ownership and co-sharing of the cloud infrastructure components between customers. The last cloud deployment model, recently introduced to the CCSP CBK, is the **multi-cloud** deployment model, which is a result of the fact that many cloud customers utilize cloud services from multiple cloud service providers simultaneously.

NIST Special Publication 800-145, **The NIST Definition of Cloud Computing**, provides the definitions of the four cloud deployment models:

- **Public cloud**: The cloud infrastructure is provisioned for open use by the general public. It may be owned, managed, and operated by a business, academic, or government organization, or some combination of them. It exists on the premises of the cloud provider.

- **Private cloud**: The cloud infrastructure is provisioned for exclusive use by a single organization comprising multiple consumers (e.g., business units). It may be owned, managed, and operated by the organization, a third party, or some combination of them, and it may exist on or off premises.

- **Community cloud**: The cloud infrastructure is provisioned for exclusive use by a specific community of consumers from organizations that have shared concerns (e.g., mission, security requirements, policy, and compliance considerations). It may be owned, managed, and operated by one or more of the organizations in the community, a third party, or some combination of them, and it may exist on or off premises.

- **Hybrid cloud:** The cloud infrastructure is a composition of two or more distinct cloud infrastructures (private, community, or public) that remain unique entities but are bound together by standardized or proprietary technology that enables data and application portability (e.g., cloud bursting for load balancing between clouds).

> **Note**
>
> The definitions are from **The NIST Definition of Cloud Computing-Recommendations of the National Institute of Standards and Technology** in NIST Special Publication 800-145 by Peter Mell and Timothy Grance.

- **Multi-cloud:** The (ISC)2, in the most recent updates to the CCSP CBK, introduced the fifth cloud deployment model, i.e., the multi-cloud model. This cloud infrastructure is provisioned with two or more public cloud hosting providers, allowing cloud customers to utilize multiple cloud services from various providers often in combination with private, community, or hybrid cloud deployment models.

> **Note**
>
> This newly introduced multi-cloud deployment model is not defined in NIST Special Publication 800-145 but is expected to be introduced in the next revision of this special publication. You can access the cloud deployment models in NIST Special Publication 800-145 at `https://csrc.nist.gov/publications/detail/sp/800-145/final`.

For now, you will learn how the multi-cloud deployment model compares with the other four deployment models, as shown in *Figure 2.2*:

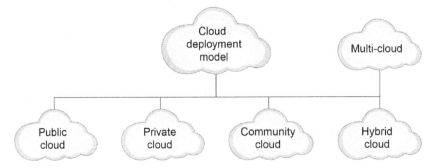

Figure 2.2 – Cloud deployment models

Shared Responsibility Model

As discussed earlier, various cloud deployment models offer different levels of control over the computing resources that are held by the cloud provider or cloud consumer. Unlike traditional IT environments—where one organization controls the full stack of computing resources and the entire lifecycle of the systems—in cloud computing, cloud providers and cloud consumers often collaborate to jointly design, build, deploy, and run cloud-based systems. Due to the varying levels of ownership and control over cloud resources, and different service models being provisioned and consumed in many aspects, both parties share responsibility for ensuring that the cloud-based systems are adequately protected.

In cloud computing, security is a shared responsibility. Selection and adoption of security controls in cloud computing requires analysis of which party is able to implement these protections and countermeasures based on the level of ownership and control over the cloud resources. This analysis must take into consideration service models and deployment models, as these imply various levels of control between cloud providers and cloud consumers.

At a high level, one way of looking at the boundaries of responsibility between cloud provider and cloud consumer is as follows:

- The cloud service provider is responsible for the security of the cloud
- The cloud customer is responsible for security in the cloud

The cloud service provider owns and controls both the physical and logical infrastructure components that run cloud services, including facilities, hardware, networking, and software, and therefore is responsible for the protection of these components.

The cloud customer's level of responsibility, in relation to the selection and implementation of security controls in cloud computing, depends on the specific cloud service(s) consumed by the customer. Depending on the cloud service model being utilized—IaaS, PaaS, SaaS—cloud consumer responsibilities can range from configuring and managing the guest operating system (including patching and other OS updates) to configuration of access control lists in firewalls and patching consumer software applications.

Figure 2.3 shows an overview of shared responsibility models in cloud service models:

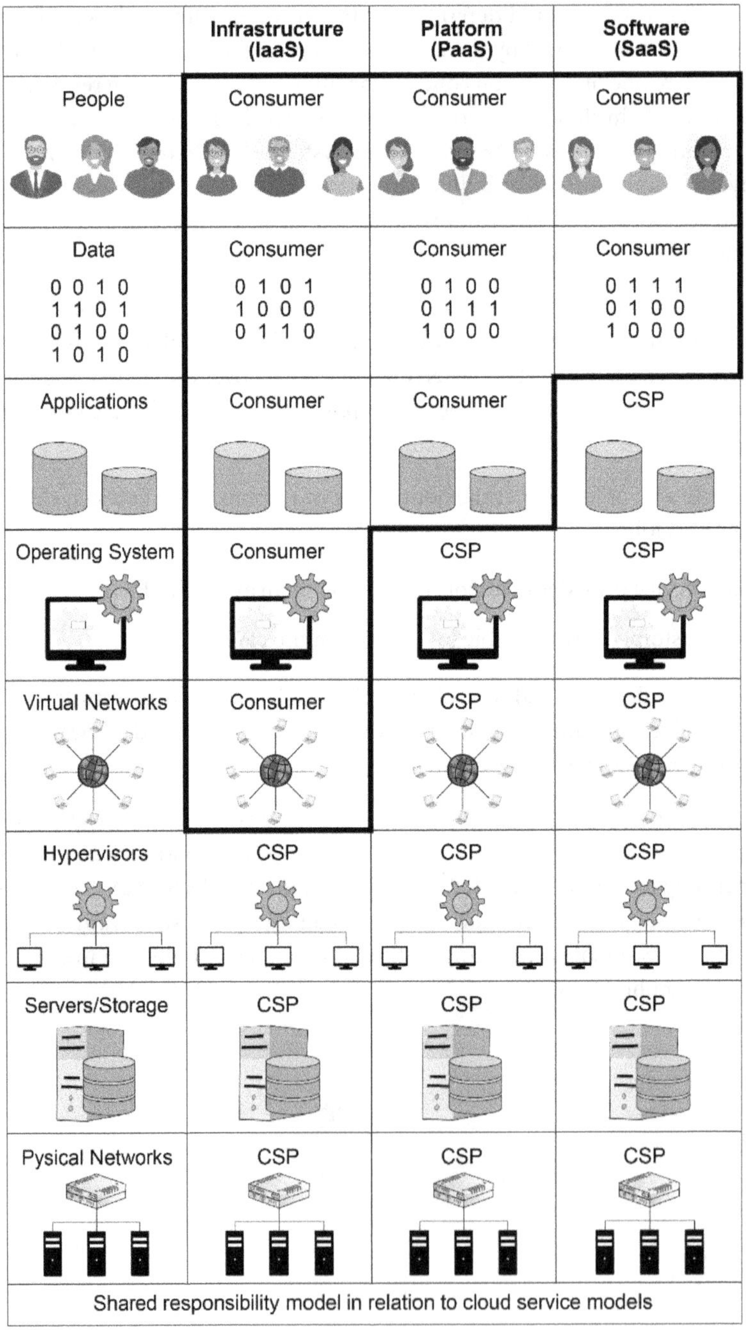

	Infrastructure (IaaS)	Platform (PaaS)	Software (SaaS)
People	Consumer	Consumer	Consumer
Data	Consumer	Consumer	Consumer
	0 0 1 0 1 1 0 1 0 1 0 0 1 0 1 0	0 1 0 1 1 0 0 0 0 1 1 0	0 1 0 0 0 1 1 1 1 0 0 0	0 1 1 1 0 1 0 0 1 0 0 0
Applications	Consumer	Consumer	CSP
Operating System	Consumer	CSP	CSP
Virtual Networks	Consumer	CSP	CSP
Hypervisors	CSP	CSP	CSP
Servers/Storage	CSP	CSP	CSP
Pysical Networks	CSP	CSP	CSP
Shared responsibility model in relation to cloud service models			

Figure 2.3 – Shared responsibility model in relation to cloud service models

You will revisit the cloud computing shared responsibility model several times in this book from the perspectives of data security, resiliency, and regulatory compliance. For your CCSP exam, it is critical that you learn how the level of responsibilities shifts based on the service model being consumed.

The key aspect to remember is that although the customer and cloud provider share the risks and responsibilities, the customer will ultimately bear the legal responsibility for unauthorized and illegal data disclosures since they are the data owner.

> **Note**
>
> You will learn more about legal responsibilities in *Chapter 22, Legal Challenges and the Cloud*, which discusses the legal challenges in relation to privacy and data protection in the cloud.

Shared Considerations for Cloud Deployments

The adoption of cloud computing involves several considerations that will lead to a combination of technical and business decisions for both cloud service providers and cloud customers. In this section, you will explore several key considerations for cloud deployments:

- **Interoperability**: In cloud computing, interoperability can be defined as the capacity of cloud ecosystem components to operate together in achieving desired outcomes. The components of a cloud computing ecosystem may come from several sources, including traditional and cloud sources, as well as public and private cloud implementation (known as hybrid cloud).

- **Portability**: This is the capability of cloud components to be moved or reused in different cloud deployment models, providers, platforms, infrastructures, and locations, with little or no modification (e.g., without converting the program to a different language). When choosing cloud providers, portability is an important factor to consider as it can help decrease the chance of vendor lock-in and provide various business benefits by enabling identical cloud deployments to be provisioned by different cloud providers, either for resiliency or to distribute service components to geographically dispersed deployments.

- **Reversibility**: Reversibility, in relation to cloud computing, can be looked at from two perspectives: cloud service provider and cloud consumer. For cloud service providers, reversibility is seen as a set of actions by which workload deployments or entire cloud environments can be returned to a previous state (rolled back). In a similar way, reversibility allows cloud service consumers to roll back the changes to return to a previous state, be it via the ability to retrieve deleted data or by rolling back application updates and patches.

- **Availability**: The availability of a cloud service is determined by the success or failure of underlying resources or components that are used to deliver services. The availability of cloud services is defined by the customer's ability to access target assets or resources when needed. The inability to access target assets or resources is described as the loss of availability. Cloud service providers describe the expectations in relation to the availability of their cloud services using **Service-Level Agreements** (**SLAs**), in which cloud service providers offer availability percentages (e.g., 95% or 99% availability). Failure to deliver services at the agreed availability levels may result in a refund of fees, service credits, and so on, and in the long term can lead to a loss of confidence that the cloud service provider is able to deliver services. This would result in reputational loss and in some cases loss of business for cloud service providers.

- **Security**: Security is a paramount component of secure cloud deployments, as it protects the **confidentiality, integrity, and availability** (**CIA**) of data and services hosted in the cloud. The CIA triad is an internationally recognized model that encompasses three core principles of information security: confidentiality, integrity, and availability. Here is how the CIA triad can be applied to cloud deployments:

  - **Confidentiality**: This principle guarantees that sensitive information is only accessible to authorized users, preventing unauthorized access or disclosure. In a secure cloud deployment, confidentiality can be achieved through various measures, such as data encryption, user authentication, and access controls. For instance, encrypted data can be encrypted both at rest and during transit, while role-based access controls manage user access rights.

  - **Integrity**: Integrity refers to the accuracy and consistency of data and the ability to guarantee that it will remain unchanged unless altered by authorized users. In cloud deployments, integrity can be maintained through data integrity checks, checksums, digital signatures, and version control systems. These mechanisms help detect any unauthorized modifications to information so they are prevented.

  - **Availability**: Availability is the principle that guarantees data and services are accessible to authorized users when needed. In secure cloud deployments, this may involve using redundant systems, load balancing, and disaster recovery plans to guarantee resource availability. This could include deploying applications across multiple geographic regions or using autoscaling for fluctuations in demand.

Cloud service providers also look at security from the perspective of the physical security of their data centers and hardware, including perimeter security, physical access to servers, and so on:

  - **Privacy**: Information privacy is the guarantee of correct and consistent collection, processing, communication, use, and disposal of **Personally Identifiable Information** (**PII**) throughout its lifecycle.

  - **Resiliency**: The capacity to minimize the effects and/or duration of disruptions to an organization's critical infrastructure. A resilient infrastructure or enterprise is measured in its capacity for anticipating, absorbing, adapting to, and quickly recovering from such events.

- **Performance**: In cloud computing, performance can be looked at as the ability of cloud services and individual cloud components to meet contractual obligations (e.g., the SLA) in relation to compute, networking, and storage. Cloud service providers use a combination of methodologies and metrics that enable monitoring and management of cloud resources performance from the perspective of network bandwidth, processing throughput and availability, and downtime of services.

- **Governance**: This is the process by which an enterprise makes sure that stakeholder drivers, needs, and goals are assessed to establish balanced, agreed-upon business objectives. Governance includes establishing direction through decision-making and prioritization, as well as monitoring performance and compliance with established business objectives.

- **Maintenance and versioning**: **Maintenance** describes actions made to a cloud service or cloud computing components to prevent failure or malfunction of equipment or to restore its operating capability. **Versioning** is the process of applying proper labeling of a service or application so both the cloud services provider and cloud consumer can identify the service and application version currently in use.

- **Service levels and SLA**: An SLA is a contractual agreement between cloud service providers and their customers that sets expectations regarding the performance, availability, and reliability of provided services. SLAs play a critical role in cloud computing as they set expectations and define the terms and conditions of the agreement; guaranteeing that both parties know their responsibilities as well as the quality of service that will be provided. Typically, these SLAs include components such as performance metrics, availability, reliability, and security; support, remediation, and change management; and disaster recovery and business continuity plans and termination clauses.

- **Auditability**: Auditability can be defined as the ability to successfully deliver an accurate audit exam report. It depends heavily on obtaining access to necessary information within the scope and the objective of the audit. Information used in the audit as evidence should be relevant, well organized, and accurate.

- **Regulatory considerations and compliance**: The use and delivery of cloud services may be impacted by a wide range of regulatory requirements. Some requirements have their roots in industry-specific regulations or laws. Many of the regulatory requirements exist to stop personal information from falling victim to theft or fraud. Location-based laws occur at a variety of levels, from national laws to state and local laws. For example, within the United States, several states have stand-alone security and privacy laws that apply to operations within their borders or to consumer interactions within those states. Similar granularity occurs elsewhere in the world, such as in Europe and Asia.

> **Note**
>
> You will learn about legal and regulatory requirements and their impact on the provisioning and consumption of cloud services in *Chapter 22, Legal Challenges and the Cloud*.

- **Outsourcing**: In IT, outsourcing can be defined as the use of external service providers (suppliers) to enhance or replace IT-enabled business processes, infrastructure solutions, or application services. In cloud computing, **outsourcing** can range from utility services (power, HVAC) and physical security (perimeter security) to SaaS, and it enables clients to create the best sourcing strategies and forge long-lasting, win-win relationships with outside suppliers. By outsourcing, businesses can cut costs, improve time to market for the delivery of products and services, and benefit from outside experience, resources, and/or intellectual property.

> **Note**
>
> You can access the cloud deployment models in NIST Special Publication 800-145 at `https://csrc.nist.gov/publications/detail/sp/800-145/final`.

You have now reviewed the shared considerations for cloud deployments that are required for the CCSP exam. Responsibility and accountability for addressing these considerations and making relevant decisions will often depend on the cloud deployment model and cloud service models in use. Many of the considerations mentioned here will come up frequently in the following chapters of this book.

Emerging Technologies in Cloud Computing

Cloud computing characteristics such as resource pooling, automation, scalability, and the innovative use of virtualization allowed for the development of many new and emerging technologies that provide businesses around the world with new capabilities. You will review some of the most notable new technologies that were developed as a result of growing cloud adoption, or which are taking cloud computing into new emerging areas.

Data Science

Data science has revolutionized how businesses analyze, process, and utilize vast amounts of data in order to gain insights and make decisions based on that knowledge. With cloud computing's continuing evolution in offering more services and platforms for data scientists to collaborate on their work, innovate within their field, and scale operations, data scientists now have an ideal environment in which they can collaborate, innovate, and scale operations.

Cloud computing is seeing the emergence of serverless computing, which hides the underlying infrastructure from users. By not having to worry about managing servers, networks, or storage for data scientists to focus on creating and deploying **Machine Learning** (**ML**) models instead, they free up time to focus on their primary responsibilities such as feature engineering, model selection, and evaluation—freeing up considerable time and energy otherwise spent managing infrastructure.

Data lakes and data warehouses hosted in the cloud are essential for storing, organizing, and managing vast amounts of structured and unstructured data. Data scientists now have access to centralized and scalable storage solutions that make it simple for them to gain relevant data for analysis and modeling. Moreover, these storage solutions offer sophisticated data processing and querying capabilities that make it simpler for data scientists to gain useful insights from complex datasets.

Data scientists now have the capacity to work with live data and gain valuable insights more quickly thanks to modern cloud computing technologies that enable real-time processing and streaming. These capabilities make it simpler to construct real-time analytics models that are flexible enough to adjust according to changing environmental conditions, giving businesses the capacity to respond promptly and make timely decisions.

JupyterHub and Databricks, for instance, provide cloud-based collaborative environments where data scientists can work together on projects, share code, and streamline workflows. These platforms facilitate communication and expedite the development of data science solutions by offering a centralized workspace with access to necessary tools, libraries, and other resources.

Cloud computing technologies have had a profound effect on data science, providing an environment that is both adaptable and scalable for data scientists to innovate within and collaborate. Data scientists can focus on what matters most to them, access more cutting-edge resources, and complete their work more efficiently when they utilize these advanced tools.

Artificial Intelligence and Machine Learning (AI/ML)

Artificial intelligence, or **AI** for short, refers to computer systems capable of performing tasks previously thought impossible. These responsibilities include learning new knowledge, solving issues, comprehending language, and recognizing spoken words; AI systems may either be rule-based with pre-defined algorithms and knowledge or be data-driven (this latter type learns through experience and adjusts itself according to changes made).

Machine Learning (ML), also known as **Machine Intelligence (MI)**, is a subfield of AI that develops algorithms and models to enable computers to learn from data. ML doesn't involve explicitly programming the system; rather, it relies on statistical techniques for pattern recognition and making predictions. With more processing power over time, ML's performance improves, leading to increasingly accurate conclusions and forecasts.

Cloud computing continues to develop and offer an ever-increasing number of services and platforms, and AI and ML have become integral parts of this field. Businesses now have the capacity to take advantage of powerful analytics, automate tedious tasks, and create intelligent applications thanks to cloud computing's integration of AI and ML. **AI as a Service** (**AIaaS**), for example, is an emerging trend in cloud computing that offers AI and ML capabilities through cloud-based services. With this model, businesses can gain access to cutting-edge AI tools and technologies without needing to invest in costly hardware or develop in-house expertise. AIaaS platforms offer pre-trained models, natural language processing, computer vision, and speech recognition services so businesses can quickly and cost-effectively incorporate AI capabilities into their applications. Technologies in distributed ML and AI make it possible to train and deploy models across a multitude of cloud-based nodes or clusters. This strategy provides improved scalability, fault tolerance, and faster processing speeds to data scientists so they can work more efficiently with large datasets and complex models. The adoption of distributed AI and ML frameworks by cloud service providers, such as TensorFlow, PyTorch, and Apache MXNet, is a crucial element in the advancement of cloud-based AI and ML solutions and an integral factor in their overall success.

Edge computing is a relatively recent technology that brings computation and data storage closer to the source of the information, leading to reductions in both latency and bandwidth consumption. Cloud providers can enable real-time processing at their network's edge through the integration of AI and ML with edge computing. This combination is especially advantageous for applications such as autonomous vehicles, smart cities, and **Internet of Things** (**IoT**) devices that require fast data processing with low latency to run optimally.

AI and ML technologies are being employed not only for data analysis and prediction but also to enhance cloud infrastructure and services. Cloud service providers utilize AI/ML algorithms to automate resource allocation, load balancing, and security management, leading to cloud operations that are more efficient and cost-effective.

Platforms for analytics in the cloud that utilize AI allow businesses to gain actionable insights from vast amounts of data quickly and efficiently. These platforms use sophisticated ML algorithms to analyze each piece of information in order to spot patterns, trends, and correlations. As a result, businesses gain the capacity to make informed decisions and enhance their operations.

Blockchain

Blockchain is a decentralized and distributed ledger technology that has recently gained prominence due to its potential to revolutionize various industries, such as healthcare and finance. To enhance safety, transparency, and operational effectiveness, blockchain technology is increasingly being integrated into cloud computing services and platforms as the cloud computing industry matures. The following are some primary examples of how blockchain technology can benefit cloud computing services and platforms:

- **Blockchain as a Service**: An emerging trend in cloud computing that gives companies subscription-based access to blockchain infrastructure and related services is **Blockchain as a Service (BaaS)**. Businesses can build, deploy, and manage applications using BaaS without investing in costly underlying infrastructure or possessing the specialist skillsets required for doing so. Major cloud providers such as **Amazon Web Services (AWS)**, Microsoft Azure, and IBM offer BaaS platforms, so it becomes simpler for businesses to implement and integrate blockchain technology into their operations, such as smart contracts and **decentralized applications (dApps)**.

- **Decentralized cloud storage**: Traditional centralized storage services can be replaced with blockchain-based decentralized cloud storage solutions that offer both security and transparency. These solutions take advantage of the decentralized nature of blockchain technology to store data across multiple nodes, increasing redundancy and fault tolerance. Taking this approach reduces the likelihood of data loss or tampering, helping ensure sensitive information remains secure. Filecoin (`https://filecoin.io/`), Storj (`https://www.storj.io/`), and Sia (`https://sia.tech/`) are examples of projects offering decentralized storage services that can be integrated into existing cloud computing infrastructure.

- **Identity and access management**: Cloud computing environments are increasingly utilizing decentralized ledger technology known as **blockchain** to improve identity and access management. Cloud providers can now generate secure, verifiable digital identities for users and devices by taking advantage of the immutable ledger and cryptographic features provided by blockchain technology. Authentication and authorization processes have both been simplified due to this improvement in cloud service security, which also increases their own level of protection.

- **Smart contracts and decentralized applications**: Smart contracts are automated contracts that take action when certain conditions are met, bypassing the need for intermediaries. dApps running on peer-to-peer networks have become possible due to the integration of smart contracts with cloud computing technology. When compared with traditional centralized apps, decentralized apps offer greater levels of security, transparency, and efficiency.

Internet of Things (IoT)

IoT is a network of interconnected devices, sensors, and actuators that collect, exchange, and process data to enhance operational effectiveness, decision-making capabilities, and user experience. IoT devices generate vast amounts of data that can be managed and analyzed with cloud computing's powerful infrastructure, which continues to expand at an incredible rate. The following are capabilities that IoT can provide:

- **Edge computing and fog computing**: New cloud technologies known as edge computing and fog computing bring computation and data storage closer to IoT devices, helping reduce latency and bandwidth consumption. With these methods, real-time processing, decision-making, and analytics can be done at the network's edge—perfect for time-sensitive IoT applications such as autonomous vehicles, industrial automation, and smart cities.

- **IoT device management and security**: Cloud-hosted IoT platforms offer comprehensive device management and security solutions. These programs help businesses remotely monitor, update, and secure their IoT devices. Features such as device provisioning, firmware upgrades, and access control guarantee that these IoT devices are always protected against potential threats with the most up-to-date software versions available.

- **Data storage and analytics in IoT**: Cloud computing offers scalable solutions for data storage and analytics that can effectively handle the vast amounts of information generated by IoT devices. Cloud-based data lakes and data warehouses offer centralized storage for structured and unstructured data generated by IoT devices. Advanced analytics tools and ML algorithms give businesses actionable insights to boost productivity, while cloud-based data lakes and data warehouses store the generated information from IoT devices. These solutions enable organizations to process and analyze IoT data in real time, allowing them to make faster decisions with more efficiency and maximize resource usage.

- **Integration and interoperability**: Recent cloud technologies enable IoT devices and other types of cloud services to integrate and communicate with one another in a seamless manner. APIs and data integration tools make it simple for businesses to link IoT devices with cloud platforms, ensuring data flows freely between different systems. As a result, organizations are able to create interconnected ecosystems that maximize the potential generated by data generated by IoT.

Containers

Containers have become a key element in cloud computing, revolutionizing the process of application development, deployment, and management. Although still relatively new to this space, containers hold great promise for increasing scalability, optimizing resource use, and streamlining workflows within cloud environments.

Virtual machines, or **VMs**, have long been the go-to method for simultaneously running multiple software programs on a single physical server. Unfortunately, VMs consume lots of resources since each requires its own operating system in addition to dedicated resources for it. Containers make use of the kernel of the host OS and distribute resources more effectively leading to improved performance, quicker boot-up times, and reduced overhead. Here are some key use cases where containers can be beneficial:

- **Microservices architecture**: Containers have made it simpler to transition to a microservices architecture, in which applications are divided into smaller, more autonomous components. This strategy enhances scalability, resilience, and flexibility since each component can be independently developed, deployed, and updated.

- **Kubernetes and container orchestration**: As containers became more commonplace, it became evident that efficient management and orchestration were required. Kubernetes has quickly become the go-to container orchestration platform due to its automatic deployment, scaling, and management capabilities for containerized applications. As a result, innovation accelerated exponentially; now it's possible to run more complex applications as well as large-scale deployments in the cloud!

- **Serverless computing**: Containers have been a driving force behind the rise of serverless computing—an approach to cloud computing in which code execution is triggered by events and users are charged according to data usage. Platforms such as AWS Lambda and Google Cloud Functions leverage containers as part of their environment for providing flexible, cost-effective, scalable event-driven application use cases.

Containers have made adopting multi-cloud and hybrid cloud strategies much simpler. This frees organizations from being locked into one vendor while optimizing their infrastructure costs.

Containers have proven their potential as a disruptive force in cloud computing, enabling many new technologies and revolutionizing how applications are built and deployed.

Quantum Computing

Quantum computing research seeks to use the principles of quantum mechanics to solve problems and perform calculations that are too difficult for traditional computers to handle. With cloud computing advancing at an exponential rate, the integration of quantum computing into cloud services and platforms could revolutionize data processing as well as security and optimization processes.

Quantum computers differ from classical computers in that they use quantum bits (qubits), which can exist simultaneously in multiple states due to superposition and entanglement. Classical computers use bits (0s and 1s) as storage and processing units for information. Quantum computers utilize **qubits**—a unique property that allows them to process vast amounts of data simultaneously and thus solve difficult problems and perform specific calculations much more quickly than their classical counterparts. However, practical quantum computing is still in its early stages and researchers are currently striving to overcome obstacles such as error correction and stability in order to create quantum systems that are both more powerful and dependable. With this in mind, here are some early use cases that make good use of cloud computing's capabilities.

Quantum as a Service (QaaS)

Quantum as a Service, or **QaaS** for short, is an emerging cloud computing trend that provides businesses with subscription access to quantum computing resources and services on demand. Through QaaS, businesses gain access to cutting-edge quantum hardware and software without investing in either infrastructure or expertise. Major cloud service providers such as IBM, Microsoft, and Google are conducting research into QaaS platforms so businesses can experiment and take advantage of quantum computing capabilities.

Quantum-Enhanced Optimization and ML

Optimization and ML algorithms have the potential to benefit from quantum computing's application of this paradigm. By drawing upon quantum mechanics' unique properties, quantum-enhanced algorithms are able to solve complex optimization problems more quickly and more accurately, as well as being able to integrate them with cloud-based machine learning platforms for data processing, pattern recognition, and decision-making more quickly and more accurately.

Quantum Simulation and Modeling

Quantum computing has the potential to transform the simulation and modeling of complex systems such as chemical reactions, material properties, and financial markets. With this advanced technology, these complex processes can be more accurately predicted.

Integrating quantum computing capabilities into cloud-based simulation platforms can enable more accurate and efficient simulations, leading to deeper comprehension and more accurate prediction of complex phenomena. This has applications across various industries such as pharmaceuticals, materials science, and finance, all of which stand to benefit from improved modeling, which in turn drives innovation and enhances decision-making.

Quantum Cloud-Based Software Development

As quantum computing advances, so too does the demand for special software tools and platforms. Cloud service providers are exploring developing quantum-aware programming languages, libraries, and development environments so businesses can easily create and deploy quantum applications without needing extensive in-house expertise or infrastructure. Through this approach, businesses will be able to take advantage of quantum computing without needing costly in-house resources or expertise.

Edge Computing

Edge computing, a relatively recent technology in cloud computing, is rapidly gaining traction due to its potential to revolutionize data processing, storage, and consumption. Edge computing attempts to address traditional cloud computing's shortcomings such as latency or bandwidth restrictions by decentralizing processing closer to its source of creation. Edge computing will become an indispensable element of today's advanced technological landscape as the IoT experiences explosive growth and real-time processing demands continue to increase.

Edge computing is a distributed computing paradigm that shifts processing and storage functions away from centralized data centers to the periphery of the network. At its core, edge computing is simply another type of cloud computing. It enables data processing to take place close to where it originates by offloading computational tasks to devices or servers located at this edge location. Not only does this reduce latency, conserve bandwidth, and maximize resource utilization efficiency, but it also addresses privacy and security concerns associated with data transmission.

Edge computing is opening up an array of innovative application possibilities across a range of business sectors. Here are some of the most exciting potential uses:

- **Smart cities**: Edge computing can aid smart cities in streamlining traffic management, public safety, and resource allocation by processing data locally.

- **Healthcare**: Computing at the edge can enhance healthcare by providing real-time monitoring and analysis of patient data, leading to more timely diagnoses and decisions regarding treatment plans.

- **Manufacturing and industrial automation**: With edge computing, manufacturing and industrial equipment can process data locally to optimize production processes and boost efficiency. This is especially beneficial in industrial automation applications.

- **Augmented reality and virtual reality**: Edge computing can reduce latency in **Augmented Reality (AR)** and **Virtual Reality (VR)** applications, providing a smoother, more immersive user experience.

Edge computing is an innovative cloud technology that offers various advantages, such as reduced latency and enhanced data privacy, and will have a major influence on how computing infrastructure evolves in the future, especially as IoT devices continue to become more prevalent and real-time data processing becomes more important.

Confidential Computing

Cloud computing is seeing the emergence of **confidential computing**, which addresses privacy and safety concerns regarding data stored in cloud environments. Confidential computing protects information against prying eyes, including those of cloud service providers—by encrypting it using specialized hardware and isolating sensitive information at each stage of processing. As businesses increasingly turn to cloud services for storing and processing sensitive information, confidential computing will become an increasingly important component for maintaining trust in digital spaces.

Confidential computing utilizes hardware-based security mechanisms such as **Trusted Execution Environments (TEEs)** and secure enclaves to protect data while it's being processed. In the traditional cloud computing model, data is encrypted both during transit and storage; however, this must be decrypted before processing, leaving sensitive information vulnerable to interceptions by hackers or unauthorized access. Confidential computing solves this flaw by encrypting data at all times—even while being processed—keeping sensitive info safe throughout its entire lifecycle.

Computing in private settings is especially advantageous for businesses and applications that prioritize the confidentiality and safety of their data. Here are some potential applications of confidential computing:

- **Financial industry**: Confidential computing allows for the secure processing of sensitive financial data such as credit card transactions, bank records, and investment portfolios, thus helping ensure compliance with applicable regulations.

- **Healthcare**: Confidential computing can assist healthcare organizations in maintaining **Health Insurance Portability and Accountability Act (HIPAA)** observance and safeguarding patient privacy by encrypting patient data during processing.

- **Sensitive data**: Confidential computing can safeguard sensitive data in government applications, such as voter registration and tax records, along with information related to national security and defense. This technology is also utilized within the public sector.

- **Encrypted data**: Confidential computing allows organizations to train ML models on encrypted data, protecting proprietary algorithms and sensitive datasets from unauthorized access. ML and AI confidential computing also enables organizations to train ML models using encrypted data.

You have now grasped the new and emerging technologies that relate to cloud computing. Many of the technologies mentioned in this section result either directly or indirectly from the broadening adoption of cloud computing worldwide. Some of these related technologies, such as containers and edge computing, are now an integral part of cloud computing services offered by many cloud service providers (e.g., Azure AKS, AWS EKS, GCP GKE, AWS CloudFront, and Microsoft Azure IoT Edge).

You will continue to see these technologies being mentioned in the chapters to follow. For the CCSP exam, ensure that you learn about these technologies, how they relate to each other, and why they are important from the perspective of cloud computing. You can refer to additional information and study references here:

- NIST Cloud Computing Standards Roadmap, NIST Special Publication 500-291, Version 2: `http://www.datascienceassn.org/sites/default/files/NIST%20 Cloud%20Computing%20Reference%20Architecture%20-%202013.pdf`

- Protecting Controlled Unclassified Information in Nonfederal Systems and Organizations: `https://csrc.nist.gov/CSRC/media/Publications/sp/800-171b/ draft/documents/sp800-171B-draft-ipd-with-line-nums.pdf`

Summary

In this chapter, you learned about cloud service capabilities and how they relate to cloud service models. You also learned about the different types of cloud deployment models and how responsibilities change depending on the cloud deployment models and cloud service models being offered and used. Many new and emerging technologies which are related to cloud computing were also introduced.

For your CCSP exam, make sure you grasp the definitions that are introduced in this chapter. Many CCSP exam questions focus on key cloud terms and definitions. You need to be able to describe different cloud service models. It is very important that you know the differences between the three cloud service models—IaaS, PaaS, and SaaS—and the different features and characteristics associated with them. It is vitally important that you grasp the features of the five cloud deployment models—public, private, community, hybrid, and multi-cloud—as well as being able to describe their differences.

In the next chapter, you will turn your attention to the top threats that cloud deployments are facing every day. The chapter will also discuss how the application of the key cloud security concepts and controls can help mitigate risks.

Exam Readiness Drill – Chapter Review Questions

Apart from a solid understanding of key concepts, being able to think quickly under time pressure is a skill that will help you ace your certification exam. That is why working on these skills early on in your learning journey is key.

Chapter review questions are designed to improve your test-taking skills progressively with each chapter you learn and review your understanding of key concepts in the chapter at the same time. You'll find these at the end of each chapter.

> **How to Access These Materials**
>
> To learn how to access these resources, head over to the chapter titled *Chapter 25, Accessing the Online Resources.*

To open the Chapter Review Questions for this chapter, perform the following steps:

1. Click the link – `https://packt.link/CCSPE1_CH02`.

    Alternatively, you can scan the following **QR code** (*Figure 2.4*):

Figure 2.4 – QR code that opens Chapter Review Questions for logged-in users

2. Once you log in, you'll see a page similar to the one shown in *Figure 2.5*:

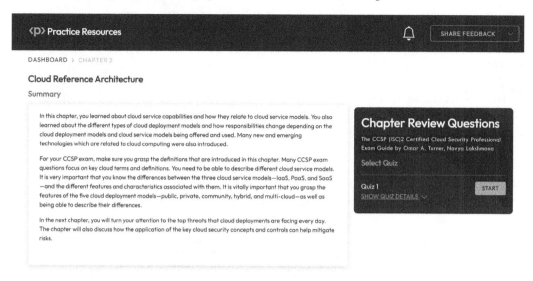

Figure 2.5 – Chapter Review Questions for Chapter 2

3. Once ready, start the following practice drills, re-attempting the quiz multiple times.

Exam Readiness Drill

For the first three attempts, don't worry about the time limit.

ATTEMPT 1

The first time, aim for at least **40%**. Look at the answers you got wrong and read the relevant sections in the chapter again to fix your learning gaps.

ATTEMPT 2

The second time, aim for at least **60%**. Look at the answers you got wrong and read the relevant sections in the chapter again to fix any remaining learning gaps.

ATTEMPT 3

The third time, aim for at least **75%**. Once you score 75% or more, you start working on your timing.

> **Tip**
> You may take more than **three** attempts to reach 75%. That's okay. Just review the relevant sections in the chapter till you get there.

Working On Timing

Target: Your aim is to keep the score the same while trying to answer these questions as quickly as possible. Here's an example of how your next attempts should look like:

Attempt	Score	Time Taken
Attempt 5	77%	21 mins 30 seconds
Attempt 6	78%	18 mins 34 seconds
Attempt 7	76%	14 mins 44 seconds

Table 2.1 – Sample timing practice drills on the online platform

> **Note**
> The time limits shown in the above table are just examples. Set your own time limits with each attempt based on the time limit of the quiz on the website.

With each new attempt, your score should stay above **75%** while your "time taken" to complete should "decrease". Repeat as many attempts as you want till you feel confident dealing with the time pressure.

3

Top Threats and Essential Cloud Security Concepts and Controls

In the last chapter, you studied cloud reference architectures, cloud service models, cloud deployment models, and their inherent capabilities. In this chapter, you will discuss security concepts that are relevant to cloud computing and review some of the common threats to cloud deployments. The chapter will introduce the most common control categories and control types necessary to secure data, network, and virtualization layers for cloud computing.

By the end of this chapter, you will be able to confidently answer questions on the following topics:

- The CIA triad—confidentiality, integrity, and availability
- Common threats to cloud deployments
- Enterprise security control categories and types

The CIA Triad—Confidentiality, Integrity, and Availability

This section will describe some of the fundamental information security concepts that form the basis of all information security initiatives and are required for your CCSP exam.

> **Note**
> Although some of you may already be aware of these concepts, this section is included to establish key terminology used in the chapters to follow.

The fundamental information security concepts include the following:

- **Confidentiality**: This can be defined as the protection of data to ensure that it can only be accessed by people with the right level of permissions, that is, those who are authorized to view the data. Theft of credit card information is an example of a breach of confidentiality. Various confidentiality controls can be used in order to enforce the required level of secrecy within the data processing system and prevent the unauthorized disclosure of confidential information. Confidentiality should be ensured for data in transit, in use, and while data is stored on systems and devices.

- **Integrity**: This relates to data accuracy and authenticity. Loss of integrity includes instances where an attacker modifies the contents of a file or a database, or defaces a website. Integrity safeguards guarantee the reliability, accuracy, and authenticity of data and protects against unauthorized or unexpected changes to the data. To maintain data integrity, hardware, software, and other IT systems must work together and validate data integrity while data is transported from one system to another and while data is being stored for processing or for archival purposes.

- **Availability**: This relates to ensuring that data and computing resources are accessible to authorized users when needed. This component of the triad focuses on preventing disruptions to service, such as hardware failures, software issues, and cyberattacks, and ensuring reliable access to data systems and data in a timely manner. For example, a **Denial of Service (DoS)** attack that prevents access to a website is an example of a loss of availability. Applications, platforms, and infrastructure should be designed and built to ensure the availability of information. Security measures should be put in place to guard against internal and external threats that could reduce IT systems' availability.

Figure 3.1 shows the three elements of the CIA triad and their key focus area—data:

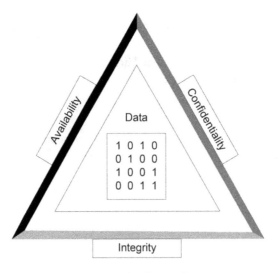

Figure 3.1 – The CIA triad showing the three information security concepts

This concludes the review of the core information security concepts that every CCSP candidate should be familiar with. It is important to remember all these core concepts while preparing for the CCSP exam. The next step is to identify the risks and impacts of cyber-attacks on the confidentiality, availability, and/or integrity of information.

So, the next section will discuss the most common threats that aim to breach the confidentiality of data, cause a loss of integrity, or prevent authorized users from accessing the data.

Common Threats to Cloud Deployments

You will now cover the top five common threats to cloud deployments. These threats are universal, apply to all cloud service models, and can impact public, private, hybrid, community, and multi-cloud deployments.

> **Note**
>
> In the chapters that follow, you will engage in a detailed review of more specific threats to IaaS, PaaS, and SaaS.

The following are some of the most common threats to cloud deployments (in no particular order).

Data Breaches

Data breaches are one of the greatest dangers to cloud deployments. They happen when unauthorized individuals gain access to sensitive information stored in the cloud. As a result, businesses suffer severe consequences, such as loss of customer trust, financial penalties, and reputational harm.

Thus, cloud security practitioners should help organizations take the following steps to prevent data breaches:

- Implement strong access control mechanisms to restrict access to sensitive data.
- Encrypt sensitive data both at rest and in transit.
- Regularly monitor and audit cloud environments in order to detect suspicious activity and take appropriate responses.
- Utilize **Intrusion Detection and Prevention Systems (IDPSs)** to detect and block unauthorized access attempts.

Misconfiguration

Misconfiguring cloud resources is another common security risk that can leave your system vulnerable to attack. An inadequate or improper configuration of security settings, network access controls, and storage permissions may expose sensitive data to unauthorized users or allow attackers to exploit the system.

To mitigate the risk of misconfiguration, cloud security experts should help organizations do the following:

- Establish and adhere to stringent configuration management policies.
- Regularly review and upgrade security configurations in accordance with industry standards and best practices.
- Implement automated tools that monitor configuration changes and deviations from the defined baseline.
- Conduct periodic security audits and assessments to detect, remediate, and prevent misconfigurations.

Insecure APIs

Application Programming Interfaces (**APIs**) facilitate communication between cloud services and applications. While APIs are essential for the smooth running of cloud environments, they can also present a significant security risk if not adequately protected. Insecure APIs could allow attackers to gain unauthorized access to data, manipulate applications, or launch **Distributed Denial of Service** (**DDoS**) attacks.

Cloud security practitioners should help organizations take the following steps to minimize the risk of insecure APIs:

- Implement strong authentication and authorization processes to limit API access to authorized users and applications.
- Encrypt data transmitted over APIs for protection against interception or modification.
- Regularly monitor and audit API usage to detect any abuse or malicious activity.
- Implement rate limiting along with other robust security measures, such as API gateways.

Insider Threats

Insider threats are security incidents caused by individuals within an organization, either intentionally or unintentionally. These risks could stem from employees, contractors, or business partners with authorized access to cloud resources for the organization. Insider breaches can result in data leaks, system sabotage, and unauthorized access to sensitive information.

Cloud security practitioners should help organizations take the following steps to prevent insider threats:

- Establish stringent access control policies, restricting sensitive data and resources to those who need to know.

- Keep an eye on what users do in the cloud environment to spot suspicious behavior or possible insider attacks.

- Give your employees regular security training, focusing on how important it is to protect sensitive data and what will happen if security is compromised.

- Implement comprehensive incident response processes in order to quickly detect and mitigate insider threats.

Account Hijacking

Account hijacking occurs when attackers gain unauthorized access to cloud user accounts through phishing attacks, password breaches, or exploiting insecure authentication mechanisms. Once in control of an account, they can view sensitive data, manipulate applications, and even launch attacks against other users or systems within the organization's cloud environment.

Cloud security practitioners should help organizations take the following steps to prevent account hijacking:

- Adopt strong **Multi-factor Authentication (MFA)** for all cloud user accounts to reduce the risk of unauthorized access, even if a password is compromised.

- Regularly update and enforce password policies that require users to create strong, unique passwords that are periodically changed.

- Warn users about phishing attacks and the importance of not sharing sensitive account information (e.g., login credentials) with unauthorized parties.

- Monitor user account activities and implement anomaly detection to identify potential account hijacking attempts.

Cloud deployments continue to increase in popularity, so organizations must stay vigilant against potential threats that could compromise their data and systems. With the knowledge of the five common risks—data breaches, misconfiguration, insecure APIs, insider threats, and account hijacking—businesses can develop and implement robust security measures to protect their cloud environments. Combining strong access control, encryption, continuous monitoring, and user education helps organizations greatly reduce the risks associated with cloud deployments.

In the chapters to follow, you will discuss many more threats and concerns for IaaS, PaaS, and SaaS. The next section will explore the leading cloud security controls, methods, and tools that both CSPs and cloud customers can use to respond to threats. It will also throw light on how organizations approach the selection, design, and implementation of security controls for a given asset.

Security Control Categories and Types

An **asset** in the context of information security is data, an information system, a facility, or any IT infrastructure component (hardware, software, or data) perceived to have either tangible or intangible value and used to support enterprise operations.

Security controls, also known as **safeguards** or **countermeasures**, are mechanisms, processes, or technologies put in place to mitigate risk and protect the CIA of an asset. Implementing encryption for data at rest or in transit, a strong password policy, and logging and monitoring techniques are just a few examples of security controls.

Enterprises can choose to implement hundreds of controls for a given information system or asset based on their risk appetite and risk thresholds. Many organizations develop their own definitions of control categories and classes; however, in this section, you will use the most common industry definitions and categorizations of controls.

You will also learn fundamental details about information security controls, including control categories, also referred to as **classes**, control functionalities, and types of controls.

Security Control Categories/Classes

Controls can be grouped into classes or categories, the names of which vary depending on the framework or standard being used. The three most common, established enterprise/industry categories of controls are as follows:

- Administrative
- Technical
- Physical

It is worth mentioning that the US' **National Institute of Standards and Technology (NIST)** organizes controls into the following three primary classes:

- Management
- Operational
- Technical

The most common enterprise control classes (administrative, technical, and physical) do not map one-to-one to the NIST control classes (management, operational, and technical), except for the technical category. Some controls, such as risk management, are administrative and management controls. Other controls, such as employee termination, are administrative and operational controls.

Different organizations use different terminology for controls. Here is a description of the terms used most often that you can expect to see on your CCSP exam:

- **Administrative controls**: These controls are a group of management-oriented controls such as policies, procedures, guidelines, risk management procedures, and human resource management practices (hiring, performance management, and employment termination, for example). Administrative controls are also referred to as **soft controls** or **managerial controls**.

- **Technical controls**: These are a group of hardware or software components that provide security control functionality to protect the CIA of an asset. Some examples of technical controls are encryption mechanisms, authentication and authorization techniques (such as MFA and **Role-Based Access Control (RBAC)**), **Intrusion Detection Systems (IDSs)** and **Intrusion Prevention Systems (IPSs)**, and network firewalls. Technical controls are also referred to as logical controls.

- **Physical controls**: These are tangible controls put in place to protect people, assets, and facilities against physical threats. This category of controls includes walls, fences, gates, doors, physical locks, bollards, intrusion and fire alarms, and security guards.

As mentioned, NIST defines a different set of classes. NIST special publication 800-53 Rev. 5 (Security and Privacy Controls for Information Systems and Organizations) lists the following classes of controls:

- **Management**: These are the security controls (i.e., safeguards or countermeasures) for an information system that focus on the management of risk and the management of information system security.

- **Operational**: These are security controls for an information system that are primarily implemented and executed by people (as opposed to systems).

- **Technical**: They are security controls for an information system that are primarily implemented and executed by the information system through mechanisms contained in the hardware, software, or firmware components of the system.

You will next look at some of the important elements of security control types and how they assist in securing cloud infrastructure.

Security Control Types and Functionality

The classes/categories of enterprise controls—administrative, technical, and physical group controls—are based on the framework or standard being used but they do not address how control operates and functions. Security controls can be broken down further based on their functionality and the intended objectives. The functionality of controls describes what the controls allow organizations to achieve to secure an asset.

The following is a list of control types based on their functionality:

- **Directive controls**: This type of control typically belongs to the administrative class, and the objective is to communicate expected behavior by specifying what actions are permitted or prohibited in the organization. Security policies, along with standards and procedures, are examples of directive controls.

- **Deterrent controls**: The goal of this type of control is to discourage or deter a potential bad actor from carrying out an attack or engaging in unwanted behavior. Examples include physical posters, warning signs, and IT system warning banners and warning messages displayed on the software user interface.

- **Preventative controls**: The objective of preventative (or preventive) controls is to prevent an adverse event or incident from happening. Examples include mandatory background checks for employees and contractors in the administrative class of controls; door locks, fences, and bollards in the physical class of controls; and IPSs and firewall **Access Control Lists** (**ACLs**) in the technical class of controls.

- **Detective controls**: These controls, as indicated by their name, are designed and implemented to discover, detect, and/or identify a potential intrusion, bad actor, security event, or incident. An example of detective controls in cloud environments is the use of IDSs or **security information and event management** (**SIEM**) systems. These systems monitor network traffic and system activities for suspicious behavior, logging potential security threats and alerting administrators to anomalous or unauthorized activities. This is done by reviewing security logs obtained from IDS. Another example is mandatory vacations of at least 10 consecutive working days for employees of many financial institutions. The absence of an employee for two weeks requires another employee to conduct their tasks and potentially uncover illicit activities conducted in the past.

- **Corrective controls**: The main functionality of this type of control is to correct adverse events that have occurred by fixing an IT system, a process, or an activity. An example of a corrective control in a cloud environment is the implementation of automated incident response systems. These systems are designed to respond to security breaches or policy violations by taking predefined action to mitigate damage. For instance, if unauthorized access to a sensitive cloud resource is detected, an automated incident response system can immediately revoke access permissions or shut down affected systems to prevent further compromise.

- **Recovery controls**: The goal of recovery types of controls is to recover from an incident and restore the environment or operations to their original functionality. Many similarities exist between corrective and recovery controls; however, recovery controls typically have an advanced level of capability. Examples of recovery controls include the use of backups or snapshots to restore an IT system to its most recent version, rolling back installation of the software patches to restore the application to its previous version, and utilizing an IT system performance management tool to detect a corrupt process and restart it to restore a desired level of performance or functionality.

- **Compensating controls**: Compensating controls are security measures implemented to offset deficiencies in primary controls, serving as alternative or secondary safeguards. These controls are often adopted when primary measures are deemed infeasible due to high costs, complexity, or the challenges associated with implementation in a cloud setting. For example, if updating or patching a cloud-based legacy application is not possible due to compatibility issues, a compensating control might involve segmenting this application into a separate, secure cloud environment to isolate it from other systems.

It is worth noting that a specific control can be associated with more than one control functionality. For example, a CCTV camera installed near a perimeter fence could be considered not only a detective control but also a preventive and deterrent control at the same time.

> **Note**
>
> During your CCSP exam, the correct answer to the question about controls will often depend on the context of the question and how the specific scenario is being presented.

For your CCSP exam, remember that the selection of security controls starts with the knowledge of each asset's importance to the business. Organizations can then use a tailored approach to determine, design, and implement a set of controls to secure a given asset. After implementing the security controls, organizations need to test them, and then monitor the operational effectiveness of controls and changes to the asset's importance to the business to achieve ongoing compliance throughout the asset's life cycle.

> **Note**
>
> Here are links to some additional information and study references for cloud threats:
>
> **Top Threats**: `https://cloudsecurityalliance.org/research/topics/top-threats/`
>
> **11 Top Cloud Security Threats**: `https://www.csoonline.com/article/3043030/top-cloud-security-threats.html`
>
> **NIST Security and Privacy Controls for Information Systems and Organizations – SP 800-53**: `https://csrc.nist.gov/publications/detail/sp/800-53/rev-5/final`

Summary

In this chapter, you learned about some fundamental security concepts that are relevant to securing confidentiality, integrity, and availability of cloud resources. You also reviewed the five most common threats to cloud deployments that many cloud service providers and cloud customers face every day. This chapter also dealt with the most common control categories, types, and control functionalities many organizations use to secure information assets. In the next chapter, you will focus on service model security considerations for **Infrastructure as a Service (IaaS)**, **Platform as a Service (PaaS)**, and **Security as a Service (SaaS)**.

Exam Readiness Drill – Chapter Review Questions

Apart from a solid understanding of key concepts, being able to think quickly under time pressure is a skill that will help you ace your certification exam. That is why working on these skills early on in your learning journey is key.

Chapter review questions are designed to improve your test-taking skills progressively with each chapter you learn and review your understanding of key concepts in the chapter at the same time. You'll find these at the end of each chapter.

> **How to Access These Materials**
>
> To learn how to access these resources, head over to the chapter titled *Chapter 25, Accessing the Online Resources*.

To open the Chapter Review Questions for this chapter, perform the following steps:

1. Click the link – `https://packt.link/CCSPE1_CH03`.

   Alternatively, you can scan the following **QR code** (*Figure 3.2*):

Figure 3.2 – QR code that opens Chapter Review Questions for logged-in users

2. Once you log in, you'll see a page similar to the one shown in *Figure 3.3*:

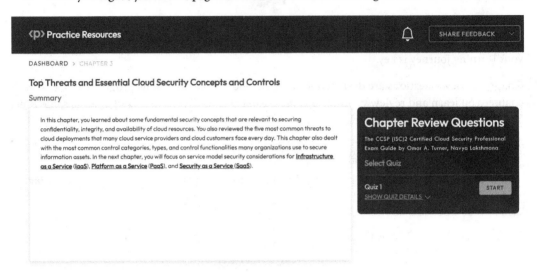

Figure 3.3 – Chapter Review Questions for Chapter 3

3. Once ready, start the following practice drills, re-attempting the quiz multiple times.

Exam Readiness Drill

For the first three attempts, don't worry about the time limit.

ATTEMPT 1

The first time, aim for at least **40%**. Look at the answers you got wrong and read the relevant sections in the chapter again to fix your learning gaps.

ATTEMPT 2

The second time, aim for at least **60%**. Look at the answers you got wrong and read the relevant sections in the chapter again to fix any remaining learning gaps.

ATTEMPT 3

The third time, aim for at least **75%**. Once you score 75% or more, you start working on your timing.

> **Tip**
>
> You may take more than **three** attempts to reach 75%. That's okay. Just review the relevant sections in the chapter till you get there.

Working On Timing

Target: Your aim is to keep the score the same while trying to answer these questions as quickly as possible. Here's an example of how your next attempts should look like:

Attempt	Score	Time Taken
Attempt 5	77%	21 mins 30 seconds
Attempt 6	78%	18 mins 34 seconds
Attempt 7	76%	14 mins 44 seconds

Table 3.1 – Sample timing practice drills on the online platform

> **Note**
>
> The time limits shown in the above table are just examples. Set your own time limits with each attempt based on the time limit of the quiz on the website.

With each new attempt, your score should stay above **75%** while your "time taken" to complete should "decrease". Repeat as many attempts as you want till you feel confident dealing with the time pressure.

4

Design Principles for Secure Cloud Computing

In the previous chapter, you studied the common threats to cloud deployments and attack vectors and also introduced the control frameworks and control types necessary to secure data, network, and virtualization layers for cloud computing. This chapter covers the fundamental principles for secure cloud computing design, an essential subject for CSSP. To fully comprehend these challenges and best practices in cloud security, you must become well versed in the realities of the shared responsibility model, data protection measures, and robust access control techniques, just to name a few. With this solid foundation in place, organizations can confidently utilize their resources on the cloud with security in mind.

If an organization is using the capabilities of IaaS, PaaS, or SaaS for its digital transformation journey, it must be concerned and informed of the security risks and threats associated with each. By doing so, they can increase their assurance that their infrastructure and applications adhere to governance standards, are secure, and can take advantage of cloud scalability and flexibility. This chapter offers an overview of the security implications that must be considered when adopting any cloud service paradigm.

By the end of this chapter, you will be able to confidently answer questions on the following topics:

- Security considerations for IaaS
- Security considerations for PaaS
- Security considerations for SaaS
- Shared responsibility model for different cloud service models

In *Chapter 2, Cloud Reference Architecture*, you learned about **Infrastructure as a Service (IaaS)**, **Platform as a Service (PaaS)**, and **Software as a Service (SaaS)**. Each provides different levels of control, flexibility, and responsibility to cloud consumers. Here's a brief review of these models before you delve deeper:

- **IaaS**: This service allows organizations to consume virtual computer resources such as storage, networking, and virtual machines from a **Cloud Service Provider (CSP)**. The provider manages the physical infrastructure while your organization takes responsibility for operating systems, middleware, applications, and data. AWS, Azure, and GCP are some examples of IaaS providers.

- **PaaS**: This service is ideal for developers who need to construct and manage applications quickly and efficiently. The provider takes care of the operating system, middleware, and runtime environment, while your organization focuses on application development and maintenance. Examples of PaaS providers include Microsoft's Azure SQL Database or AWS's **Relational Database Service (RDS)**.

- **SaaS**: With this service, organizations can access entire application suites through the internet. The provider is responsible for infrastructure to software development and updates, so users don't need to install or update anything themselves. Examples of SaaS providers include Salesforce, Microsoft Office 365, and Google Workspace.

This model framework is important for organizations to get right and essential for you, as a CCSP candidate, to know the nuances of each framework. Take an example: since IaaS is a cloud computing framework that lets you consume computing resources such as storage, networking, memory, and servers using a dedicated connection such as AWS Direct Connect or Azure Express Route, this eliminates the need to manage your own physical infrastructure. In other words, the provider takes care of any physical issues, such as data centers or hardware, while you manage only virtual ones, such as operating systems and software.

As an organization that uses IaaS resources, you will have access to computing resources without worrying about maintaining physical infrastructure. You have complete control over scaling up or down your resources without investing in equipment, facilities, or personnel. Plus, you get to choose the type and quantity of infrastructure, including compute, storage, memory, and networking—all from one place.

You are accountable for safeguarding all information above the physical infrastructure, such as operating systems, applications, and internal networks.

But how do you do that exactly?

The next section will delve into the security considerations for IaaS, SaaS, and PaaS. Think about the examples and scenarios that you face in your current role (if you are a cloud architect for example).

Security for IaaS, SaaS, and PaaS

Security risks can negatively impact the data and applications on your platform. These dangers put confidential information, integrity, and accessibility at risk which could result in financial losses, legal troubles, or damage to your reputation.

Security Considerations for Infrastructure as a Service (IaaS)

IaaS is like having your own virtual computer and storage space online to consume. You do not need to purchase or manage hardware such as servers or data centers as the cloud provider bears that responsibility. With IaaS, all the underlying infrastructure is taken care of for you so you can focus on developing on top of that infrastructure.

IaaS uses various deployment strategies that are determined by location, control, and vendor engagement. Different IaaS models require different deployment strategies that depend on several factors, such as location, control, and vendor engagement. No matter the model used, it's essential to really know your responsibilities and apply necessary security and compliance measures.

While IaaS providers provide a secure environment, it's still essential to be aware of potential weaknesses or security risks. The following are the security risks:

- **Shared responsibility model**: Security responsibilities are split between the cloud provider and the customer. While the provider is accountable for safeguarding the fundamental infrastructure, customers are accountable for safeguarding operating systems, middleware, and applications. Failure to recognize or address the appropriate responsibilities may lead to security gaps.

- **Misconfigurations**: Misconfigured virtual machines, storage, and networking components can leave systems vulnerable to security threats. Common misconfigurations include overly permissive access controls, open ports, and unpatched software.

- **Data breach risks**: Storing sensitive data in the cloud increases the potential risk of data breaches due to weak access controls, lack of encryption, or misconfigurations that allow unauthorized access.

- **Unauthorized access**: Weak authentication mechanisms or improper management of access credentials can allow unauthorized entry to critical infrastructure components such as management consoles and APIs.

- **Compliance challenges**: Maintaining compliance with industry regulations and standards can be more complex in IaaS environments. Organizations should know their obligations under these laws and put in place suitable controls within their cloud infrastructure to ensure these outcomes.

- **Resource hijacking**: Attackers can exploit vulnerabilities in IaaS environments to gain unauthorized access to resources, such as compute power, and use them for malicious activities, such as crypto-mining or launching attacks against other targets.

By being aware of these potential hazards, you can take necessary precautions and implement measures to safeguard your IaaS environment effectively. It is essential to set up your virtual machines, storage, and networking correctly to avoid security risks. Conduct regular checks of these settings to guarantee you're adhering to industry best practices.

If you're storing sensitive data in the cloud, you need strong access controls, encryption, and data loss prevention techniques. Encrypt data both during transmission and at rest. Make sure your identity management system is functioning properly as well. Furthermore, consider preventing unauthorized access to administration consoles or APIs by implementing **Multi-Factor Authentication (MFA)** and following the principle of least privilege.

To ensure your IaaS environment is secure, adhere to the following best practices:

- **Know the shared responsibility model**: Be clear on what tasks you and the cloud provider are responsible for. Have it all written down so there can be no confusion later.

- **Implement robust access controls**: Utilize an **Identity and Access Management (IAM)** solution and adhere to the principle of least privilege.

- **Encrypt data at rest and in transit**: Make your sensitive information safe with encryption. Utilize standard encryption algorithms and key management practices to guarantee that your information remains confidential and secure.

- **Regularly monitor and audit**: Utilize monitoring and logging solutions to stay abreast of security incidents. Review your cloud environment, settings, and access controls regularly for best results.

- **Secure network configurations**: Implement firewalls, dedicated connections such as AWS Direct Connect, and network segmentation to keep your cloud network secure. Additionally, restrict public internet access where necessary for additional protection.

- **Maintain security patches and upgrades**: Stay current with the latest security patches, and establish a systematic process for regularly conducting vulnerability assessments using automated tools.

- **Prepare for disaster recovery and business continuity**: Create a strategy to cope with disasters and unexpected events. Regularly assess and update your plan to guarantee its efficacy remains high.

Core Elements of Security for IaaS

To solve the IaaS security-related questions in the CCSP exam, you should be well versed in the following competencies:

- Recognize and classify IaaS cloud service models such as public, private, hybrid, and community.

- Recognize the distinctions between cloud deployment models and their security implications.

- Be acquainted with the shared responsibility model in IaaS, including the roles of CSPs and customers.

- Be knowledgeable about common IaaS providers such as AWS, Azure, and Google Cloud Platform.

- Implement secure access controls, such as IAM.

- Encrypt IaaS data storage with encryption both at rest and during transit.

- Gain insight into network security concepts such as VPNs, firewalls, IDS/IPSs, and security groups.

- Implement secure network architecture and segmentation within IaaS environments.

- Monitor and audit IaaS environments to ensure security and compliance.

- Guarantee data security and privacy by adhering to regulations such as GDPR, HIPAA, and PCI-DSS.

- Conduct vulnerability assessments as per the CSP's guidelines.

- Develop incident response and disaster recovery plans tailored specifically for IaaS.

- Recognize the role of containerization and orchestration technologies in enhancing IaaS security.

- Be familiar with DevSecOps principles and best practices for secure development and deployment in IaaS environments.

- Evaluate and select suitable security controls and tools tailored toward IaaS environments.

This section focused on the main security considerations that you must be aware of while implementing IaaS services. Now let's look at what's needed from a security perspective for PaaS.

Security Considerations for Platform as a Service (PaaS)

PaaS is like having your own virtual toolbox in the cloud that developers can use to develop apps without worrying about all technical details. It comes equipped with pre-made tools and services, so developers can focus on coding features while the cloud provider takes care of maintaining servers, networks, and databases, thus providing an effortless way to develop and maintain apps.

PaaS offerings differ depending on the development and deployment environments they support, or the services they offer. Here are some popular types of PaaS solutions:

- **General purpose**: These platforms provide a comprehensive set of tools and services to make it effortless to design, deploy, and maintain various types of applications. Examples include Google App Engine, Microsoft Azure App Service, and IBM Cloud Foundry.

- **Integration PaaS**: The **Integration PaaS (iPaaS)** platforms facilitate seamless integration of applications, data, and systems across various environments such as on-premises, cloud, or hybrid. Popular examples include MuleSoft Anypoint Platform, Dell Boomi, and Informatica Cloud.

- **Communications PaaS**: The **Communications PaaS (cPaaS)** platforms enable developers to incorporate real-time communication features, such as audio, video, and messaging, into their applications. Examples include Twilio, Vonage (formerly Nexmo), and Plivo.

- **Business Process Management PaaS**: The **Business Process Management PaaS (BPM PaaS)** platforms offer tools and services for creating, implementing, and managing business processes, workflows, and rules within applications. Popular options include Appian, Pega Platform, and Oracle Process Cloud Service.

- **Data Analytics PaaS**: The **Data Analytics PaaS (dPaaS)** platforms offer tools and services for data storage, processing, and analysis, as well as machine learning and artificial intelligence capabilities. Examples include Google Cloud Dataflow, Databricks, and Snowflake.

- **Internet of Things Platform as a Service**: The **Internet of Things Platform as a Service (IoT PaaS)** platforms make the development and deployment of IoT applications much simpler by providing tools and services for managing devices, processing data, and performing analytics. Examples include Amazon IoT Core and Microsoft Azure IoT Suite.

- **Function PaaS or Serverless PaaS**: The **Function PaaS (fPaaS)** platforms enable developers to run code or functions without worrying about the underlying infrastructure. They scale automatically based on workload, with customers charged according to actual consumption. Examples include Amazon Lambda, Microsoft Azure Functions, and Google Cloud Functions.

When selecting a provider for your organization's cloud resources, it's essential to know its individual security requirements and objectives. With so many choices, selecting an appropriate PaaS solution shouldn't be a difficult decision; it all comes down to what makes sense for your organization's overall strategy.

The following are essential concepts that will be critical as you utilize PaaS:

- Be familiar with the shared responsibility model, which outlines the roles and responsibilities of both your organization and PaaS provider. Your organization must secure applications and data while the provider takes care of managing the underlying infrastructure and platform.

- Enhance the security of your application development process by employing secure development practices to reduce vulnerabilities in applications. This includes code reviews, **Static Application Security Testing (SAST)** and **Dynamic Application Security Testing (DAST)**, and adhering to secure coding standards such as the **Open Worldwide Application Security Project (OWASP)** Top 10.

- Control application access by implementing strong IAM technologies to restrict who has access to your applications. Ensure that only authorized personnel have access, adhering to the principle of least privilege.

- Protect data security and privacy by encrypting sensitive information stored in databases, as well as transmission between applications and users. Comply with data protection standards such as GDPR or HIPAA, and follow best practices for data privacy and security.

- Monitor and log your PaaS environment regularly for security events and unusual activity. Utilize logging tools to monitor user actions and application performance, then review these logs regularly for security vulnerabilities.

- Implement robust API security by using authentication and authorization systems such as OAuth or OpenID Connect. Regularly audit and test your APIs for vulnerabilities and adhere to recommended security practices.

- Protect third-party components such as apps and plugins. If your applications rely on third-party libraries, plugins, or frameworks, make sure these elements are up to date and secure. Regularly assess third-party components for security flaws and update them if necessary.

- Adhere to industry-specific regulations and standards applicable to your PaaS environment. Implement necessary controls and conduct regular audits to guarantee ongoing observance. Ensure that your **Service-Level Agreements** (**SLAs**) include language that requires cloud vendors(s) to adhere to your governance policies.

- Create a comprehensive incident response plan that addresses potential PaaS security risks. Verify and upgrade the plan periodically to guarantee its efficacy in case of an actual security incident.

Core Elements of Security for PaaS

Before answering questions related to PaaS concepts in the exam, you should be familiar with the following topics:

- Know cloud computing concepts and PaaS architecture.

- Recognize and categorize PaaS cloud service models such as public, private, hybrid, and community in order to be successful.

- Be familiar with the various cloud deployment models and their security implications.

- Learn about the PaaS shared responsibility model, including the roles of CSPs and customers.

- Know about popular PaaS providers such as AWS, Azure, Google Cloud Platform, and Heroku.

- Implement secure access controls, such as IAM, for PaaS environments.

- Protect application data storage in PaaS applications by using encryption both at rest and during transit.

- Gain insight into network security concepts relevant to PaaS such as VPNs, firewalls, and security groups.

- Create secure network architecture and isolation within PaaS environments.

- Monitor and audit PaaS environments for security risks and compliance violations.

- Guarantee data protection and privacy by adhering to regulations such as GDPR, HIPAA, and PCI-DSS.

- Conduct vulnerability assessments and penetration testing on PaaS applications and platforms.
- Know how to craft incident response and disaster recovery plans tailored specifically for PaaS users.
- Be familiar with containerization, orchestration, and serverless technologies in PaaS security measures.
- Learn about the DevSecOps principles and best practices for secure development and deployment within PaaS environments.
- Evaluate and select appropriate security controls and tools suitable for PaaS environments.

By mastering these skills and concepts, you'll be well prepared for questions that would probably confuse you if you were not prepared. The next section will help you learn about the risks associated with SaaS environments and their security considerations.

Security Considerations for Software as a Service (SaaS)

Software as a Service (SaaS) is a cloud-based application delivery model where the software is hosted and managed by a third-party provider. Users do not need to purchase, install, or manage the software locally; instead, they can access the application directly through a web browser. The SaaS provider is responsible for all the underlying infrastructure, maintenance, updates, and bug fixes. This arrangement allows users to focus solely on using the software to achieve their business objectives without worrying about technical management. SaaS is one of the most convenient ways to deploy software, offering easy scalability, accessibility, and minimal IT overhead.

SaaS environments have the following risks associated with them:

- **Shared responsibility model**: Both cloud providers and customers have distinct security responsibilities. It is essential to comprehend these distinctions so breaches or other security issues do not take place.
- **Unauthorized access**: Inadequate access controls or weak authentication processes might allow unauthorized individuals to view sensitive data stored in SaaS applications.
- **Data breaches**: Data breaches can occur due to weak access controls, lack of encryption, misconfigurations, or vulnerabilities in the application.
- **Insider threats**: Malicious insiders with access to your SaaS environment pose a significant security risk.
- **Account hijacking**: Attackers may compromise user accounts and gain unauthorized access to SaaS applications through phishing, social engineering, or credential theft techniques.
- **Data loss and leakage**: In SaaS environments, there is an increased risk of data loss or leakage since data is often stored and processed remotely.

- **Compliance challenges**: Adhering to relevant industry regulations and standards can be more complex in SaaS environments. It's essential that you comprehend your obligations under applicable legislation, and then implement necessary controls within your applications.

- **Insecure APIs**: Cloud providers often provide APIs to integrate and manage SaaS applications. However, insecure APIs or vulnerabilities in API implementations can expose an application to security risks.

- **Service availability**: Unexpected issues such as technical difficulties or DDoS attacks can disrupt service availability and disrupt business operations.

Core Elements of Security for SaaS

When considering security for SaaS solutions, there are some essential elements to consider, so please review the following competencies:

- Recognize and categorize SaaS cloud service models such as public, private, hybrid, and community.

- Recognize the distinctions between cloud deployment models and their security implications.

- Acquaint yourself with the shared responsibility model in SaaS, including the roles of CSPs and customers.

- Be familiar with common SaaS providers and applications such as Salesforce, Microsoft 365, and Google Workspace.

- Enforce secure access controls such as IAM for SaaS environments.

- Protect SaaS application data storage by encrypting it both at rest and during transit.

- Be familiar with network security concepts applicable to SaaS, such as secure web gateways and CASB solutions.

- Construct secure network architecture and isolation within SaaS environments.

- Monitor and audit SaaS environments for security risks and compliance obligations.

- Assure data security and privacy by adhering to regulations such as GDPR, HIPAA, and PCI-DSS.

- Conduct vulnerability assessments and penetration testing on SaaS applications.

- Create incident response and disaster recovery plans tailored specifically for SaaS operations.

- Be familiar with APIs and integrations as they relate to SaaS security.

- Learn best practices for safeguarding user access, such as strong authentication and authorization methods.

- Assess and select suitable security controls and tools appropriate for SaaS environments.

You have covered a fair bit of information so far, as you have read, there are many components of the various cloud deployment models that you need to understand. You have also touched on terms such as APIs, DevSecOps, and GDPR. If you are not familiar with them don't worry, you will cover them in more detail in subsequent chapters. For now, just know that each cloud deployment model is dependent on a variety of critical processes that, when working correctly, can help ensure a secure and reliable architecture. Let's now look at an overview of the shared responsibility model.

Shared Responsibility Model for Cloud Service Models

Overall, cloud security can be daunting, but the shared responsibility model provides some clarity. It outlines who is accountable for what when it comes to security, reliability, and resiliency in IaaS, PaaS, and SaaS cloud services. *Table 4.1* shows the responsibility in each model:

Iaas	Paas	SaaS
With IaaS, the cloud provider takes care of securing data centers, networks, and hardware.	In the PaaS model, the cloud provider secures the platform—from operating systems to middleware and runtime components—on your behalf.	With SaaS, the cloud provider takes care of everything—infrastructure, platform, and apps.
You are ultimately responsible for safeguarding your operating systems, applications, and data; maintaining software updates; managing your network settings; and controlling who has access.	You remain accountable for safeguarding your apps and data. It is your responsibility to set up appropriate security measures in your applications, manage access rights, and ensure data is sent securely.	The CSP keeps the app secure and up to date; however, you must take control of your data and access. It is up to you to create strong identity management policies, protect your information, and remain abreast of industry standards and regulations.

Table 4.1: Accountability in IaaS, PaaS, and SaaS

When it comes to the cloud, you need to ensure your users are secure by providing security training on potential risks such as phishing or data leaks. Unfortunately, there are a lot of misconceptions about the shared responsibility model which may cause misunderstandings and create security gaps.

It is time to debunk some of the misconceptions about the shared responsibility model. They are as follows:

- **Cloud providers are responsible**: Cloud providers have their own security obligations, but customers also have their own responsibilities depending on which cloud service they utilize (IaaS, PaaS, or SaaS).

- **The CSP is a leader in the cloud space**: Customers must manage user access, safeguard their data, and ensure the security of their applications, among other responsibilities, regardless of who the CSP is.

- **Compliance is the cloud provider's burden**: Customers must also guarantee their use of cloud services complies with regulatory requirements and includes the implementation of essential controls.

- **Data security is the cloud provider's responsibility**: Customers **MUST** protect their data using encryption, access controls, and other security practices.

- **Security responsibilities remain constant**: The shared responsibility model differs based on which cloud service (IaaS, PaaS, or SaaS) is being utilized. This is especially true as each provider releases a myriad of features on a semi-frequent basis.

- **Customers don't need to worry about user behavior**: Customers must educate their users on security best practices and offer training to reduce risks related to user actions.

By comprehending the shared responsibility model and dispelling these myths, organizations can take proactive measures to safeguard their cloud environments.

While preparing for CSSP, keep in mind that security is paramount with cloud services. While they offer many advantages, such as flexibility and scalability, cloud technologies also come with potential security risks that could compromise the confidentiality, integrity, and availability of your data and applications.

Therefore, it's essential to implement strong security measures such as access management, data encryption, network security, physical security, and incident response and recovery when using cloud services. By taking a proactive approach to security, you can help minimize the likelihood of cyber-attacks, data breaches, or other security risks that could negatively affect your business operations and reputation.

Additionally, knowing the shared responsibility model for cloud services is paramount. While the CSP is responsible for infrastructure security, you have a role in data and application security.

Review of Your Responsibilities

To succeed on the CCSP exam's shared responsibility model questions, you must be familiar with the following skills and concepts:

- Recognize the significance of this concept within cloud security.

- Acquaint yourself with the differences between IaaS, PaaS, and SaaS models and how responsibilities are shared between CSPs and customers.

- For IaaS:

  - CSPs are responsible for safeguarding the underlying infrastructure, such as physical data centers, networking equipment, and virtualization layers.

  - You must take responsibility for protecting your own virtual machines, applications, data, and network configurations.

- For PaaS:

  - CSPs manage security for the underlying infrastructure and platform, such as operating systems, runtime environments, and middleware.

  - You are accountable for protecting your own applications, data, and custom configurations on the platform.

- For SaaS:

  - CSPs handle security for the underlying infrastructure and platform including operating systems, runtime environments, and middleware.

  - CSPs are responsible for securing the entire software stack, including infrastructure, platform, and applications.

  - You are accountable for managing access control and permissions as well as safeguarding your own data within a SaaS application.

  - You must acquire the knowledge of CSP security capabilities and the division of responsibilities in SLAs.

  - You must familiarize yourself with tools and services provided by CSPs to assist you in fulfilling your responsibilities.

  - You must be able to implement security best practices and controls in each model for a secure cloud environment.

- Recognize how to monitor and audit the shared responsibility model in order to guarantee security and compliance.

- Recognize the significance of communication and collaboration between CSPs and customers in a shared responsibility model.

By mastering these skills and concepts, you'll be well prepared for the shared responsibility model portion of the CCSP exam regarding IaaS, PaaS, and SaaS domain topics.

Summary

For the CCSP exam, it is essential that you comprehend cloud security concepts such as IaaS, PaaS, and SaaS. Each service model has distinct security challenges and requirements that must be learned in order for you to succeed.

When it comes to IaaS, PaaS, and SaaS, you and the cloud provider share responsibility for security. With IaaS, you must protect your applications, data, and operating systems on virtual machines while the provider takes care of the physical security of the infrastructure. With PaaS, the provider will take care of platform and middleware security so all that remains for you to focus on is keeping apps and data safe. While SaaS providers do most of the heavy lifting when it comes to security and access control, data protection always remains your responsibility.

The following is a summary of the *Core Elements of Security for IaaS*, *Core Elements of Security for PaaS*, *Core Elements of Security for SaaS* sections:

- Recognize and classify XaaS cloud service models such as public, private, hybrid, and community.

- Recognize the distinctions between cloud deployment models and their security implications.

- Acquaint yourself with the shared responsibility model in XaaS, including the roles of CSP and customers.

- Enforce secure access controls such as IAM.

- Encrypt data storage with encryption both at rest and during transit.

- Implement secure network architecture and segmentation within XaaS environments.

- Monitor and audit XaaS environments to ensure security and compliance.

- Guarantee data security and privacy by adhering to regulations such as GDPR, HIPAA, and PCI-DSS.

- Conduct vulnerability assessments and penetration testing in IaaS environments.

- Develop incident response and disaster recovery plans tailored specifically for IaaS.

- Evaluate and select suitable security controls and tools tailored towards XaaS environments.

By mastering these topics, you'll be prepared to design, set up, and manage secure cloud environments like an expert. Becoming familiar with IaaS, PaaS, and SaaS security protocols is essential for passing the exam and ensuring your company's cloud resources remain safe and secure.

Designing a secure cloud computing environment necessitates taking a proactive and ongoing approach to security. Organizations must adhere to the shared responsibility model, acknowledging their individual security responsibilities in the cloud. They must implement strong access controls, encryption, and data loss prevention techniques to safeguard their data and applications. Regular monitoring and auditing of the cloud environment are key for detecting potential security risks early on and taking appropriate measures to address them. Additionally, organizations must abide by relevant industry regulations and standards, as well as have an established incident response plan for handling security incidents. By adhering to these design principles for secure cloud computing, companies can reduce the risks associated with cloud computing while ensuring the confidentiality, integrity, and availability of their data and applications.

In the next chapter, you will learn about the key cloud service contractual documents from the perspective of cloud service consumers. You will also learn about the best practices of how to evaluate CSPs based on a set of criteria.

Exam Readiness Drill – Chapter Review Questions

Apart from a solid understanding of key concepts, being able to think quickly under time pressure is a skill that will help you ace your certification exam. That is why working on these skills early on in your learning journey is key.

Chapter review questions are designed to improve your test-taking skills progressively with each chapter you learn and review your understanding of key concepts in the chapter at the same time. You'll find these at the end of each chapter.

How to Access These Materials

To learn how to access these resources, head over to the chapter titled *Chapter 25, Accessing the Online Resources*.

To open the Chapter Review Questions for this chapter, perform the following steps:

1. Click the link – `https://packt.link/CCSPE1_CH04`.

    Alternatively, you can scan the following **QR code** (*Figure 4.1*):

Figure 4.1 – QR code that opens Chapter Review Questions for logged-in users

2. Once you log in, you'll see a page similar to the one shown in *Figure 4.2*:

Figure 4.2 – Chapter Review Questions for Chapter 4

3. Once ready, start the following practice drills, re-attempting the quiz multiple times.

Exam Readiness Drill

For the first three attempts, don't worry about the time limit.

ATTEMPT 1

The first time, aim for at least **40%**. Look at the answers you got wrong and read the relevant sections in the chapter again to fix your learning gaps.

ATTEMPT 2

The second time, aim for at least **60%**. Look at the answers you got wrong and read the relevant sections in the chapter again to fix any remaining learning gaps.

ATTEMPT 3

The third time, aim for at least **75%**. Once you score 75% or more, you start working on your timing.

> **Tip**
>
> You may take more than **three** attempts to reach 75%. That's okay. Just review the relevant sections in the chapter till you get there.

Working On Timing

Target: Your aim is to keep the score the same while trying to answer these questions as quickly as possible. Here's an example of how your next attempts should look like:

Attempt	Score	Time Taken
Attempt 5	77%	21 mins 30 seconds
Attempt 6	78%	18 mins 34 seconds
Attempt 7	76%	14 mins 44 seconds

Table 4.2 – Sample timing practice drills on the online platform

> **Note**
>
> The time limits shown in the above table are just examples. Set your own time limits with each attempt based on the time limit of the quiz on the website.

With each new attempt, your score should stay above **75%** while your "time taken" to complete should "decrease". Repeat as many attempts as you want till you feel confident dealing with the time pressure.

5

How to Evaluate Your Cloud Service Provider

In the previous chapter, you reviewed the important security considerations for IaaS, PaaS, and SaaS cloud models. In this chapter, you will learn about key cloud service contractual documents from the perspective of cloud service consumers. You will also learn about the best practices for evaluating **Cloud Service Providers (CSPs)** based on a set of criteria. This is another important topic that both CCSP candidates and cloud architects should be familiar with.

By the end of this chapter, you will be able to confidently answer questions on the following topics:

- Key cloud service contractual documents
- **Cloud Service Agreement (CSA)**
- **Acceptable Use Policy (AUP)**
- **Service-Level Agreement (SLA)**
- Evaluation of the CSP services

Regardless of their size, businesses are increasingly migrating to the cloud. Cloud computing provides economies of scale, agility, cost savings, and other competitive advantages that give small businesses an upper hand in today's fast-paced business environment. Medium and large businesses use cloud computing to remain competitive by accelerating their digital transformation. Additionally, CSPs provide the infrastructure and services necessary to create, store, manage, and analyze vast amounts of data in real time.

To evaluate a CSP, you will first need to review the key cloud service contractual documents from that perspective. Read on to know how.

Key Cloud Service Contractual Documents

Due to the important and critical role that the cloud has for organizations, you must properly evaluate your CSP with a high level of due diligence. The right CSP can provide flexible and scalable solutions for businesses of all sizes; the wrong one could cause headaches and potential security risks. To evaluate a CSP effectively, there are several factors to consider. They are as follows:

- Firstly, you should know your business needs and ensure that the CSP can meet them
- After finding the right CSP that meets your needs, you need to review the contractual documents that make up the CSP's agreement
- You must review the CSA, AUP, and SLA because these are instrumental in selecting which CSP you will ultimately partner with
- You should not consider cost as the only determining factor here

With these legal documents, the cloud provider and the customer set expectations for service quality, security, and usage, defining their respective responsibilities. By knowing the importance of these agreements, organizations can ensure they choose a CSP aligned with their business needs and make informed decisions. Now, let's look at each of these in a little more detail.

CSA

Initially, SLAs were all that were required when signing a cloud service contract. However, as the complexity of cloud services has increased, SLAs have largely been replaced by CSAs. A CSA is a detailed contract that includes an SLA, an AUP, and a customer agreement. All these documents are designed to ensure that the CSP meets its obligations and that customers know their responsibilities when using a cloud service.

The Customer Agreement

The customer agreement highlights the various rights and obligations of both the customer and the CSP. This binding document accurately details the scope, limitations, and exclusions of liability. Before signing the customer agreement, customers should review it. It explains how the service will be used and what happens if a party fails to fulfill its obligations.

The following terms are included in the customer contract:

- **Services provided**: This section provides information about the cloud services that the CSP will provide to customers. You must read this section very carefully to learn what services are included/excluded and whether they meet your business needs.
- **Customer responsibility**: This section describes the responsibilities that a CSP has set for its customers when using its cloud service. Requirements such as compliance with laws and regulations, proper use of the service, and maintaining account security are covered in this section. Complying with the customer responsibilities listed in this section is of the utmost importance.

- **Term and termination**: In this section, the length of the agreement and the conditions for termination by either party are thoroughly explained. It contains details such as notice periods, reasons for termination, and any fees or penalties for early termination. Since contract terminations can be costly, taking the time to read and understand this section to ensure that you are comfortable with the terms before signing is imperative.

  Apart from the cost, your data protection and compliance needs should also factor into the evaluation. It has often been said that "the first and foremost thing that needs to be clear in a CSA is how the customer will egress their data out of the system." For example, it is always recommended to save a backup copy at an on-premise location to ensure data portability. It doesn't matter if they violate your contract or go bankrupt; there's not much you can do, so replicating your data nightly should be a core part of your disaster recovery and business continuity plan.

- **Confidentiality**: Here is where the requirements to maintain the confidentiality of information related to the cloud service are explained – things such as non-disclosure agreements and restrictions on using confidential information. In today's increasingly digital world, security and data privacy are essential to any business. You have all heard stories of companies losing sensitive data or facing legal liabilities due to inadequate security measures and privacy policies. By taking the time to read and understand a CSP's customer agreement, you can ensure that your company is better protected against the potential risks associated with using cloud services.

- **Limitations of liability**: This section outlines the limitations and exclusions of liability for the CSP and the customer. It includes details such as liability caps and exclusions for certain types of damages.

Overall, the customer agreement is a critical component of a CSA. It outlines the terms and conditions for using the cloud service, explaining the customer's rights and responsibilities as well as those of the CSA. Before signing, cloud customers should review the agreement carefully and leverage legal advice if needed.

The CSA from a CSP Perspective

If you were to read the CSA of a CSP, you would most likely see a format similar to the following:

- **Introduction**: This includes an overview of the agreement, including the scope and purpose of the document.

- **Definitions**: This includes the explanation of the terms and concepts used throughout the agreement.

- **Service offerings**: This option includes a detailed description of the cloud services being provided, including any additional features or add-ons.

- **Service levels**: This is the information on the expected performance levels of the services, often referring to an SLA for specific metrics and commitments.

- **Customer responsibilities**: This outlines the customer's responsibilities, such as providing accurate information, complying with the AUP, and maintaining account security.

- **Provider responsibilities**: This option provides an outline of the provider's responsibilities, such as maintaining the infrastructure, providing customer support, and ensuring data privacy and security.

- **Data management and security**: This shows the high-level details on how the provider will handle customer data, including data storage, backup, and protection measures.

- **Billing and payment**: This provides information on pricing, billing cycles, and payment terms.

- **Intellectual property rights**: These are clauses related to the ownership and usage rights of the intellectual property involved in the services.

- **Confidentiality**: This is an agreement to maintain the confidentiality of sensitive information exchanged between the provider and the customer.

- **Warranties and disclaimers**: This is a description of any warranties provided by the provider, as well as disclaimers related to the services.

- **Limitation of liability**: These are clauses defining the extent of the provider's liability if there are service issues or damages.

- **Indemnification**: This is an agreement by the customer to indemnify the provider against certain claims or losses arising from the customer's use of the services.

- **Term and termination**: This is the information on the duration of the agreement and the procedures for terminating the contract, including termination for cause and termination for convenience.

- **Dispute resolution**: This is the outline of the process for resolving disputes between the provider and the customer, such as arbitration or mediation.

- **Governing law**: This is the statement of the jurisdiction and legal framework governing the agreement.

- **Miscellaneous**: These are any additional clauses or provisions, such as force majeure, assignment, or notices.

For the exam, become familiar with the verbiage and major sections of sample CSAs.

> **Note**
> You can most likely find an example of your favorite cloud provider's CSA online for you to review.

AUP

An **AUP** is a crucial document that tells you what is acceptable and unacceptable when using the cloud. There are rules for downloading files, sending messages, and respecting copyright agreements. It helps make sure that everyone is safe and follows the same rules.

> **Note**
>
> Don't be complacent about AUPs, as they constitute an important part of any cloud computing environment.

The Purpose of an AUP

An AUP is typically a short document that outlines the expectations of the service provider regarding customer behavior when using their services. An AUP's primary purpose is to ensure that the service provider's network and resources are used responsibly and securely.

> **Note**
>
> Google Cloud Platform's AUP can be found at `https://cloud.google.com/terms/aup`.

AUP covers several critical issues. The critical issues are as follows:

- **Prohibited activities**: One of the essential components of an AUP is outlining specific activities that are not allowed – for instance, spamming, hacking, and other activities that could harm the service provider or other users. It's important to note that laws governing acceptable use may vary from country to country, so it's vital to research and include any relevant local legislation in the AUP.

- **User conduct**: Have you ever wondered what kind of behavior is expected from users when using a service? The AUP provides the answer! It provides guidelines on language, behavior, and tone so that users can interact respectfully with each other. These guidelines help protect against activities that could be offensive or cause harm to others.

- **Security**: The AUP specifies actions that users need to take to guarantee the security of the service. These actions might involve having strong passwords, performing regular software updates, and utilizing antivirus software. Other security measures, such as two-factor authentication, may be required in certain situations.

- **Consequences for violations**: An AUP outlines the consequences of violating any of the guidelines. They include suspension or termination of service, account deactivation, or other legal actions depending on the severity of the violation.

- **Enforcement**: An AUP outlines the measures that the service provider will take to enforce the policy. It may include monitoring user activity, reporting violations to authorities, or revoking user access.

Now that you have understood AUP's critical issues, the next section will showcase why AUP holds such significance.

The Importance of an AUP

An AUP is an essential component of any organization that provides services to users. It ensures that users know what constitutes unacceptable behavior when using the service. The most significant advantage of having an AUP is that it protects the service provider from any legal liability if users violate their policies.

In short, an AUP is an essential component of any service provider's infrastructure. It provides clear guidelines for users and protects the service provider from legal liability. By outlining prohibited activities, expected user behavior, security measures, consequences for violations, and enforcement procedures, an AUP ensures that the service is used responsibly and securely.

SLA

An **SLA** is another important document that outlines the terms and conditions for the quality of the services that a customer can expect from a CSP. It outlines the responsibilities of both the customer and the CSP regarding services, performance, availability, and other metrics.

The Purpose of an SLA

An SLA serves as a tool to align the expectations of both parties, the cloud service provider and the customer. Everything – from the scope of services to the performance level that a customer can expect to the metrics used to measure performance – is explained in detail. With the help of an SLA, customers can be sure that they'll get the promised level of quality from the service provider.

The Key Characteristics of an SLA

The SLA should include several essential components that make it effective. The essential components are as follows:

- **Service description**: An SLA includes a detailed description of the services provided by the service provider. It should be clearly defined and aligned with the customer's requirements to avoid any confusion. Make sure the CSP knows exactly what your needs are and that it delivers on its promises. Service levels are the minimum performance targets that the service provider must achieve for each service defined in the SLA. These service levels should be measurable and quantifiable to allow for objective measurement and evaluation of performance.

- **Performance metrics**: The SLA should specify the performance metrics used to measure the service levels. The performance metrics should be relevant, objective, and measurable to ensure that the service provider can demonstrate compliance with the agreed service levels. For example, the SLA may stipulate that your website will be available 99.5% of the time with an average response time of fewer than two seconds.

  SLAs can be updated regularly to reflect changing customer requirements or changes in the service provider's capabilities. For example, the SLA for **Microsoft Online Services** Worldwide (`https://www.microsoft.com/licensing/docs/view/Service-Level-Agreements-SLA-for-Online-Services?lang=1`), which is over 100 pages in length, has been updated at least eight times this year.

- **Remedies**: An SLA outlines the remedies that will be available to the customer if the service provider fails to meet the agreed service levels. Remedies can include credits, refunds, and termination of the contract. In certain cases, a customer may be eligible for compensation due to losses incurred from service failures. Although rare, this clause can provide an added layer of protection for the customer.

- **Reporting**: An SLA should include a reporting mechanism to provide regular feedback to the customer on service performance. This mechanism should be clearly defined, and the frequency and format of reporting should be agreed upon. The reporting can be used to assess performance and identify areas for improvement. It can also be used to track service-level violations, which can lead to the invocation of remedies if there are serious or repeated violations.

- **Review and revision**: As explained in Microsoft's SLA documentation previously, an SLA is a living document that should be reviewed and updated regularly. This review should be carried out by both parties to ensure that the SLA remains relevant and aligns with the customer's changing needs.

The Importance of an SLA

An SLA is a crucial document designed to be a clear and concise communication tool for both the service provider and the customer. It also provides a framework for measuring and evaluating service performance, essential for continuous improvement.

In short, an SLA is a critical component of any service-based business. It defines the scope of services, the expected service levels, and the performance metrics used to measure compliance with these service levels. The key characteristics of an SLA include a detailed service description, measurable service levels, objective performance metrics, available remedies, reporting mechanisms, and regular review and revision. By creating an SLA, both parties can achieve a common understanding of the services provided and the expected level of performance, which is imperative for the success of any service-based business.

In the next section, you will look at some best practices on how to evaluate CSP services.

Evaluation of the CSP Services

From shared infrastructure to specialized services, there are a variety of CSPs in the market today. In addition to large CSPs such as **Amazon Web Services (AWS)**, Microsoft Azure, and **Google Cloud Platform (GCP)**, there are also smaller and less expensive CSPs that offer specialized services.

Before selecting a CSP, you should evaluate their services to determine which provider offers the best solution for your needs. The following best practices will help you make an informed decision.

Know Your Business Needs

The first step in evaluating CSP services is to know your business needs. To do so, answer these three questions:

- What type of data do you want to move to the cloud?
- What cloud-based applications and services do you need?
- What are the compliance requirements that you must meet?

When selecting a CSP, the choice hinges significantly on the specific operational needs and business strategies of the organization. For software development teams that require rapid provisioning and dynamic scaling capabilities, a CSP that excels in offering efficient, on-demand service deployment is crucial. This includes CSPs that provide robust auto-scaling, flexible resource management, and a wide array of ready-to-use development environments. This ensures that development teams can swiftly set up test environments, scale applications in response to real-time demand, and implement changes without significant delays.

Conversely, small businesses, particularly those operating with a limited number of critical applications, should prioritize cost-effectiveness in their CSP selection. For these entities, the primary considerations often include straightforward pricing structures, low-cost setup and maintenance, and bundled services that align with their limited scalability requirements. These businesses benefit from CSPs that offer simplified management interfaces, lower-cost data storage solutions, and security features that do not require extensive customization, thus aligning their cloud investments directly with their operational scope and financial constraints.

No one-size-fits-all solution exists; each CSP provides different services and features that may or may not meet your needs. By assessing your business needs, you can select a CSP that can meet your specific requirements.

Assessing Security and Compliance

Security and compliance are critical factors to consider when evaluating CSP services. You should look for CSPs that have robust security and compliance programs in place. The following checklist can be used to assess a CSP's security and compliance:

- **Adherence to standards and frameworks**: This confirms that the CSP complies with established security standards such as the following:

  - **ISO 27001:2013**: Information technology, security techniques, information security management systems, and requirements

  - **ISO-27017:2015**: Information technology, security techniques, and a code of practice for information security controls based on ISO/IEC 27002 for cloud services

  - **ISO-27002**: Information technology, security techniques, and a code of practice for information security controls

  - **ISO-27001**: Information technology, security techniques, and a code of practice for information security controls

  Also, check ISO-27018 (Information technology – security techniques – a code of practice for protection of **personally identifiable information** (**PII**) in public clouds acting as PII processors) to learn about their policy for personal data protection.

- **Checking authentication and identity controls**: It is important to know whether the service provider has implemented **Multi-Factor Authentication** (**MFA**), strong password policies, and identity governance tools in their backend systems.

- **SLA**: You must review the SLA provided by the CSP to determine their commitment to delivering a secure environment.

- **Internal management resources**: It is important to know the service provider's internal resources and processes to ensure compliance with security standards. This includes reviewing their incident response plans, data encryption policies, and vulnerability management programs.

- **Data center/storage locations**: You must be aware of the data center and storage locations of the CSP to ensure that all your data is stored in secure environments.

- **Uptime and performance**: Check the availability and performance of the CSP to ensure that your business data is accessible when needed.

- **Backup and data recovery process**: You must know the backup and data recovery process of the CSP and make sure that it allows you to restore your data if there is an emergency.

- **Migration services and support**: Review the migration services and support offered by the CSP to ensure a successful transition to their platform.

Summary

Evaluating CSPs is a crucial process that requires careful consideration of your specific business needs and security requirements. A well-informed decision will help you select the right CSP that offers scalable, reliable, and secure services.

Critical contractual documents, including the CSA, AUP, and SLA, provide important information that should be thoroughly reviewed before signing a contract with a CSP. You must be aware of their differences to be prepared for the exam.

When evaluating CSP services, best practices include knowing your business needs, assessing security and compliance, evaluating performance and uptime, checking customer references, and reviewing pricing and contract terms. In other words, you must take a holistic approach when evaluating CSP services.

Security, risk management, and compliance in the cloud rely heavily on these agreements. A solid grasp of these concepts will not only help you succeed in the CCSP exam, but also contribute to your professional expertise in managing cloud security and mitigating risks associated with cloud services.

In the next chapter, you will explore cloud data concepts, cloud data storage architectures, data security, data classification, and cloud data security technologies – all very critical to managing data governance in the cloud.

Exam Readiness Drill – Chapter Review Questions

Apart from a solid understanding of key concepts, being able to think quickly under time pressure is a skill that will help you ace your certification exam. That is why working on these skills early on in your learning journey is key.

Chapter review questions are designed to improve your test-taking skills progressively with each chapter you learn and review your understanding of key concepts in the chapter at the same time. You'll find these at the end of each chapter.

> **How to Access These Materials**
>
> To learn how to access these resources, head over to the chapter titled *Chapter 25, Accessing the Online Resources*.

To open the Chapter Review Questions for this chapter, perform the following steps:

1. Click the link – `https://packt.link/CCSPE1_CH05`.

    Alternatively, you can scan the following **QR code** (*Figure 5.1*):

Figure 5.1 – QR code that opens Chapter Review Questions for logged-in users

2. Once you log in, you'll see a page similar to the one shown in *Figure 5.2*:

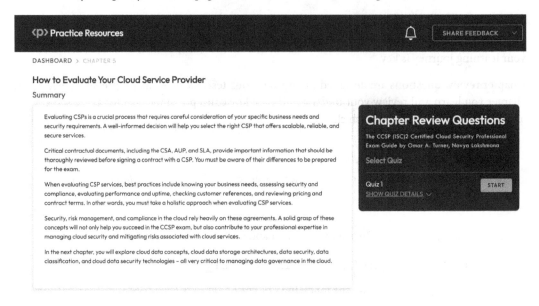

Figure 5.2 – Chapter Review Questions for Chapter 5

3. Once ready, start the following practice drills, re-attempting the quiz multiple times.

Exam Readiness Drill

For the first three attempts, don't worry about the time limit.

ATTEMPT 1

The first time, aim for at least **40%**. Look at the answers you got wrong and read the relevant sections in the chapter again to fix your learning gaps.

ATTEMPT 2

The second time, aim for at least **60%**. Look at the answers you got wrong and read the relevant sections in the chapter again to fix any remaining learning gaps.

ATTEMPT 3

The third time, aim for at least **75%**. Once you score 75% or more, you start working on your timing.

> **Tip**
>
> You may take more than **three** attempts to reach 75%. That's okay. Just review the relevant sections in the chapter till you get there.

Working On Timing

Target: Your aim is to keep the score the same while trying to answer these questions as quickly as possible. Here's an example of how your next attempts should look like:

Attempt	Score	Time Taken
Attempt 5	77%	21 mins 30 seconds
Attempt 6	78%	18 mins 34 seconds
Attempt 7	76%	14 mins 44 seconds

Table 5.1 – Sample timing practice drills on the online platform

> **Note**
>
> The time limits shown in the above table are just examples. Set your own time limits with each attempt based on the time limit of the quiz on the website.

With each new attempt, your score should stay above **75%** while your "time taken" to complete should "decrease". Repeat as many attempts as you want till you feel confident dealing with the time pressure.

Cloud Data Security Concepts and Architectures

In the last chapter, you looked at how to evaluate your CSP. Reviewing and understanding the key cloud service contractual documents, such as acceptable use policies and service-level agreements, is a critical part of selecting and working with a cloud provider. As readers of this book are most certainly aware, cloud computing is one of the most visible and impactful tech trends of the past few decades. Thousands of organizations worldwide have adopted it. In the coming years, more businesses will embrace it to become **cloud-native**.

However, to maximize the return on their cloud investments in a secure way, businesses must first understand how their data will be stored, managed, and protected in the cloud. As an individual studying for the CCSP certification, having a thorough understanding of data protection measures, data lifecycle management, and other important data protection topics is vital to pass the exam.In this chapter, you will review the following topics:

- Structure and unstructured data
- The cloud data lifecycle phases and data flows
- Various storage types and common threats based on the storage type
- Data classification and discovery
- Cloud data security technologies and common strategies
- Critical cloud data security strategies
- Best practices for data retention, archival, and deletion

Knowledge of how data is stored, managed, and protected in the cloud enables candidates to design secure systems, manage risks, ensure compliance with regulatory requirements, and implement advanced security measures. This expertise not only prepares you for the CCSP exam but also enhances your professional credibility, equipping you with the necessary skills to advise organizations on best practices for securing data in the cloud.

> **Note**
>
> You should pay immense attention to this chapter as it contains important topics with respect to the CCSP exam.

Structured and Unstructured Data

Before covering the main topics, it's important to understand the difference between types of data. You may already be familiar with **structured** and **unstructured data**. If so, you can skip to the next section.

Structured and unstructured data are two primary types of data that organizations handle, and they differ significantly in terms of their formats, storage, and processing methods.

Structured data is highly organized and formatted in a manner that enables simple and direct search operations via search engine algorithms or other search mechanisms. It is usually stored in **relational databases** (**RDBs**) and managed using **Structured Query Language** (**SQL**). Examples of structured data include information stored in relational databases, spreadsheets, and CSV files. This type of data is typically numerical and can easily be mapped into predefined fields. For example, details such as names, dates, and credit card numbers in an e-commerce transaction are considered structured data.

On the other hand, unstructured data is data that does not conform to traditional row-and-column databases and lacks a predefined format. It is often text-heavy but may also include elements such as dates, numbers, and facts.

Unstructured data encompasses various types, including text files, social media posts, email messages, word processing documents, videos, photos, audio files, web pages, and other types of business documents. It is important to note that while these types of data can be categorized and analyzed, they do not fit neatly into a structured database.

Key Differences

Structured data is highly organized and adheres to a consistent format, making it straightforward to search and analyze using traditional data analytics tools and techniques. It is typically stored in relational databases that utilize SQL. In contrast, unstructured data lacks a predefined format and organization, which complicates its processing and analysis. It requires more sophisticated methods such as **natural language processing** (**NLP**), text analytics, data mining, or machine learning, and is often stored in various ways, including data lakes, NoSQL databases, or file systems such as Hadoop. Although both types of data can be voluminous, unstructured data generally makes up the majority of data in the digital universe, primarily due to the explosion of data from sources such as social media, IoT devices, and other digital channels. This vast amount of unstructured data presents significant challenges in data management and utilization.

The Cloud Data Lifecycle

Today, more and more enterprise data is in the cloud. This includes customer and employee data, financial information, business intelligence, intellectual property, trade secrets, and sensitive personal healthcare data.

As cloud storage consumption increases, data growth and the management of that data have become ever-increasing challenges for organizations. Fortunately, organizations (with the coordinated efforts of data governance professionals and business stakeholders) can minimize these challenges by understanding the cloud data lifecycle. By understanding the different phases that data goes through in the cloud, companies can simplify data management and security, and maximize its benefits.

Grasping the data lifecycle extends well beyond a basic understanding; it forms an integral component of an organization's data governance strategy. Recognizing the roles, processes, policies, procedures, and metrics that ensure efficient and effective data usage is an ongoing effort. Furthermore, a thorough understanding of the cloud data lifecycle within your organization enables the implementation of precise security controls and measures at each stage. For instance, selecting the appropriate encryption method during the data creation and storage phases is crucial for effectively preventing unauthorized access.

Let's review the data lifecycle with examples of systems or technologies used to help you understand what happens at each stage. The cloud data lifecycle consists of the following six phases.

Data Creation or Procurement

In the initial phase of data lifecycle management, data is generated or collected from both internal and external sources. For instance, within one organization, the primary source of data might be the outputs from research and development activities, while in another, the bulk of data could stem from manual data entry. Additionally, many organizations gather data from a variety of other sources, such as devices, customer engagements, financial transactions, and enterprise applications such as CRM and ERP systems. These systems are integral for providing detailed insights into various aspects of business operations. Internally, aside from R&D and manual inputs, significant data comes from CRM and ERP systems that track and record customer and transactional data, enhancing operational insights. Externally, data collection extends to include information from devices and sensors, diverse customer interaction points, financial transactions, and extensive market research, which may also encompass publicly available data. This comprehensive approach to data collection from multiple sources is typical of modern organizations, enabling them to capture a broad spectrum of data crucial for informed decision-making and strategic planning.

The acquired data may exist in many different forms and formats, such as Word or PDF files, SQL databases, and email content. Also, some data may be structured, but it is increasingly common for most enterprise data to be unstructured.

Data Storage

Once data is created or procured, it can be stored and secured in the cloud with appropriate controls. Some organizations store all their data in the public cloud, while others store it only in private clouds. Most organizations commonly store their data both in on-premises storage and in the cloud (public and/or private).

As discussed in *Chapter 4, Design Principles for Secure Cloud Computing*, **cloud service providers (CSPs)** follow a "shared responsibility model" for cloud data security. In this model, CSPs take care of the security *of* the cloud, meaning they protect the cloud infrastructure that houses enterprise data and applications. Customers are responsible for security *in* the cloud, so they must implement appropriate controls to store and secure it. They must also classify their data and implement backup and recovery processes to ensure it is recoverable.

Data classification and its relation to security are explored in further detail in later sections of this chapter.

Data Usage

In this phase, authorized users can use the data to support the organization's business activities. For example, they may use data to understand their customers, market, or competitors, guide decision-making, inform new innovations, improve existing processes, and improve their marketing, sales, or financial strategies. They may simply view or access the data, or they may process, modify, and re-save it.

Organizations are vulnerable to data breaches and cyberattacks in this phase since the data may be stored insecurely or in an insecure location, making it accessible to unauthorized or malicious users. To ensure that data use is restricted only to authorized users, organizations must implement data governance strategies. They should define roles and permissions around data access and use and maintain an audit trail, especially for sensitive or critical data, to ensure that all modifications are traceable and controllable.

Data Sharing

In this phase the cloud centralizes data, making it easy for users to share it with others, both inside and outside the organization. For example, colleagues can share data between themselves as they collaborate on a project; mid-level managers can leverage data to prepare reports that they then share with senior leadership; and the company can share data with its customers and other stakeholders.

But just like the usage stage, the sharing stage also creates security gaps for enterprises. A least-privileged access approach can help close these gaps.

The goal of this approach is to provide authorized users with access only to data that is critical for their role. This minimizes the possibility of unauthorized data access, use, and sharing.

Data Archiving

Data archival involves determining which inactive data is relevant to the organization. Most CSPs provide archival services for long-term data retention and digital preservation. One example is AWS, with its Amazon S3 Glacier storage classes that support low-cost data archival and flexible retrieval.

In the archival stage, inactive but relevant data is moved from an active production environment to long-term storage. No maintenance or general usage occurs in the archival environment. However, data can be transferred to the production environment when required.

It's important to consider legal and regulatory requirements when archiving data. Other factors including the costs of archival, privacy and data availability requirements, and the time needed to keep data should also be considered when moving data into deep storage.

Data Destruction

Sometimes enterprise data stops being useful or valuable. At other times, data retention costs can overshadow its benefits. Either way, organizations may choose to destroy the data instead of simply archiving it. Privacy laws such as the GDPR, which give individuals the "right to be forgotten", also create a need for the permanent destruction of data.

Methods of data destruction vary based on the sensitivity of the data and include physical destruction techniques such as shredding, degaussing, and incineration, which physically destroy the media. Digital methods include overwriting, where software is used to replace old data with random information, and cryptographic erasure, which makes data inaccessible by deleting the encryption keys. Additionally, disk formatting can be employed, ranging from standard formatting, which removes data pointers, to secure formatting, which overwrites the entire disk multiple times. Choosing the right data destruction method depends on the data's sensitivity and the specific compliance requirements governing the data.

Regardless of the purpose, organizations must follow commonly accepted best practices, internal governance policies, and legal standards when deleting or destroying data. Any failure to adhere to these practices may result in improper or incomplete data destruction. Such data may also end up in the wrong hands, creating a serious risk of a breach or leak.

A thorough understanding of these stages is essential for correctly answering questions given to you on this topic.

As a reminder (and quick summary), the typical phases of the cloud data lifecycle are as follows:

1. **Create**: This initial phase involves the generation of data by users, applications, or devices.

2. **Store**: Once created, data is stored in a database or another storage system. The format and location of the data may vary depending on the cloud service model.

3. **Use**: During this phase, applications or users access and process the data. This may involve tasks such as reading, updating, or transforming the data.

4. **Share**: Data can be shared with other users, applications, or services. This may entail transferring data between different cloud environments or sharing it with third parties.

5. **Archive**: Inactive or infrequently accessed data is moved to long-term storage. Archiving helps reduce storage costs and enhances system performance.

6. **Destroy**: Finally, data is permanently deleted when it is no longer required. It is crucial to carry out this process securely to prevent unauthorized access to the deleted data.

> **Note**
>
> To aid your memory for the CCSP exam, you can utilize mnemonics or acronyms. For instance, you could use the acronym **CSUSAD**, representing **Create**, **Store**, **Use**, **Share**, **Archive**, **Destroy**. Alternatively, you can create a sentence or story that incorporates these phases.
>
> I have used: "Computer Systems Undergo Software Adjustments Daily"
>
> You probably won't forget this version:
>
> "Cats Start Using Skateboards At Dusk"

Visualizing the data lifecycle as a cycle or flowchart is another effective strategy. It helps us to grasp the flow of data and the interactions between different phases. Relating each phase to real-world scenarios or use cases further facilitates comprehension and memorization.

Ultimately, practice is key. Apply these concepts to practice questions or scenarios to reinforce your understanding of the phases and their application, which is crucial for this exam.

Understanding the data lifecycle is crucial as it provides a framework for managing the security and accessibility of data from creation to destruction, ensuring compliance and efficient data management. With this foundational knowledge, you can now transition to exploring various storage types and the common threats associated with each, enabling us to better prepare and protect our data across different storage environments.

Various Storage Types and Common Threats

The cloud allows organizations to store vast amounts of data and access it from anywhere and at any time. The cloud also provides a more cost-effective and scalable storage alternative than expensive on-premises data centers. Since they don't have to buy and manage their data storage infrastructure, businesses can focus on deriving value from their data at a near-unlimited scale. Data redundancy is another advantage of cloud storage. The largest CSPs store the same data on multiple offsite servers, ensuring high data availability and reliability, even if one server suffers an outage.

Enterprises have three main options for cloud data storage types, discussed in the following sections.

Object Storage

Object storage is suitable for unstructured data including video, photos, audio, ML data, IoT sensor data, web pages, and data lakes. It uses a flat address space that scales horizontally, making it perfect for big data applications and **content delivery networks (CDNs)**.

In the object storage architecture, data is managed as objects and placed in secure repositories instead of rigid file/folder hierarchies. Each object stores data in the same format it arrives in. It includes an identifier and metadata (data about data) to identify and access each data point. Users can customize the metadata to simplify data access and analysis.

Object repositories deliver almost unlimited scalability. Also, large data volumes can be stored at an affordable and controllable cost. For these reasons, object storage is ideal for storing unstructured data, which is growing at an unprecedented scale for many organizations. Object storage is also suitable for data lakes and for archiving static files, i.e., data that does not change frequently.

File Storage

File storage is suitable for structured data, development platforms, and home directories. It is designed for the hierarchical organization of files in folders, similar to traditional file servers, supporting standard protocols such as NFS or SMB/CIFS, making it easy to integrate with existing applications.

A file storage mechanism in the cloud stores data in hierarchical files, folders, directories, and sub-directories on a **network-attached storage (NAS)** server. All data is retained in its original format.

The hierarchical storage format provides an easy and intuitive way to find and retrieve files as needed. However, retrieval can become cumbersome as the number of files increases and if they contain more web-based digital content or unstructured data. In such situations, object and block storage are better options.

Block Storage

Block storage is suitable for large databases, virtual machines, and business-critical applications. In cloud block storage systems, all data is organized into equally sized chunks or blocks, where each block represents a separate hard drive and is stored under a unique address. These blocks don't follow a rigid hierarchical storage structure. Instead, they can be stored anywhere in the system and easily retrieved using the assigned unique address.

Block storage provides low I/O latency (delay) for easy retrieval and better performance over a network. It also enables more granular access to data. For these reasons, the architecture is highly suited for large or transactional databases, such as databases that store large volumes of website visitor data. Block storage is also ideal for virtual machines and critical business applications that require low-latency data access.

Like storage, security is also an equally important aspect of data in the cloud. All organizations that store their data in the cloud must be aware of the threats to their data and the security measures available to protect it from bad actors.

Prevalent Threats to Cloud Data

Identifying the threats associated with the different types of storage solutions is essential when choosing a suitable option based on your risk profile. It is critical to implement the appropriate security controls and strategies to protect your valuable assets in the cloud from potential vulnerabilities.

Cloud storage security is a serious concern for organizations, especially sensitive data. To launch attacks targeting cloud storage, as many do, attackers leverage malicious software tailor-made to target cloud services and infrastructure that encrypts data and prevents authorized users from accessing it. The attackers demand payment for a decryption key to unlock the encrypted data.

Different cloud storage services come with their own set of risks, which can vary depending on factors such as data sensitivity, access management policies, and encryption methods used. Some common threats include unauthorized access, data breaches, and cyber-attacks targeting infrastructure or applications hosted by cloud providers such as **Amazon Web Services** (**AWS**), Microsoft Azure, or **Google Cloud Platform** (**GCP**).

Malware injection attacks against cloud data are devastating. In these attacks, threat actors inject malicious software, such as viruses, into cloud resources or infrastructure. Attackers either exploit vulnerabilities in these resources or add a malicious module/VM to a PaaS or IaaS system. These attacks enable them to compromise the target resources and steal or compromise enterprise data.

In addition to malware, some other severe threats to cloud storage and data are the following:

- **Denial-of-service (DoS) attacks**: Attackers flood a cloud service or application with a large volume of traffic, overwhelming the system and making it unavailable for legitimate users.

- **Account hijacking**: An attacker gains unauthorized access or control over an enterprise cloud account and exploits its resources or data.

- **User account compromise**: Adversaries gain access to an authorized user's account by tricking the user into revealing their login credentials or exploiting a security vulnerability in a cloud service or application.

- **Cookie poisoning**: A threat actor modifies a website's or web application's cookie or injects malicious content into it to steal users' credential data.

- **Security misconfigurations and insecure APIs**: Improperly configured cloud computing resources and vulnerable APIs create pathways that allow attackers to compromise enterprise cloud systems and data.

Security Measures for Cloud Object Storage, File Storage, and Block Storage

To mitigate these risks and maintain customer trust in your business's ability to keep their data safe, it's essential to implement a comprehensive cloud security strategy. Some key measures involved in this are as follows:

- **Encryption**: Encrypting sensitive data at rest and in transit can help protect against unauthorized access. Ensure that encryption keys are securely managed and stored separately from the encrypted data.

- **Access management**: Implement strict user access controls, such as **multi-factor authentication (MFA)** and role-based permissions, to minimize the risk of unauthorized access.

- **Patching and updates**: Regularly update your cloud systems with the latest patches and software updates to address potential security vulnerabilities.

- **Data backup and recovery plans**: Develop robust backup strategies for critical information stored in the cloud along with disaster recovery plans to ensure business continuity during unexpected events such as cyber-attacks or natural disasters.

A competent CSP will leverage robust governance processes, implement best practices, and use layered security solutions to protect its core architecture. As stated in *Chapter 4, Design Principles for Secure Cloud Computing,* the customer of a CSP is ultimately responsible for what is stored in their cloud environment, but protections still exist. In general, block and object storage provide more reliable security in the cloud than file storage. Block storage, for example, includes two critical safeguards: data encryption and mirroring.

Encryption translates data into another form so that only users with a secret decryption key can decrypt and read the data. Mirroring ensures that all data in the organization's cloud account gets automatically replicated at a secondary site, allowing for its recovery in case of a cyberattack or other disruptive event.

In the case of object storage, CSPs may use **Information Dispersal Algorithms (IDAs)** to secure data and ensure private transmission. Also, data at rest is encrypted using secure per-object generated keys. Also, the storage is accessed over HTTPS instead of HTTP, and internal storage devices communicate with each other using TLS.

Object storage also provides robust API support with solid security capabilities (such as **write once, read many (WORM)**, which protects object data from ransomware attacks), multiple critical management systems, and encryption options. For the encryption and security of block storage systems, CSPs usually use AES-256 encryption and manage keys in-house using the industry-standard **Key Management Interoperability Protocol (KMIP)**.

Additionally, cluster-to-cluster traffic is encrypted using TLS, and a **Redundant Array of Independent Disks (RAID)** may be implemented to maintain data backups and ensure redundancy and recovery.

CSPs that provide file storage also encrypt data at rest by default, usually with AES-256 encryption and KMIP for key management.

Other security safeguards include the following:

- Endpoint protection
- Content filtering
- Email filtering
- Threat analysis
- Regular patches
- Robust access controls
- User authentication

Despite these security measures implemented by CSPs, organizations must not rely solely on them. All cloud-reliant businesses must look beyond a CSP's baseline security measures and default controls, and implement further controls and practices to minimize the risks of attacks and data breaches.

These practices should include the following:

- Encrypting data at rest and data in transit (assuming that capability is available).
- Use dedicated private connections to connect with the CSP instead of the public-facing internet.
- Implement **identity and access management (IAM)** solutions to control access to storage resources and data.
- Take regular data backups to ensure they can be recovered in case of a cyberattack, human error, or natural catastrophe.
- Ensure that no complete copy of the data resides in a single storage node.
- Enable **two-factor authentication (2FA)** to make it harder for threat actors to access or compromise cloud data.
- Maintain secure copies of encryption keys to prevent access by unauthorized or malicious parties.

Data classification is a vital first step in data security. Let's explore this concept next.

Data Classification and Discovery

Data classification helps with data discovery and access for authorized users. Properly classified data in the cloud tends to be more usable and valuable. It also enables enterprises to determine and mitigate the risks associated with their cloud data—risks that, if left unaddressed, may impact data confidentiality, integrity, and availability (the CIA Triad).

Mishandling sensitive information has severe consequences that go beyond immediate financial losses. These consequences include a breach of trust, legal penalties, and significant reputational damage, each of which can deeply impact an organization:

- **Breach of trust**: When sensitive data is compromised, it fundamentally erodes the trust that customers place in an organization. This loss of trust can lead to a decrease in customer loyalty and retention, as clients may feel their personal information is no longer secure with the company.

- **Legal penalties**: Organizations are required to adhere to various data protection regulations, such as the **General Data Protection Regulation (GDPR)**, **Health Insurance Portability and Accountability Act (HIPAA)**, and **Payment Card Industry Data Security Standard (PCI DSS)**. Failing to protect sensitive information can lead to substantial fines, legal action, and other regulatory penalties. These can not only affect the financial standing of the company but also consume valuable resources in legal battles and compliance efforts.

- **Reputational damage**: A publicized data breach can significantly tarnish a company's public image. The loss of reputation is particularly damaging in the digital age where news spreads rapidly online. This damage can lead to lost business opportunities, a decrease in stock value, and challenges in attracting both customers and talent in the future.

Understanding these risks underscores the importance of robust data management and security measures to prevent the mishandling of sensitive information and safeguard the organization's longevity and trustworthiness.

Once all data is classified, it becomes easier to secure, govern, and manage for day-to-day and long-term/strategic use.

As per the shared responsibility model of cloud security, data classification and security are an organization's responsibility. However, they can use a CSP's security controls as a baseline to guide their data protection strategies.

There are many ways to classify cloud data. One way is to adopt the NIST's three-tier scheme (`https://www.nccoe.nist.gov/publication/1800-26/VolA/index.html`) that categorizes data based on the potential impact of a cyberattack or breach on the data's CIA Triad:

- **Low**: Limited adverse effect on organization operations or assets.

- **Moderate**: Serious adverse effect on organization operations or assets.

- **High**: Possibly severe or catastrophic adverse effect on organization operations or assets.

Organizations can also categorize their cloud data as follows:

- **Public**: Data that can be made available for public consumption.

- **Confidential**: Data that must not be publicly shared because doing so can harm the business.

- **Highly confidential or secret**: Data can only be shared with certain approved parties (e.g., senior leadership).

- **Sensitive**: Data whose loss can severely harm the company or its stakeholders (e.g., customers) and, therefore, requires reliable security/privacy controls.

One effective way to implement data classification is to understand its intended use and determine how its loss could adversely impact the organization. For instance, the loss of sensitive data could affect a business's customer relationships and reputation, whereas the loss of public data may have a minimal visible effect. Understanding the loss impact makes it easier to categorize data.

Once data is classified, organizations can determine the risks that affect it and implement appropriate controls to mitigate them. A baseline set of controls must be implemented for each data type to protect it against vulnerabilities and threats.

Recommended Data Classification Process

The general recommendation is a five-step data classification process for the cloud:

1. **Establish a data catalog:** Prepare an inventory of the various data types the organization uses and then group the types into classification levels (public, sensitive, restricted, etc.).

2. **Conduct an impact assessment**: Identify all business-critical functions and assess the criticality of data type to each function.

3. **Label information:** Label all datasets in their respective classification buckets.

4. **Set handling guidelines for each data type**: Formalize and document the handling guidelines for each data set based on its classification tier.

5. **Monitor data:** Monitor data usage and security (ideally through an automated system) to maintain normal operations and identify (and mitigate) discovered threats.

Most cloud providers also recommend a three-tier system security categorization based on data classification levels. For example, data classified as "unclassified" or "general" can be assigned a "low" security category. In contrast, official, secret, and other critical data types should be categorized as "moderate" or "high" security and therefore protected with more robust controls.

Cloud Data Security Technologies and Common Strategies

The rapid migration of organizations and businesses to cloud-based services has led to an increased focus on understanding cloud data security technologies. These technologies are essential in ensuring that sensitive information stored, processed, or transmitted within a cloud environment remains protected against unauthorized access, breaches, and other cybersecurity threats.

One example of a core technology used for securing data in the cloud is encryption. By effectively encrypting data both at rest (storage) and during transit (transmission), organizations can maintain control over their sensitive assets by making them unreadable to anyone without the necessary decryption keys.

Access control mechanisms such as IAM tools also play a vital role in protecting enterprise resources from unauthorized access.

In addition to implementing robust security controls such as encryption and IAM solutions, it's crucial for businesses utilizing cloud infrastructure to regularly update their risk management strategies based on emerging threat intelligence.

Employing **Security Information and Event Management** (**SIEM**) systems enables real-time monitoring for early detection of potential threats or anomalies while **Data Loss Prevention** (**DLP**) tools help protect against accidental exposure or loss through strict authorization settings.

To ensure comprehensive protection of sensitive information stored in the cloud, organizations must adopt best practices such as implementing a robust security policy backed by regular audits and assessments to identify vulnerabilities within systems before they can be exploited by malicious actors.

Training employees on cybersecurity awareness coupled with MFA goes a long way to prevent unauthorized access through stolen credentials or targeted phishing attacks.

Let's look at some of the primary technologies in more detail.

Encryption in Cloud Data Security

Encryption is a fundamental technology in cloud data security that provides end-to-end protection of customer data. It involves the process of converting plaintext data into ciphertext to protect sensitive information, which can only be decrypted by authorized users with access to the encryption keys.

In cloud computing, storage providers encrypt user data and provide encryption keys for safe decryption when needed. This helps prevent unauthorized access to sensitive company or personal information such as financial records or **protected health information** (**PHI**).

It's essential to ensure comprehensive encryption both in transit and at rest for vital business assets such as intellectual property and PII.

IAM

IAM is a vital aspect of cloud data security technology that focuses on managing user identities and access permissions within an organization. IAM products offer **Role-Based Access Control (RBAC)** to regulate employee access to critical information systems and data stored in the cloud.

One example of IAM capability is MFA, which requires more than one credential for secure logins, such as using biometrics or SMS codes alongside passwords.

RBAC is another widely used feature of IAM technology that limits access requests based on the roles given to users, rather than individual employees' usernames.

IAM security emphasizes protecting user identities while ensuring only authorized individuals can perform specific actions with their accounts, such as opening certain files required for their jobs or increasing account privileges from within the system.

SIEM

One key technology used in cloud data security is SIEM. SIEM solutions, especially cloud-native ones, allow organizations to gain valuable visibility by aggregating telemetry from various elements of their architecture.

This way, they can identify and respond to potential security threats more effectively.

SIEM tools are designed to gather and analyze logs and events from operating systems, applications, servers, network devices, and intrusion detection systems, among others. This allows analysts to track activities across the entire IT infrastructure proactively.

The collected events are then analyzed using various threat intelligence feeds that help to detect suspicious patterns or anomalies.

In conclusion, utilizing SIEM technologies is crucial for detecting emerging cyber threats before they escalate into full-blown attacks on cloud infrastructure. Today's modern companies must adopt proactive measures such as integrating the cutting-edge technology of SIEM tools to create an effective, agile cybersecurity posture.

Firewall and Intrusion Detection Systems

Firewall and **Intrusion Detection Systems (IDSs)** are critical components of cloud data security.

A firewall is a network component (either physical or virtual) that is responsible for filtering and blocking network traffic based on an organization's rules or policies. IDSs can also be on-premises or cloud-based appliances, responsible for monitoring network traffic for malicious or anomalous behavior.

An effective IDS can be crucial in detecting potential cyber threats. For instance, if malware somehow makes it past the firewall, an IDS can spot unusual behavior within the network and alert administrators so they can respond quickly to the event or incident.

Overall, implementing firewalls and IDSs is fundamental to secure networks from attackers who use various techniques such as vulnerability exploits or phishing attacks where gaining the information needed to break into targets originates through social engineering rather than technical hacking methods associated with vulnerability exploitation. Those familiar with the deployment of IDSs and firewalls in the traditional data center will find that leveraging these controls in the cloud allows for greater management and scalability. For example, in an on-premises environment, these controls are generally deployed at the network perimeter, whereas in a cloud environment, they can be deployed in front of virtually any resource or service. With respect to micro-segmentation, one could provide firewall rules and IDS policies at a granular level.

Data Loss Prevention Tools

Data Loss Prevention (**DLP**) tools are critical for protecting sensitive information from unauthorized access, misuse, and loss. These tools ensure that data is protected in real time by continuously tracking the flow of data across networks and devices.

DLP solutions offer advanced features such as **Endpoint Detection and Response** (**EDR**), incident management workflows, and web application protection capabilities.

Users should choose a software solution based on their business requirements and risk assessment needs.

Critical Cloud Data Security Strategies

A cloud security policy is a critical component of an organization's overall security framework, specifically designed to manage the risks associated with cloud computing. This policy provides a comprehensive set of controls and procedures that guide the deployment, operation, and management of cloud services and resources. The importance of a cloud security policy stems from its role in safeguarding sensitive data and maintaining operational integrity within the cloud. This policy not only addresses security concerns but also aligns cloud operations with business objectives and regulatory compliance requirements.

Importance of a Cloud Security Policy

The significance of implementing a cloud security policy lies in its ability to provide clear guidelines and standards for securing cloud environments. As organizations increasingly rely on cloud solutions for scalable and efficient operations, the need to protect critical assets and data from threats such as unauthorized access, data breaches, and other cyber threats becomes paramount. A well-defined cloud security policy helps with the following:

- **Risk management**: Identifying and mitigating risks associated with cloud deployments
- **Compliance**: Ensuring adherence to legal, regulatory, and security standards
- **Consistency**: Standardizing security practices across all cloud services and providers
- **Incident response**: Providing strategies for prompt and effective action in the event of a security breach

Implementing a Cloud Security Policy

Implementing such a policy requires careful planning, execution, and ongoing management to adapt to new security challenges and changes in the business environment. The creation of a cloud security policy in an organization is typically a collaborative effort involving several key roles. The **Chief Information Security Officer** (**CISO**) leads the initiative, setting the overall security direction and ensuring alignment with the organization's broader security goals. The IT security team provides essential technical expertise, detailing specific security controls, tools, and practices. Cloud security architects design and maintain the security architecture, ensuring integration with existing systems. Legal and compliance officers ensure the policy adheres to relevant laws and regulations, while the risk management team evaluates potential risks and proposes mitigation strategies. Operational and business unit leaders also contribute to ensuring that the policy supports daily business processes without compromising security. Additionally, while not directly involved in creating the policy, coordination with CSPs is crucial to align their offerings with the organization's security needs. Once the policy is drafted, it requires approval from senior management to guarantee organizational support and resources for effective implementation. This wide-ranging collaboration ensures the cloud security policy is comprehensive, practical, and attuned to both technical and business needs.

Now that you have defined the importance of those responsible for policy creation, let's now look at the steps of developing and implementing the policy in an organization:

1. **Define the scope of the policy**: Determine which cloud models (IaaS, PaaS, or SaaS), services, and providers the policy covers. This scope should align with the organization's overall IT and business strategy.

2. **Assess risks**: Conduct a thorough risk assessment of the organization's cloud environment to identify specific security risks associated with the chosen cloud service models and deployment types. This assessment should consider the data criticality and regulatory requirements specific to the organization's industry.

3. **Specify security requirements**: Based on the risk assessment, define security requirements tailored to the cloud services in use. These requirements should address aspects such as data encryption, IAM, network and application security, and incident response.

4. **Define roles and responsibilities**: Clearly outline the security responsibilities of all stakeholders involved, including the cloud service provider and the organization's internal teams. This clarity is crucial for maintaining security governance and operational effectiveness.

5. **Develop control measures**: Establish security controls to mitigate identified risks. These controls may include technical measures, such as the use of security tools and services offered by cloud providers, and managerial measures, such as regular security training for employees.

6. **Implement data protection measures**: Define how data is classified, handled, and protected in the cloud. Include measures for data at rest, in transit, and during processing, incorporating the use of encryption, tokenization, and other protective technologies.

7. **Create incident response plans**: Develop comprehensive incident response plans that specify how to respond to security incidents in the cloud. This plan should be integrated with the organization's broader incident management framework.

8. **Regular audits and revisions**: Schedule regular audits to ensure compliance with the cloud security policy. Revise the policy periodically to address emerging threats, technological changes, and business needs.

From there implementation should typically follow the following steps:

1. **Communication and training**: Communicate the policy across the organization and conduct training sessions to ensure that all employees are aware of the policy and understand their respective roles and responsibilities.

2. **Integration with existing policies**: Integrate the cloud security policy with existing IT and security policies to create a cohesive governance framework. This integration helps in addressing security comprehensively across all IT environments.

3. **Monitoring and enforcement**: Use security monitoring tools to continuously monitor the cloud environment for compliance with the policy. Enforce the policy strictly to maintain security standards and discipline.

4. **Continuous improvement**: Use feedback from audits, incident response activities, and new threat intelligence to continuously improve the cloud security policy. This iterative process helps the organization stay ahead of potential security challenges.

Regular Security Audits and Assessments

Regular security audits and assessments are crucial components in maintaining the integrity and security of cloud data. To ensure comprehensive coverage and adherence to best practices, the following are some detailed considerations for effectively managing cloud data security:

- **Frequency of audits**: It's recommended that security audits be conducted at least annually to confirm alignment with evolving industry standards and stringent regulatory compliance requirements. For industries dealing with highly sensitive data, such as finance or healthcare, more frequent audits may be necessary.

- **Scope of audits**: These audits should encompass every facet of the cloud infrastructure, including the following:

  - **Physical access controls**: Ensuring that physical access to data centers and hardware is strictly regulated and monitored.

  - **Logical access controls**: Review of authentication mechanisms, authorization processes, and other cybersecurity practices that control access to data and services.

  - **Network security**: Examination of network configurations, firewall policies, intrusion detection systems, and other defenses that protect against unauthorized network access.

  - **Data protection measures**: Verification of data encryption practices, data loss prevention strategies, and backup solutions to safeguard data integrity and availability.

- **Preparation for audits**: To facilitate thorough and effective audits, organizations must maintain an up-to-date inventory of all cloud resources and data assets. This inventory should be complemented by ready access to documented security policies, procedures, and any relevant forensic data that can support the auditing process.

- **Audit techniques**: Auditors should employ a strategic mix of manual review techniques and advanced automated tools to thoroughly identify vulnerabilities and potential weaknesses within the system. This dual approach allows for both granular inspection and broad coverage of the cloud environment.

- **Ongoing assessments**: Beyond regular audits, continuous assessments are vital for detecting emerging threats or shifts in the risk landscape that might necessitate immediate remedial actions. These assessments help in staying ahead of potential security threats by dynamically adjusting to new vulnerabilities and attack vectors.

- **Strategic insights**: Regular assessments provide critical insights into operational and security practices, highlighting areas that may benefit from enhanced training or updates to existing policies and protocols. This ongoing feedback mechanism is essential for the continual improvement and adaptation of security strategies.

Ultimately, the objective of regular security audits and assessments is to ensure the persistent protection of cloud data. By proactively identifying and addressing potential risks, organizations can significantly mitigate the likelihood of data breaches or other security incidents, thereby protecting their reputation and the trust of their customers.

Employee Training and Awareness

Employee training and awareness in cloud data security are essential for safeguarding organizations against cyberattacks and protecting integral business operations. Educating employees about security protocols significantly reduces risks associated with human error—one of the largest vulnerabilities in any security system. Such training includes teaching best practices for creating strong passwords, thus mitigating risks from one of the most common hacker entry points. Additionally, as employees increasingly use public Wi-Fi networks, they learn to navigate these insecure environments safely, often through the use of VPNs to secure data transfers. The growing reliance on mobile devices to access corporate resources also underscores the importance of mobile device security training, which includes managing security settings and understanding app-associated risks.

Moreover, training programs highlight the importance of identifying phishing attempts—common and effective tactics used by cybercriminals—as well as the necessity of regular data backups to ensure business continuity in the event of data loss or corruption. Regular updates and refresher courses are vital, as cyber threats are constantly evolving, and ongoing education helps maintain a security-conscious culture within the organization. This continuous educational effort is especially crucial in environments reliant on cloud technologies, where data breaches can have extensive repercussions. By instilling a comprehensive understanding of cybersecurity best practices, organizations not only enhance their security posture but also foster a culture where security is viewed as a collective responsibility.

Regular Data Backups and Recovery Planning

Routine data backup and recovery planning are critical components of a robust cloud data security strategy, essential for any cloud security practitioner focused on safeguarding organizational data. Effective backup and recovery measures ensure that data remains accessible and secure, even in the face of system failures, cyberattacks, or natural disasters. Here's a detailed approach to establishing a reliable backup and recovery framework:

- **Regularly scheduled backups**: It's crucial to establish a backup schedule tailored to the needs of the organization. Depending on the volatility and importance of the data, backups might be scheduled daily, weekly, or at another appropriate frequency. This regularity ensures that all critical data is duplicated systematically, minimizing data loss between backups.

- **Off-site backups**: Storing backup copies in a secure, off-site location is vital for protecting against physical threats such as fires, floods, or theft that might affect the primary site. Off-site backups provide an additional layer of security and ensure data availability if the main storage environment is compromised.

- **Testing backups**: Regular testing of backup systems is essential to confirm that data can be successfully restored from backup copies. This step verifies the effectiveness of the backup strategy and the integrity of the data within the backups, ensuring that recovery processes are operational and effective when needed.

- **Disaster recovery plan**: Every organization should have a comprehensive disaster recovery plan that outlines specific actions to be taken in response to various disaster scenarios, including cyberattacks, hardware failures, or natural events. This plan should detail procedures for securing data against unauthorized access during disruptions and strategies for quick recovery to maintain business continuity.

- **Access control**: Limiting access to data backups is fundamental to prevent unauthorized retrieval or alteration of sensitive information. Access should be restricted to authorized personnel only, with stringent authentication mechanisms in place to enforce this policy.

- **Integration with the security strategy**: Backup and recovery systems should be an integral part of the organization's overall security and disaster recovery strategies. By incorporating these systems into broader security planning, organizations can ensure a unified approach to data protection and disaster response.

For cloud security practitioners, understanding and implementing these backup and recovery practices is not just about compliance or operational necessity; it's about assuring stakeholders of the organization's resilience against data loss and its capacity to recover swiftly and effectively, thus maintaining trust and safeguarding the organization's reputation and operational capabilities in a cloud environment.

Best Practices for Data Retention, Archival, and Deletion

In addition to data classification and security, it's important to follow proven and accepted best practices for data retention, archival, and deletion. In doing so, businesses can ensure that only required data is retained and all inactive data is stored in low-cost secondary locations or securely deleted.

Implementing a sound data retention policy is essential for organizations to comply with legal requirements, regulatory guidelines, and industry standards. An effective data management plan includes identifying what data needs to be retained or deleted, the retention period, and ensuring that the right tools are leveraged in the process.

Retaining sensitive data beyond the compliance regulations' required time frame exposes a company to increased risks of cyber attacks or litigation.

One solution that organizations can explore is leveraging cloud-based archiving solutions that offer long-term data storage as well as quick access to archives through various devices regardless of location globally.

Developing a comprehensive data retention plan requires a clear understanding of the compliance regulations relevant to your organization, outlining the different types of electronic records to be retained or deleted entirely based on their value. These records could be customer-facing emails or internal employee communication, for instance.

Organizations should therefore define and enforce a data retention policy that clarifies the data types to be retained, the reasons for retention (e.g., to comply with regulatory requirements), and the required retention periods. They should also create archival and deletion policies to clarify what data should be archived or disposed of, when, and how.

Some other practices that can help streamline enterprise data retention, archival, and deletion processes:

- Map policies and processes to business needs and applicable legal regulations.
- Create different policies for different data types.
- Automate the processes based on business and regulatory requirements.

Understanding Data Lifecycle Management

Data Lifecycle Management (DLM) is a fundamental aspect of cloud data security, crucial for managing information from its creation to its destruction within a business. It involves the meticulous planning and execution of policies related to data creation, storage, access, sharing, retention, and deletion, ensuring that all stages comply with strict security standards. DLM starts by mapping the data flow across an organization, pinpointing how data is stored and accessed and by whom, thus securing sensitive or confidential information against unauthorized access while ensuring its availability for legitimate use.

Ensuring compliance with regulatory standards is another critical aspect of DLM, as these rules vary significantly across industries and geographies. Regularly reviewing and adapting to these regulations is essential to prevent legal issues and maintain operational legitimacy. Implementing effective DLM not only helps in meeting these regulatory demands but also in mitigating the risks associated with data breaches and cyberattacks by enforcing robust access controls and secure data handling practices.

The benefits of a well-implemented DLM strategy are manifold. It ensures regulatory compliance across cloud or hybrid environments, reduces the risk of security breaches, and enhances operational efficiency by minimizing downtime and optimizing resource utilization. This operational resilience translates into cost savings and supports business continuity. Furthermore, DLM enables scalability and future-proofing of business operations, making it an invaluable strategy for organizations looking to sustain long-term success in today's dynamic technological and regulatory environment. For cloud security practitioners, mastering DLM is essential to effectively secure data throughout its lifecycle and navigate the complexities of modern data governance and protection.

Implementing Data Retention Policies

Ensuring compliance with regulatory standards and maintaining data security necessitates the implementation of robust data retention policies. These policies dictate how long various types of data should be kept and outline procedures for securely disposing of data that is no longer needed. The following are some detailed best practices for creating and effectively implementing data retention policies:

- **Prioritize data security**: When developing data retention policies, security should be the foremost consideration. This involves not only protecting data while it is stored but also ensuring it is securely deleted when no longer needed.

- **Identify data retention requirements**: Carefully determine which types of data need to be retained and for how long based on both legal and business requirements. Different types of data may have different retention periods mandated by laws such as the **Federal Information Security Management Act (FISMA)**, **North American Electric Reliability Corporation (NERC)** standards, HIPAA, and the **Sarbanes-Oxley Act (SOX)**.

- **Develop a data purging plan**: Establish a systematic plan for regularly purging data that has exceeded its required retention period. This helps in minimizing risks associated with data breaches and reduces storage costs while ensuring that the organization does not retain unnecessary data.

- **Employee training**: Educate employees about the importance of adhering to the data retention policy. Training should cover how to handle data securely, the significance of compliance, and the potential consequences of policy violations.

- **Regular policy review and updates**: Continually review and update the data retention policy to reflect changes in regulatory requirements, technological advancements, and shifts in business strategy. This ensures that the policy remains relevant and effective in addressing current compliance and security challenges.

Regulatory compliance requirements such as FISMA, NERC, HIPAA, and SOX provide frameworks and set standards for how certain data should be handled and protected. For example, HIPAA requires healthcare organizations to retain medical records and other related health information for a minimum period, typically six years from the date of its creation or the date it was last in effect, depending on the given state's law. This retention period ensures the availability of medical records for patient care, billing purposes, and potential audits. FISMA requires federal agencies to develop, document, and implement an information security and protection program. While FISMA itself does not specify exact retention periods, it mandates that agencies' policies must ensure the effectiveness of security controls over information throughout its lifecycle, which includes retention. By implementing and adhering to strong data retention policies, organizations can ensure they meet these legal obligations while protecting sensitive information from unauthorized access or breaches. Effective data retention is not only a compliance issue but also a critical component of an organization's data governance and security strategy, helping to safeguard its information assets and maintain trust with clients and stakeholders.

Secure Data Deletion Techniques

Effective data deletion protocols are critical for maintaining robust security within cloud environments, ensuring that sensitive information is irretrievably erased when no longer needed. These practices are pivotal for protecting against unauthorized access and data leakage, especially in cloud settings where data management is handled by external service providers. By implementing stringent data deletion strategies, organizations can prevent potential security vulnerabilities associated with outdated or unnecessary data retention. This not only aids in regulatory compliance by adhering to prescribed data retention schedules but also optimizes cloud resource management by clearing up storage space. Thus, secure data deletion is integral to upholding an organization's integrity, ensuring data privacy, and maintaining customers' trust in the organization's cloud security measures. The following are some techniques that can be used:

- **Overwriting**: This technique involves writing over the existing data with random characters, making it difficult to recover the original data.

- **Cryptographic techniques**: This technique involves encrypting the data before deleting it, ensuring that even if it's recovered, it won't be readable without the encryption key.

- **Data wiping software**: This technique involves using specialized software to erase all traces of the data on a storage device.

These techniques can help ensure that deleted data is completely removed and cannot be accessed by unauthorized individuals.

Effective Data Archiving Solutions

Having a comprehensive data archiving solution is crucial for optimizing cloud security, ensuring that data that is no longer actively used but still valuable for future reference or regulatory compliance is securely stored. These solutions help organizations manage the volume of data in cloud environments by relocating outdated (but necessary) data to more cost-effective storage solutions. By implementing robust archiving strategies, organizations can enhance data accessibility and retrieval processes, while also ensuring comprehensive protection against data breaches. This practice supports compliance with legal retention requirements and reduces the burden on primary storage systems, ultimately improving overall system performance and reducing costs. Therefore, investing in efficient data archiving is fundamental for maintaining operational efficiency, compliance, and security in the cloud.

To ensure secure and reliable data archiving, organizations should consider these effective solutions:

- **Implementing a tiered storage approach**: This approach ensures that frequently accessed data is stored in high-performance storage while rarely accessed data is archived in lower-cost storage.

- **Utilizing cloud-based archives**: Cloud-based data archiving solutions offer significant cost savings and enable remote access to archived data.

- **Automated archival policies**: Automated policies ensure that only the necessary data is included in archives, reducing costs and improving the accuracy of long-term retention schedules.

- **Encryption for secure archiving**: Data encryption provides an extra layer of protection against unauthorized access to archived data, ensuring that sensitive information remains secure.

By utilizing these effective data archiving solutions, organizations can enhance their cloud data security by reducing the risk of loss or theft of important digital assets.

Compliance with Data Protection Regulations

Compliance with data protection regulations is a fundamental component of cloud data security currently. Various international, regional, and industry-specific regulations mandate stringent measures to ensure that organizations respect user privacy and secure private data effectively. These regulations encompass a broad range of requirements, including privacy laws, data governance standards, and personal information protection statutes.

For example, the **General Data Protection Regulation (GDPR)** in the European Union, the **California Consumer Privacy Act (CCPA)** in the United States, and the **Personal Information Protection and Electronic Documents Act (PIPEDA)** in Canada set comprehensive rules for data handling and consumer privacy. Organizations must adhere to these laws to avoid hefty fines and legal repercussions.

Moreover, compliance with these regulations often involves more than merely following legal mandates. Many companies choose to exceed baseline legal requirements by adopting additional best practices and standards such as ISO/IEC 27001 for information security management. These measures not only bolster the organization's data protection frameworks but also enhance trust among consumers by demonstrating a commitment to safeguarding their data.

Implementing robust compliance policies is not only a legal obligation but also a critical business strategy. Strong compliance helps prevent data breaches and other security incidents, thereby reducing potential financial losses and preserving the company's reputation. In essence, rigorous compliance with data protection regulations and standards is indispensable for maintaining customer trust and ensuring the long-term success of cloud-based operations.

You will cover privacy regulations and international laws in *Chapter 22, Legal Challenges and the Cloud* and *Chapter 23, Privacy and the Cloud*.

Lessons from Cloud Data Security Breaches

Recent cloud data security breaches have provided valuable lessons for organizations, emphasizing the importance of a strong security posture.

A notable example of a recent cloud data security breach that has offered critical lessons for organizations is the Capital One breach in 2019. In this incident, an engineer exploited a misconfigured web application firewall to access the data of approximately 106 million Capital One customers stored on AWS. The breach exposed sensitive information, including names, addresses, phone numbers, email addresses, dates of birth, and self-reported income, as well as credit scores and transaction data.

This breach highlighted several important lessons for organizations relying on cloud services:

- **Configuration and vulnerability management**: The breach was primarily due to a misconfigured firewall, underscoring the need for meticulous configuration management and regular security reviews to identify and rectify such vulnerabilities. Organizations must invest in automated tools to monitor configurations and compliance continuously.

- **Least-privilege access**: The extensive access allowed by the misconfiguration demonstrated the importance of implementing a least-privilege access policy, where users are granted only the permissions necessary to perform their job functions. Regular audits of access policies and practices are crucial to prevent exploitation.

- **Incident response and monitoring**: Capital One was notified of the breach by a third party, not through its internal systems, indicating a need for improved monitoring and detection strategies. Effective security monitoring and an agile incident response plan are vital to identify and mitigate threats promptly.

- **Employee training and insider threat programs**: The fact that the breach was carried out by a former employee of a cloud service provider highlights the risk posed by insiders. Organizations must strengthen their security training programs and implement robust insider threat prevention strategies.

- **Vendor risk management**: Since the data was hosted on a third-party service (AWS), the incident underscores the importance of comprehensive vendor risk management, including understanding the security measures and compliance policies of all third-party services used.

By learning from such incidents, organizations can enhance their cloud security posture, implement stronger safeguards, and ensure better protection against potential data breaches in the future.

The following table highlights key takeaways and best practices to improve cloud data security:

Lesson	Best Practice
Avoid leaving storage containers and databases exposed.	Regularly review and update security configurations, ensuring that storage containers and databases have proper access controls in place.
Address misconfiguration of cloud security settings.	Implement a cloud security posture management strategy to monitor, identify, and remediate security misconfigurations.
Understand the differences between cloud data breaches and on-premises attacks.	Adapt security strategies to address the unique risks of cloud environments, focusing on securing access to sensitive information rather than exclusively on malware prevention.
Recognize the shared responsibility model for cloud security.	Work closely with cloud vendors to understand and fulfill the organization's responsibilities in securing data and transactions in the cloud.
Focus on disaster recovery and employee education.	Establish a comprehensive disaster recovery plan and invest in employee training to raise awareness of cloud security risks and best practices.

Table 6.1 – High-level lessons learned and best practices

Summary

In this chapter, you understood that cloud data security is a broad domain with many important topics. Each stage of the cloud data lifecycle is an important area that you should fully understand. You also covered storage types and common threats based on the storage type leveraged as well as the importance of creating cloud security policies and best practices as they relate to what can be learned from data breaches.

Leverage the examples provided to remember the stages and what happens in each stage to help with questions in this area. In addition, understanding the typical data storage types used in the cloud, and the threats commonly associated with each storage type, are also important to understand for the exam. This chapter also provided the opportunity for you to learn the importance of data classification and the overall best practices with regard to an organization's data governance strategy.

In the next chapter, you will continue delving into data governance with a review of information rights management and the best practices for the auditability, traceability, and accountability of data in cloud environments.

Exam Readiness Drill – Chapter Review Questions

Apart from a solid understanding of key concepts, being able to think quickly under time pressure is a skill that will help you ace your certification exam. That is why working on these skills early on in your learning journey is key.

Chapter review questions are designed to improve your test-taking skills progressively with each chapter you learn and review your understanding of key concepts in the chapter at the same time. You'll find these at the end of each chapter.

> **How to Access These Materials**
>
> To learn how to access these resources, head over to the chapter titled *Chapter 25, Accessing the Online Resources.*

To open the Chapter Review Questions for this chapter, perform the following steps:

1. Click the link – `https://packt.link/CCSPE1_CH06`.

   Alternatively, you can scan the following **QR code** (*Figure 6.1*):

Figure 6.1 – QR code that opens Chapter Review Questions for logged-in users

2. Once you log in, you'll see a page similar to the one shown in *Figure 6.2*:

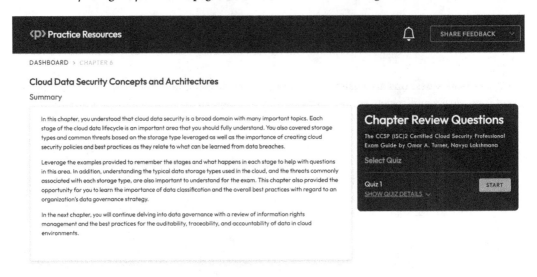

Figure 6.2 – Chapter Review Questions for Chapter 6

3. Once ready, start the following practice drills, re-attempting the quiz multiple times.

Exam Readiness Drill

For the first three attempts, don't worry about the time limit.

ATTEMPT 1

The first time, aim for at least **40%**. Look at the answers you got wrong and read the relevant sections in the chapter again to fix your learning gaps.

ATTEMPT 2

The second time, aim for at least **60%**. Look at the answers you got wrong and read the relevant sections in the chapter again to fix any remaining learning gaps.

ATTEMPT 3

The third time, aim for at least **75%**. Once you score 75% or more, you start working on your timing.

> **Tip**
>
> You may take more than **three** attempts to reach 75%. That's okay. Just review the relevant sections in the chapter till you get there.

Working On Timing

Target: Your aim is to keep the score the same while trying to answer these questions as quickly as possible. Here's an example of how your next attempts should look like:

Attempt	Score	Time Taken
Attempt 5	77%	21 mins 30 seconds
Attempt 6	78%	18 mins 34 seconds
Attempt 7	76%	14 mins 44 seconds

Table 6.2 – Sample timing practice drills on the online platform

> **Note**
>
> The time limits shown in the above table are just examples. Set your own time limits with each attempt based on the time limit of the quiz on the website.

With each new attempt, your score should stay above **75%** while your "time taken" to complete should "decrease". Repeat as many attempts as you want till you feel confident dealing with the time pressure.

7

Data Governance Essentials

In the last chapter, you learned about the essentials of cloud data security. You explored the cloud data life cycle, data flows, data classification, and common threats based on the various cloud storage types.

You will continue to look at data in this chapter, but from the perspective of data governance. Cloud data security and governance are not only important practices for your organization but also essential areas of study for the CCSP exam. You will review the key concepts of governance relating to the data life cycle phases in cloud environments, including **Information Rights Management** (**IRM**) as well as best practices for auditability, traceability, and accountability.

By the end of this chapter, you will be able to answer questions on the following topics:

- Data governance
- IRM and its implementation
- How to define, design, and implement auditability, traceability, and accountability for the cloud data life cycle phases

Data Governance

Over the past several years, industries worldwide have increasingly recognized the immense value of data as a critical business asset. This transformation has driven organizations to harness data for strategic decision-making, innovation, and competitive advantage. However, with this shift comes a set of formidable challenges, particularly the need to comply with stringent laws related to **Governance, Risk, and Compliance** (**GRC**). These regulations are evolving rapidly, and they vary significantly across different regions and sectors.

Today, large organizations across diverse sectors—including corporations, universities, hospitals, and government agencies—are under immense pressure to establish robust policies that govern data sharing, privacy, and compliance. These policies must not only meet the regulatory requirements of their home country but also adhere to international standards, making the task even more complex. For example, a multinational corporation must navigate the intricacies of GDPR in Europe, CCPA in California, and various other local data protection laws.

Effective governance is not just a box-ticking exercise; it must be ingrained in the organization's DNA from the moment data is collected. This means implementing stringent measures to ensure data accuracy, secure storage, and controlled access right from the outset. Moreover, understanding data lineage — knowing where data comes from, how it is transformed, and where it is stored — is crucial for maintaining compliance and ensuring data integrity.

Organizations must deploy comprehensive data governance frameworks that include data classification, **Role-Based Access Control (RBAC)**, and encryption to protect sensitive information. Regular audits and continuous monitoring are essential to identify and mitigate risks proactively. Additionally, fostering a culture of data stewardship within the organization can ensure that employees at all levels understand the importance of data governance and comply with established policies.

Two key governance models outline data governance principles and best practices:

- **Confidentiality, Integrity, and Availability (CIA)** is a conceptual model that will be familiar to many readers experienced in cybersecurity roles.

- **Authentication, Authorization, and Accounting (AAA)** is the model with more direct applicability in system design and implementation.

Another key element as it relates to data governance is the roles that are responsible for handling data. In an organization, various roles are essential for effective data management, governance, and protection. Each role has specific responsibilities that contribute to the overall data strategy and compliance with regulatory requirements. Here are key data role types:

- **Data owner**: The data owner is typically a senior business leader or manager who si responsible for a particular set of data within the organization. This role entails ultimate accountability for the data's integrity, security, and usage. Data owners make decisions about who has access to the data, how it should be managed, and what security measures are necessary. They also ensure that data governance policies are followed and that the data meets regulatory and business requirements.

- **Data controller**: The data controller is the person or entity that determines the purposes and means of processing personal data. In regulatory terms, such as under the GDPR, the data controller has significant responsibilities for ensuring that data-processing activities comply with data protection laws. This includes implementing appropriate technical and organizational measures to safeguard data and ensuring that data subjects' rights are upheld.

- **Data custodian**: Data custodians manage the technical environment where data is stored and processed. They are responsible for the day-to-day maintenance and protection of data, including database management, implementing security controls, and ensuring data backups and recovery processes are in place. Data custodians work closely with data owners to ensure that data governance policies are effectively implemented at the technical level.

- **Data processor**: A data processor is any person or entity that processes data on behalf of the data controller. This role includes performing operations such as the collection, storage, retrieval, and deletion of data as directed by the data controller. Data processors must adhere to the data controller's instructions and are responsible for implementing appropriate security measures to protect the data. Under regulations such as GDPR, data processors have specific legal obligations to protect personal data and can be held liable for breaches.

- **Data steward**: Data stewards are responsible for managing and overseeing the quality and life cycle of data within their domain. They ensure that data is accurate, consistent, and used appropriately. Data stewards work closely with data owners, custodians, and processors to enforce data governance policies and standards. They play a critical role in data quality management, metadata management, and ensuring that data is used in compliance with regulatory requirements and organizational policies.

- **Data subject**: The data subject is the individual to whom personal data relates. Under data protection laws such as GDPR and CCPA, data subjects have specific rights, including the right to access their data, request corrections, and demand the deletion of their data under certain conditions. Organizations must ensure that data subjects' rights are respected and that their data is processed lawfully, transparently, and for legitimate purposes.

From a CCSP exam perspective, remember the following:

- **Data owner**: Entity responsible for data integrity, security, and usage decisions

- **Data controller**: Entity determining purposes and means of processing personal data, ensuring compliance with data protection laws

- **Data custodian**: Manages technical aspects of data storage and protection, implementing security controls

- **Data processor**: Processes data on behalf of the data controller, adhering to their instructions and protecting the data

- **Data steward**: Oversees data quality and life cycle, enforcing governance policies and standards

- **Data subject**: Individual to whom personal data relates, with rights under data protection laws

Each of these roles plays a vital part in ensuring that data is managed responsibly, securely, and in compliance with relevant regulations, supporting the organization's overall data governance framework.

Data Dispersion in the Cloud

Data governance in the cloud presents unique challenges due to the distributed nature of cloud data. In cloud environments, data is often stored across multiple geographic locations, each subject to its own set of privacy laws, regulatory requirements, and region-specific regulations. This geographical dispersion necessitates a global view of all data and centralized control mechanisms to ensure comprehensive compliance and effective management.

To tackle these challenges, technologies for data governance must seamlessly integrate with the existing cloud infrastructure. This integration is essential for providing visibility and control over data, regardless of where it is stored. Advanced governance solutions should offer automated data classification, robust encryption, access management, and continuous monitoring to enforce policies consistently across the entire cloud environment.

The distribution of cloud data across various storage locations and service providers means that data governance policies must be enforced at the data level rather than the system level. Data-level governance ensures that the same standards of data protection, privacy, and compliance are applied universally, regardless of the underlying infrastructure or the physical location of the data. This approach involves tagging and tracking data assets, applying encryption to protect data at rest and in transit, and implementing RBAC to restrict data access to authorized users only.

The Importance of Data Governance for Cloud Security

Three of the five key characteristics of a cloud computing system as defined by NIST—broad network access, rapid pooling, and rapid elasticity—require data and its copies to be available in geographically spread locations that are maintained by the CSP. However, the threat surface fragments and expands when data exists in multiple locations outside the organization's boundaries, thereby enhancing the role of data governance as a key component of overall cloud security.

The main points to grasp about how data governance and cloud security impact one another are the following:

- **Data homogeneity**: Data silos resulting from organizational growth, new databases, or acquisitions lead to similar or related data being stored in disparate databases across the organization, in turn complicating data integration and governance.

- **Regulatory compliance**: Regulatory compliance (privacy, security, geographic requirements, etc.) must be built into data-handling practices within the cloud. Data located across geographic boundaries leads to various concerns with respect to regulatory compliance:

  - Complex data privacy challenges due to data movement across jurisdictions with varying regulations.

  - **Data Loss Prevention (DLP)** is a crucial practice for restricting data access to authorized users, reducing risk, and ensuring compliance across data states (at rest, in transit, and in use).

  - Verification by cloud security professionals to ensure compliance with the data-related regulation of a specific region.

Leveraging Cloud-Specific Tools and Services

At a high level, the following are features to look for in tools and services for data governance in the cloud:

- Data-driven governance technology that enforces governance policies at the data level, allowing data to remain compliant as it travels

- Technology that marries policies such as CIA (Confidentiality, Integrity, and Availability) and AAA (Authentication, Authorization, and Accounting) to data, enabling automated enforcement across clouds

- Fine-grained access controls (role-based, attribute-based, behavior-based, etc.)

- Data management methods, such as data lakes and warehouses, to help overcome the problems caused by silos

- Flexible frameworks incorporating metadata for compliance across jurisdictions, which is important for multinational companies dealing with different regulations in each country of operation

- Centralized data access and governance through platforms that offer a single pane of glass through which data cataloging and data intelligence tools are available

IRM

Controlling individuals' level of access to data within an organization is a crucial part of data governance. The set of techniques used to protect sensitive data through control of access, usage, and distribution privileges is known as **IRM**. IRM enables centralized control over who is authorized to access and manipulate data, even when the data lives outside an organization's four walls, for example, across multiple cloud environments. This secures data from potential unauthorized or malicious use and ensures access to authorized users based on their individual privileges.

IRM is a vital framework that allows organizations to protect sensitive information by managing how data is accessed and used, both within and outside the organization. By embedding security policies directly into data files, IRM ensures that data remains protected regardless of its location or who attempts to access it.

IRM helps answer fundamental questions about data security: Who does the data belong to? What rights come with it? How are those rights protected? By establishing clear ownership of data, IRM ensures that data originators or owners maintain control over their information, even when shared with others. It defines the rights and permissions associated with data, specifying who can view, edit, print, copy, or forward the information, and enforces these rights through encryption, access controls, and other security measures.

Several key IRM controls are essential for comprehensive data protection. Access control mechanisms restrict who can view or edit the data, often using RBAC to assign permissions based on user roles. Encryption ensures that data remains unreadable without proper decryption keys, protecting it from unauthorized access. Usage policies dictate how data can be used, such as allowing a document to be viewable but not printable by certain users. Expiration policies enforce time-based restrictions, making data inaccessible after a specific date. Watermarking deters unauthorized sharing and provides traceability for leaked information. Auditing and monitoring capabilities track data access and usage, providing visibility into potential misuse and aiding in compliance reporting. Additionally, IRM enables the revocation of access rights even after data has been distributed, allowing data owners to maintain control.

In cloud environments, where data is frequently shared across various platforms and devices, IRM is even more critical. CSPs offer integrated IRM solutions to help organizations manage data rights in a distributed environment, ensuring that data remains secure and compliant with organizational policies and regulatory requirements.

For CCSP candidates, understanding IRM is important as it forms the backbone of comprehensive data security strategies. Understanding IRM equips professionals with the skills to protect sensitive information and implement robust data governance frameworks. By leveraging IRM controls such as access control, encryption, usage policies, and auditing, organizations can maintain control over their data, mitigate risks, and comply with legal and regulatory obligations.

The term **DRM** stands for **digital rights management** or **data rights management** and is also commonly used in the computing industry to refer to this concept. Usually, though, DRM applies to consumer media protections, for example, music or video files to avoid piracy, while IRM applies to the protection of information within systems, such as digitized files—think Word or PowerPoint documents. You may also see variants such as **Enterprise Rights Management** (ERM) and E-DRM utilized. Note that ISC2 generally prefers the term IRM for discussing this concept. Throughout this book, the term IRM will be used for clarity and consistency.

The central aspect of IRM is the handling of information with reference to the rights of its owner. Notably, there is no universally accepted international or industry standard spelling out these terms or their application in specific processes or technical approaches. Nevertheless, the overarching concept of "rights management" involves deploying specialized controls that complement or augment an organization's existing access control mechanisms to safeguard various asset types, typically at the file level. A plausible scenario may look like this: an organization institutes an access control protocol requiring users to authenticate when accessing systems for job-related tasks. Once past initial authentication, users may encounter additional "rights management" controls that protect access to certain files, such as sensitive financial records, preventing unauthorized users from reading, distributing, deleting, altering, or copying them.

The Role of IRM in Data Governance

IRM is crucial to effective data governance. It ensures that data is managed and protected according to preassigned rights and permissions. IRM gives organizations control over sensitive data even when stored and processed in remote cloud environments, mitigating the risk of unauthorized access, misuse, or data breaches.

Key IRM policies and technologies include the following:

- Data access controls

- Data ownership and accountability

- Monitoring of data usage

- Encryption of data in transit, in use, and at rest

These measures help CSPs safeguard data through its entire life cycle.

Additionally, IRM helps organizations comply with regulatory requirements and industry standards governing data privacy and security, including GDPR, HIPAA, and PCI DSS. IRM provides the means to control, track, and protect data, supporting its CIA in cloud environments. These capabilities make IRM indispensable to a robust data governance strategy.

Key Components of an Effective IRM System

An effective IRM system is composed of several key components that work together to ensure data security and proper management of access rights. These components are essential for protecting sensitive information, maintaining compliance with regulatory requirements, and enabling secure data sharing both within and outside an organization. Here are the key components:

- **Data labeling**: Data labeling is the process of categorizing and tagging data based on its sensitivity and classification. Labels can include categories such as public, internal, confidential, and restricted. This classification helps in applying appropriate security controls and access policies. Data labeling ensures that sensitive information is easily identifiable and handled according to its classification level, thereby preventing unauthorized access and ensuring compliance with data protection regulations.

- **Access control**: Access control mechanisms are fundamental to IRM systems. They determine who can access specific data and what actions they can perform. RBAC and **Attribute-Based Access Control** (**ABAC**) are commonly used methods. RBAC assigns permissions based on user roles within the organization, while ABAC considers various attributes, such as user location, time of access, and data sensitivity. These controls ensure that only authorized users can access and interact with sensitive data.

- **Encryption**: Encryption is a critical component of IRM, providing a robust layer of security by making data unreadable to unauthorized users. IRM systems typically employ strong encryption algorithms to protect data both at rest and in transit. Encryption keys are managed securely to ensure that only authorized users can decrypt and access the data. This protects sensitive information from being intercepted or accessed by malicious actors.

- **Usage policies**: Usage policies define how data can be used once it is accessed. These policies can specify permissions such as viewing, editing, printing, copying, and forwarding. By embedding these policies directly into the data files, IRM systems ensure that the rules are enforced regardless of where the data travels. This prevents misuse and unauthorized dissemination of sensitive information.

- **Auditing and monitoring**: Continuous auditing and monitoring are essential for tracking data access and usage. IRM systems record who accessed the data, when it was accessed, and what actions were performed. This audit trail helps in detecting and responding to unauthorized activities, ensuring accountability, and maintaining compliance with regulatory requirements. Monitoring tools can also trigger alerts for suspicious activities, enabling proactive security measures.

- **Expiration and revocation policies**: Expiration policies allow data owners to set time-based restrictions, making data inaccessible after a certain date. This is particularly useful for time-sensitive information. Revocation policies enable data owners to revoke access rights even after data has been shared. This ensures that if data is accidentally shared with the wrong person or if access needs to be terminated for any reason, the data owner can maintain control.

- **Integration with existing systems**: An effective IRM system must integrate seamlessly with existing IT infrastructure, including cloud services, collaboration platforms, and enterprise applications. This integration ensures that data governance and security policies are consistently applied across all environments. It also simplifies the management of data rights and enhances the overall efficiency of the IRM system.

- **Watermarking**: Watermarking is an additional security measure that involves embedding identifiable marks into documents. These marks can include information about the document owner, creation date, and intended recipients. Watermarking helps in deterring unauthorized sharing and provides a way to trace the source of leaked information.

- **Training and awareness**: An often-overlooked component of an effective IRM system is training and awareness. Employees must be educated about data classification, usage policies, and security practices. Regular training sessions ensure that staff are aware of the importance of IRM and understand how to comply with data protection policies.

IRM and Cloud Computing

IRM is vital in cloud computing for protecting **intellectual property (IP)**, managing data access, and complying with regulations. In the cloud, as in any other computing environment, it manages the access and usage of digital assets through rights and permissions. However, IT and data security professionals implementing IRM in cloud environments face specific challenges related to replication, jurisdictional conflicts, integration with IAM, and ensuring seamless functionality across different applications and platforms.

The essential functions of IRM in the cloud include the following:

- **Persistent protection**: Continuous safeguarding of digital content
- **Dynamic policy control**: Flexible modification of access controls
- **Automatic expiration**: Cease IRM controls when legal protections expire
- **Continuous auditing**: Monitor access history and content usage
- **Replication restrictions**: Restrict unauthorized duplication
- **Remote rights revocation**: Capability to revoke IP rights, which is useful in legal cases

Some common challenges cloud security professionals run into when implementing IRM in cloud environments are as follows:

- **Replication issues**: IRM may hinder unauthorized duplication in cloud replication processes.
- **Jurisdictional differences**: Enforcing IP rights across geographical boundaries presents challenges.
- **Enterprise/agent conflicts**: Problems with integration arise with IRM solutions in cloud, virtualization, and **Bring Your Own Device (BYOD)** setups.
- **IAM and IRM integration**: Conflict between enterprise IAM and cloud IRM mechanisms. IAM is a very important prerequisite for IRM as it establishes the identity of a person accessing resources.
- **API conflicts**: IRM tools may not function consistently across different platforms.
- **IRM tool features**: Implementations vary in sophistication, including metadata, reference checks, and licensing.

Implementing IRM

Implementing IRM in real-life scenarios requires a combination of careful planning and well-chosen techniques and tools. While each organization has its own approach, some tools and best practices are vital to any successful implementation. The following is intended to provide a picture of what implementing IRM looks like for a typical cloud security professional working today.

Here are the IRM implementation steps:

- **Selection of IRM technology**: Choose an appropriate IRM solution from available tools. IRM platform solutions typically offer a suite of features for data protection, access control, and policy enforcement.

- **Integration with existing systems**: Ensure seamless integration of the chosen IRM technology with IT infrastructure already in use. This may require API integration, customization, and compatibility testing.

- **Labeling and metadata**: IRM solutions usually require some type of labeling or metadata applied to the protected material to work properly. This metadata assists in enforcing access policies, tracking data usage, and applying encryption where necessary.

- **Determine data retention policies**: This is done at the organizational level and based on specific governance and regulatory requirements

- **Access controls**: Configure granular access controls based on user roles, permissions, behavior, and content types. Fine-grained access controls can enforce security policies based on highly specific or contextual factors. Common types of access controls include RBAC, ABAC, and behavior-based access control.

- **Enable auditing**: Enable the IRM system's auditing features to track user activities, including content access, modifications, and sharing. Auditing provides a "paper trail" for both compliance audits and security incident investigations.

- **Monitoring and maintenance**: Establish regular monitoring of IRM operations. Check system performance, security compliance, SIEM, DLP and other security controls, and policy enforcement. Conduct periodic reviews and updates to stay abreast of new threat types and regulations.

> Note
>
> There are numerous IRM technologies available, varying in sophistication. Some of the most popular technologies in use today include Microsoft Azure Information Protection, Adobe LiveCycle Rights Management, Seclore FileSecure, DLP solutions, DRM platforms, **Identity and Access Management** (**IAM**) systems, encryption technologies (e.g., symmetric encryption or public-key encryption), access control mechanisms (RBAC and ABAC), auditing and logging tools, and compliance management software.

To recap, the steps to successful IRM implementation are technology deployment, policy enforcement, user education, and ongoing monitoring. With these tools and processes, organizations can ensure day-to-day reliable data protection and reduced security risk. Also, as a reminder, the CCSP exam is vendor agnostic, so don't worry about the vendor solutions listed; just know that they exist to help with your research.

Implementing IRM in a Cloud Environment

Implementing IRM in cloud environments involves specific requirements that are as follows:

- **User accessibility**: IRM policies should not hinder legitimate user access to data stored in the cloud. Implementing IRM should strike a balance between security and usability to ensure that authorized users can access information efficiently.

- **Policy flexibility**: IRM policies must be flexible and easy to configure; continually changing business requirements and compliance regulations necessitate this. Data owners must be able to adjust access controls and permissions dynamically without disrupting operations.

- **Integration with cloud services**: IRM solutions should seamlessly integrate with cloud services and platforms in use in the organization. IRM compatibility is needed with cloud storage solution collaboration tools and productivity suites for full data protection across all cloud-based services and applications.

- **Monitoring and auditing**: Continuous monitoring and auditing of data access and usage are crucial in the cloud. IRM implementations should have logging and reporting capabilities to track user activities, detect unauthorized access attempts, and ensure compliance with regulatory mandates.

- **Data classification and labeling**: Proper data classification and labeling are foundational aspects of cloud IRM. Organizations should categorize data based on sensitivity levels and apply IRM policies accordingly. Automated classification tools can streamline this process and ensure security controls are enforced consistently.

- **Disaster recovery and DLP**: Disaster recovery and DLP measures are needed for full-scale data protection. Disaster recovery measures safeguard the integrity and availability of data in the event of an unforeseen incident or system failure in the cloud environment. DLP ensures that sensitive information is not accessed, shared, or transferred inappropriately, aligning with IRM by enforcing data usage policies, monitoring data movements, and preventing unauthorized disclosures, thereby protecting data integrity and compliance with regulatory requirements.

- **Replication issues**: Cloud environments often involve replicating virtualized host instances, including user-specific content. IRM's prevention of unauthorized duplication might interfere with automatic resource allocation in the cloud.

- **Jurisdictional differences**: Cloud services extend across geographic borders, which can complicate data governance and compliance with IP laws and regulations that differ by region.

- **Enterprise/agent conflicts**: In a cloud environment, compatibility issues may arise if IRM solutions rely on local edge or endpoint-hosted software agents for enforcement. This is particularly likely with the virtualization engines and different platforms used in BYOD scenarios.

- **Integration with IAM**: IRM introduces an additional layer of access control, which needs close integration with existing IAM processes, especially in cloud environments where IAM functions may be outsourced to third-party providers such as **Cloud Access Security Brokers (CASBs)**.

- **API conflicts**: Incorporating IRM into cloud-based applications may cause inconsistencies in performance across different platforms. This can, for example, adversely impact user experience with content readers or media players.

Best Practices for IRM Implementation

There are certain practices you can follow to make your IRM implementation efficient. Here are three best practices:

- **Initial planning**: Plan the IRM strategy with the involvement of all stakeholders, including perhaps manufacturers, vendors, or data owners of content. Also, collaboration between business and IT teams is necessary to avoid misunderstandings and ensure a smooth rollout.

- **Education and training:** Conduct training sessions to educate employees on IRM policies, procedures, and best practices. Help users to appreciate their role in protecting sensitive information by adhering to IRM guidelines.

- **Continuous improvement**: Continuously evaluate the IRM implementation and look for areas for enhancement. Stay up to date on emerging IRM technologies, best practices, and industry standards to improve data protection capabilities.

Case Studies of Successful IRM Implementation

A successful IRM implementation in the context of a financial institution could look like the following.

A financial institution keeps important customer data, such as details about accounts and transaction history, in the cloud for its online banking service. Robust IRM ensures that the institution has strict control over who can view or modify this stored information. The use of access controls, encryption methods, usage policies, and other features prevents customers' private data from being disclosed without authorization. These features make certain that only permitted individuals can access sensitive information, and any efforts to share or modify data without authorization are quickly detected and stopped. If there is a security issue, IRM provides comprehensive audit paths for the institution and forensic investigators to find out why it occurred. This helps them take rapid corrective steps to mitigate security risks and keep their customer's trust intact.

Auditability in Cloud Data Governance

Auditability in cloud data governance refers to the ability to track, monitor, and analyze data-related activities, access, and changes within a cloud environment. Cloud security professionals leverage various monitoring and auditing software tools to continually monitor compliance and track data activities. Such tools detect anomalies in data access, changes, and compliance, and generate audit trails that record access and detect suspicious activities.

Auditability is an indispensable part of cloud data governance, ensuring data stays secure and compliant over the long run. Auditing can detect incidents in real time, supporting efforts for both immediate actions to avert security threats and continuous improvement of the data governance framework. Audits ensure accountability in organizations, maintaining data integrity and trust.

Implementing Auditability in the Cloud

Implementing auditability in cloud data governance can be complex. A blend of technological tools, procedural frameworks, and organizational oversight is essential to ensure reliable day-to-day auditing capabilities in the cloud environment. Implementation will look somewhat different depending on the organization. Here is a review of what implementation may look like for an organization today:

- **Tools and processes**:

  - Utilization of **Data Life Cycle Management** (**DLM**) practices to reinforce adherence to regulations and facilitate reporting processes

  - Reliance on data classification mechanisms to categorize data based on sensitivity and importance

  - Application of tailored security measures such as encryption and access controls to different data types

  - Deployment of specialized tools, such as Azure security audits and AWS Audit Manager, to continuously monitor and log data-related activities

  - Maintenance of audit trails that record data access instances, enhancing investigation and potentially improving the data governance and security program

- **People**:

  - Data stewards may enforce governance policies and oversee their implementation

  - An internal data governance committee may oversee audits and risk assessments, addressing any deficiencies

  - Regular audits of the data governance program to measure metrics and **Key Performance Indicators** (**KPIs**) and ensure alignment with regulations and internal policies

  - Third-party auditing such as the CSP assessment to evaluate the security and compliance of CSPs

- **Tools and technologies for enhancing auditability**:

  - Specialized tools such as Azure security audits and AWS Audit Manager with real-time monitoring capabilities.

  - Data classification through automated tagging or manual labeling.

  - Access management solutions such as Azure Active Directory or Google Cloud Identity offer access controls for different data types.

  - Fine-grained access controls, such as RBAC, ABAC, and **Policy-Based Access Control** (**PBAC**).

  - Systems that record audit trails to track data access and usage.

  - Audit trails provide information about data access, changes, and usage, enabling visibility into encrypted data activities without compromising security.

  - Data analytics tools.

  - Data discovery, mapping, and cataloging tools.

Let's look at how auditability could be done in an organization.

A **Software-as-a-Service** (**SaaS**) provider offers a cloud-based **Customer Relationship Management** (**CRM**) platform for businesses that operate in the **European Union** (**EU**). The SaaS provider implements certain features to manage data subject rights in accordance with GDPR requirements. These rights include the right to access, store, and erase personal data. Through auditing capabilities, the provider maintains detailed records of data subject requests, including timestamps, action taken, and communication logs. While auditing GDPR compliance, regulators examine these records to ensure appropriate responses to data subject requests, accurate record keeping, and compliance with data protection principles. By demonstrating compliance through audit records, the SaaS provider avoids regulatory sanctions. It also builds customer confidence and sets itself apart in a competitive market.

Traceability in Cloud Data Governance

Traceability in data governance involves tracking activities, changes, and access patterns throughout the data life cycle. It has much in common with auditability, which you covered earlier. While auditability primarily concerns the verification of compliance through retrospective assessments, traceability focuses on the real-time monitoring of data movements and activities. Both concepts are integral to effective cloud data governance, working in tandem to uphold regulatory requirements, mitigate risks, and foster trust in data management practices.

Traceability ensures data integrity, security, and compliance. It demonstrates accountability and readiness to remediate any data security issues that arise. As one tool in the overall data governance program, traceability is vital to data intelligence, allowing organizations to proactively manage risks and maintain compliance throughout the data life cycle.

Implementing Traceability in the Cloud

In the day-to-day implementation of traceability, organizations use a combination of tools and processes to track and trace data-related activities.

With data classification and implementation of security measures such as encryption and access controls taken care of, tools for monitoring and tracing are put in place to "watch" the data, as it were. IT professionals deploy data monitoring and auditing tools to capture real-time events related to data access, modification, and usage.

These tools generate detailed logs and audit trails, providing visibility into how data is being handled within the cloud environment. Automated validation checks ensure data quality and integrity and report data anomalies. Audit trails are maintained to aid in the detection and investigation of anomalous activities related to data. Regular assessments—internal and external—ensure ongoing compliance with changing regulations and organizational policies.

Data stewards ensure compliance by enforcing data retention and disposal policies and conducting audits to verify adherence.

Tools and Technologies for Enhancing Traceability

To help IT and data security professionals implement traceability properly, the following are tools and technologies at their disposal:

- **Security Information and Event Management (SIEM) systems**: SIEM systems aggregate security event logs from various sources, including network devices, servers, and applications. They provide centralized logging, storage, and analysis of data events, as well as real-time monitoring and detection of security incidents.

- **CASBs:** As intermediaries between CSPs and users, CASBs enhance auditability by providing visibility into cloud-based activities. They enforce security policies and access controls, monitor data access and usage, and log events with attributes such as user identity and IP address.

- **IAM solutions**: IAM solutions manage user identities, access permissions, and authentication mechanisms. They enforce the principle of least privilege, ensuring users have only the necessary access rights, and log access events and attribute data with user identities and other metadata.

- **Cloud-native logging and monitoring services**: Cloud logging and monitoring services capture and analyze data events within the cloud infrastructure. They provide scalable, resilient, and cost-effective solutions for the logging, storage, and analysis of data events. They also facilitate auditability by supporting the definition of event sources and requirements of event attributes.

- **Blockchain technology**: Blockchain technology provides a decentralized and immutable ledger for recording data transactions. It ensures a chain of custody and non-repudiation through an auditable record of data events. Tamper-proof and verifiable, blockchain can enhance auditability in cloud data governance with transparent and trustworthy data-activity logs.

- **DLP solutions**: DLP solutions monitor and police data movements within and outside the organization's network. By identifying and preventing unauthorized data transfers, they ensure data integrity and confidentiality. Additionally, they support auditability by logging data events and attributing them with relevant metadata.

- **Endpoint Detection and Response (EDR) systems**: EDR systems monitor endpoints such as desktops, laptops, and servers for signs of malicious activity. They collect endpoint telemetry data, including process execution, file modifications, and network connections, and analyze it for security threats, generating detailed event logs and attributes to aid forensic investigation and auditability.

As an example of how an organization could implement traceability, let's look at the following.

A healthcare clinic uses cloud-based **Electronic Health Record** (**EHR**) systems to store patient information. As part of regulatory compliance, the clinic undergoes periodic audits to ensure adherence to data protection laws such as HIPAA. Traceability mechanisms allow auditors to trace patient data through its life cycle in the cloud environment, including its creation, access, modification, and sharing activities. With these detailed logs and audit trails, the clinic shows compliance with regulations and increases trust among stakeholders, including patients and regulatory bodies.

Accountability in Cloud Data Governance

Accountability means the ability to hold individuals or entities accountable for their actions and decisions related to data, and the outcomes of those actions. In addition to reducing risks and preventing data breaches, accountability promotes transparency, trust, and integrity in cloud data management practices. When individuals and organizations are accountable for their actions, a culture of responsibility emerges. Confidence among stakeholders and customers increases. Also, documentation and evidence of adherence to data protection requirements and standards aid regulatory compliance.

Key Components of Accountability

Accountability involves defining roles, responsibilities, and processes to ensure that data management practices meet the highest standards of integrity, security, and compliance. The process to ensure accountability includes data collection, storage, processing, sharing, and deletion. Clear lines of responsibility and authority are established to ensure that data governance policies are enforced consistently and effectively. Here are the key components of accountability arranged in a systematic order:

- Role definition and assignment:

  - **Data owners**: Typically, senior managers or business unit leaders who are accountable for the overall quality and integrity of specific datasets. They make decisions regarding data use, access, and governance policies.

- **Data stewards**: Individuals responsible for managing and overseeing the data's life cycle, ensuring compliance with governance policies, and maintaining data quality.

- **Data custodians**: IT professionals tasked with the technical management of data storage and infrastructure. They ensure that data is stored securely and that backup and recovery procedures are in place.

- **Data processors**: Entities or individuals who process data on behalf of data controllers. They must follow the data controller's instructions and ensure data protection during processing.

- **Compliance officers**: Ensure that all data activities comply with relevant laws and regulations. They conduct audits and assessments to verify compliance.

- Policies and procedures:

  - **Data governance policies**: Clear, documented policies outlining how data should be managed, protected, and used. These policies should align with regulatory requirements and best practices.

  - **Standard Operating Procedures (SOPs)**: Detailed procedures for performing data-related tasks, ensuring consistency and compliance with governance policies.

- Monitoring and auditing:

  - **Continuous monitoring**: Implementing systems to continuously monitor data access and usage. Tools such as SIEM systems can provide real-time alerts on suspicious activities.

  - **Regular audits**: Conducting periodic audits to ensure that data governance policies are being followed and to identify any areas of non-compliance or potential risks.

- Transparency and reporting:

  - **Audit trails**: Maintaining comprehensive logs of all data-related activities to provide a clear record of who accessed data, when, and what actions were taken. This supports transparency and accountability.

  - **Reporting mechanisms**: Establishing processes for regular reporting on data governance activities, including compliance status, audit findings, and risk assessments.

- Training and awareness:

  - **Training programs**: Providing regular training for all employees on data governance policies, data protection regulations, and their specific responsibilities.

  - **Awareness campaigns**: Promoting a culture of accountability through awareness campaigns that highlight the importance of data governance and individual responsibilities.

- Risk management:

  - **Risk assessments**: Conducting regular risk assessments to identify potential threats to data security and compliance. Developing mitigation strategies to address identified risks.

  - **Incident response plans**: Establishing and maintaining incident response plans to address data breaches or other security incidents promptly and effectively.

Implementing Accountability in Cloud Data Governance

Accountability can be implemented in cloud data governance in five primary steps:

1. **Framework development**: Develop a comprehensive data governance framework that includes clear definitions of roles and responsibilities, detailed policies and procedures, and mechanisms for monitoring and enforcement.

2. **Technology integration**: Integrate advanced technologies such as AI and machine learning to enhance monitoring and compliance efforts. Use automation to streamline data management tasks and reduce the potential for human error.

3. **Stakeholder engagement**: Engage stakeholders across the organization, including executive leadership, IT, legal, and business units, to ensure a unified approach to data governance and accountability.

4. **Regulatory compliance**: Stay informed about regulatory changes and ensure that governance policies are updated accordingly. Work closely with legal and compliance teams to interpret and implement regulatory requirements.

5. **Performance metrics**: Establish KPIs to measure the effectiveness of data governance initiatives. Regularly review and adjust strategies based on performance data.

Accountability in cloud data governance is critical to ensuring that data is managed responsibly, securely, and in compliance with legal and regulatory requirements. By defining clear roles and responsibilities, establishing robust policies and procedures, and implementing comprehensive monitoring and auditing mechanisms, organizations can maintain high standards of data integrity and security. Regular training and awareness programs, coupled with advanced technological solutions, further support a culture of accountability, enabling organizations to protect their data assets and build trust with stakeholders.

Tools and Technologies for Enhancing Accountability

The following tools can be used to enhance accountability in the cloud:

- **Logging and monitoring solutions**: Platforms such as AWS CloudTrail, Azure Monitor, and Google Cloud Logging provide detailed logs of user activity within cloud environments. These logs track who accessed what data and when, aiding accountability.

- **DLP tools**: DLP tools prevent unauthorized data disclosure by monitoring and enforcing policies on data access, sharing, and storage. Popular solutions include Symantec DLP, McAfee DLP, and Microsoft Azure Information Protection. The logs generated by DLP solutions aid with the implementation of accountability in DLM.

- **Auditing and compliance tools**: Compliance management platforms such as LogicGate, Compliance Quest, and ZenGRC feature auditing, reporting, and compliance tracking capabilities, helping organizations comply with regulatory frameworks such as SOC 2, GDPR, HIPAA, and PCI DSS.

- **Blockchain**: Cloud security professionals today are seeing a rise in the use of blockchain technology for assistance with accountability, chain of custody, and nonrepudiation.

Cloud Data Life Cycle

In the previous chapter, you learned about the cloud data life cycle, even discovering a mnemonic to help you remember the steps in the CCSP exam. Let's review it again, now from the perspective of Data Governance.

The cloud data life cycle outlines six phases that data undergoes in its journey through the cloud environment. At each stage, data is subject to specific considerations. It undergoes various processes as it ages from creation or acquisition to eventual destruction or deletion. Cloud security professionals must understand this life cycle to ensure effective data management security, regulatory compliance, and risk mitigation.

The six data life cycle phases are as follows:

1. **Data creation or procurement**: Data originates from various sources, including research, manual entry, and device captures. It exists in diverse formats, such as Word or PDF files, SQL databases, and emails.

2. **Data storage**: After creation or procurement, data is stored in public, private, or hybrid clouds. Cloud customers are responsible for implementing all data security controls, including but not limited to encryption, data classification, backup, and recovery processes.

3. **Data usage**: Authorized users access and utilize data for business activities such as customer analysis, innovation, and marketing. Data governance strategies are used to restrict access, define roles, and maintain an audit trail.

4. **Data sharing**: Sharing data within and outside the organization is often necessary but introduces security risks. Policies such as **least privileged access** can limit unauthorized access and sharing.

5. **Data archiving**: Inactive but relevant data is moved to long-term storage for retention. Factors to consider include legal requirements, storage costs, and data availability.

6. **Data destruction**: Data that is no longer useful is permanently deleted. This may be to prevent unauthorized access or comply with privacy laws. Organizations must adhere to best practices and legal standards for secure data destruction.

Refer to *Chapter 6, Cloud Data Security Concepts and Architectures*, for a more in-depth explanation of the cloud data life cycle.

The Role of IRM, Auditability, Traceability, and Accountability in Each Phase of the Cloud Data Life Cycle

The cloud data life cycle comprises several phases (as listed previously), each of which necessitates specific controls and measures to ensure data security, compliance, and effective management. IRM, auditability, traceability, and accountability play crucial roles throughout this life cycle:

- **Data creation and capture**:

  - **IRM**: At the creation and capture phase, IRM controls are embedded to define who can access, edit, share, or delete the data. This ensures that data is immediately protected according to its sensitivity and classification.

  - **Auditability**: Establishing audit logs during data creation helps in recording the origin, time, and context of data capture, making future reviews and audits possible.

  - **Traceability**: Assigning unique identifiers or metadata to data at this stage facilitates tracking its origin and subsequent modifications.

  - **Accountability**: Data owners are identified, establishing responsibility for data accuracy and compliance from the moment it is created.

- **Data storage**:

  - **IRM**: IRM continues to enforce access controls and usage policies on stored data, ensuring that only authorized users can interact with it.

  - **Auditability**: Persistent logging of data access and modifications while in storage enables regular audits and compliance checks.

  - **Traceability**: Metadata management and data lineage tools track data location and movement across storage systems, providing a clear map of data residency.

  - **Accountability**: Data custodians are accountable for implementing and maintaining secure storage practices, including encryption and backup.

- **Data use**:

  - **IRM**: Usage policies enforced by IRM dictate how data can be used, shared, and processed, maintaining control over data integrity and confidentiality.

  - **Auditability**: Monitoring and logging data access and usage ensures that all actions are recorded and auditable, supporting compliance with usage policies.

- **Traceability**: Maintaining detailed logs of data interactions allows for tracking data usage patterns and identifying potential misuse.
- **Accountability**: Users accessing and utilizing the data are held accountable for their actions through RBAC and user authentication mechanisms.

- **Data sharing**:

  - **IRM**: IRM policies regulate data sharing, ensuring that sensitive data is only shared with authorized recipients and under specified conditions.
  - **Auditability**: Detailed records of data-sharing activities, including recipient information and the context of sharing context, support audits, and compliance reviews.
  - **Traceability**: Tracking mechanisms record the path of shared data, from the origin to the final recipient, ensuring full visibility of data flow.
  - **Accountability**: Individuals and entities involved in data sharing are responsible for adhering to data-sharing policies and ensuring secure transmission.

- **Data archiving**:

  - **IRM**: Archived data retains IRM controls to ensure it remains protected even when not actively used.
  - **Auditability**: Archiving systems maintain audit logs of data movement into and out of archival storage, ensuring all actions are recorded.
  - **Traceability**: Metadata and data lineage information continue to track the archived data's life cycle, maintaining a clear history of data states and movements.
  - **Accountability**: Data stewards oversee the archiving process, ensuring that data is properly categorized and securely stored according to retention policies.

- **Data deletion**:

  - **IRM**: IRM policies ensure that data deletion processes are secure and irreversible, preventing unauthorized recovery of deleted data.
  - **Auditability**: Deletion activities are logged, providing a clear record of when, how, and by whom data was deleted.
  - **Traceability**: Ensuring that all data copies, including backups and archived versions, are identified and deleted as required.
  - **Accountability**: Individuals responsible for data deletion must ensure compliance with data retention and deletion policies, maintaining accountability for the secure disposal of data.

Each phase of the cloud data life cycle requires careful attention to IRM, auditability, traceability, and accountability. IRM ensures that data access and usage are controlled according to predefined policies. Auditability supports compliance and security by providing detailed records of all data-related activities. Traceability maintains a clear view of data movements and transformations, supporting integrity and regulatory compliance. Accountability assigns responsibility for data management practices, ensuring that individuals and roles are held accountable for maintaining data security and compliance throughout its life cycle. Together, these elements form a robust framework for managing data in the cloud, ensuring its protection and effective governance.

Challenges and Solutions in Implementing IRM and Data Governance in the Cloud

Some major challenges to implementing IRM and data governance in the cloud are as follows:

- **Regulatory compliance**: Organizations must make sure their IRM and data governance practices comply with regulations such as GDPR, which can differ significantly by location or industry.

- **Data visibility and control**: Data visibility, control, monitoring, and auditing can be difficult due to the distributed and dynamic nature of cloud environments.

- **Integration**: Integrating IRM and data governance solutions with existing cloud infrastructure, applications, and services can be complex.

- **Adoption**: Training users on proper IRM and data governance practices can be difficult, but is a must for successful implementation in the long run.

The following are **practical solutions** and strategies to overcome these challenges:

- Cloud security professionals must ensure that IRM and data governance practices comply with regulations such as GDPR and HIPAA.

- Robust encryption, access controls, and authentication mechanisms can protect data from unauthorized access.

- Cloud security professionals must put techniques in place to monitor data access, usage, and movement across the entire cloud environment.

- Cloud security professionals must plan carefully for seamless integration with no adverse impact on data integrity or functionality.

- Training and support for users with diverse roles is essential to educate them on their rights and responsibilities regarding IRM and data governance.

Emerging Trends and Technologies in Cloud Data Governance and IRM

As cloud-centric technologies evolve, so does the ability to leverage new solutions and capabilities to help organizations with data governance challenges:

- **AI and machine learning** can inform more proactive and intelligent data governance decisions. Increasingly, vendors are integrating these algorithms into data governance tools for automation, anomaly detection, and predictive analytics.

- **Regulations** around rights to data and data privacy continue to roll out, making data governance trickier to navigate. Policies may come into conflict with each other. For example, it's not hard to imagine a case where regulators require retention of data for a certain period, but a customer may demand its deletion.

- **Data-driven governance** frameworks and tools will likely continue to be the choice of cloud security professionals faced with increasingly granular governance rules and a cloud environment that keeps expanding and fracturing.

- **Homomorphic encryption** enables computations on encrypted data without decryption, mitigating the risk of data exposure. It allows CSPs to process sensitive data without accessing its plaintext form, enhancing confidentiality.

- **Blockchain** can create tamper-proof records of data access and transactions, ensuring data integrity and accountability. Smart contracts can automatically create enforceable agreements, streamlining data sharing and access control.

- **Regulatory compliance automation** can make compliance with strict and intricate data privacy regulations simpler. Automated compliance tools and systems designed for the cloud can guarantee that an organization is following the rules set by regulators.

These trends are catalyzing a big change in how you manage, secure, and govern data. Together, they will make governance processes simpler, reduce manual work, and allow cloud security professionals to manage their environments proactively. They can improve transparency in how information is handled, reducing the risk of unauthorized tampering.

Removing all assumptions of trust from the management of data is a big theme as you move forward— whether through tamper-proof blockchain ledgers, zero-trust security frameworks, exposure-free encryption techniques, or automated regulatory compliance. These emerging trends will bring about data governance methods that are more efficient, secure, and successful; this way, organizations can deal with ever-changing threat types and new regulations confidently.

Summary

IRM is a crucial area of cloud data governance. Its essential components work together to keep data secure and available within the cloud for CSPs, the organizations they serve, and those organizations' customers. The correct implementation of IRM in the cloud safeguards sensitive and valuable information, building the business's and the public's trust in the cloud as a secure and reliable computing environment.

CCSP candidates should have a thorough understanding of IRM, auditability, traceability, and accountability in cloud data governance. It is vital for them to know how and why IRM mechanisms, such as access controls, encryption, data classification, monitoring, and auditing, are implemented in cloud environments. To succeed in cloud data security, CCSP candidates should have a solid understanding of the tools and techniques of IRM and understand its importance in the context of their organization, their industry, the evolving cyber-threat landscape, and relevant legislative requirements. You have covered these topics at length in this chapter, so please ensure that you understand them.

Mapping out a successful cloud data governance strategy takes careful planning, knowledgeable people, and well-chosen tools and technologies. Investing in complete IRM solutions, strong traceability methods, and sophisticated audit tools will empower organizations to get a sure handle on their data. Also, creating an environment of responsibility and encouraging knowledge about best practices for data management among workers are important to achieving success. Everyone in the organization—not just cloud security professionals—must know their roles and responsibilities regarding data governance. Additionally, they need to remain updated about developments and emerging technologies related to data governance. Staying informed can empower them to change their tactics early and gain an advantage in tackling growing issues.

In the next chapter, you will pivot a bit and look at the infrastructure and platform components of a secure data center.

Exam Readiness Drill – Chapter Review Questions

Apart from a solid understanding of key concepts, being able to think quickly under time pressure is a skill that will help you ace your certification exam. That is why working on these skills early on in your learning journey is key.

Chapter review questions are designed to improve your test-taking skills progressively with each chapter you learn and review your understanding of key concepts in the chapter at the same time. You'll find these at the end of each chapter.

> **How to Access These Materials**
>
> To learn how to access these resources, head over to the chapter titled *Chapter 25, Accessing the Online Resources.*

To open the Chapter Review Questions for this chapter, perform the following steps:

1. Click the link – `https://packt.link/CCSPE1_CH07`.

    Alternatively, you can scan the following **QR code** (*Figure 7.1*):

Figure 7.1 – QR code that opens Chapter Review Questions for logged-in users

2. Once you log in, you'll see a page similar to the one shown in *Figure 7.2*:

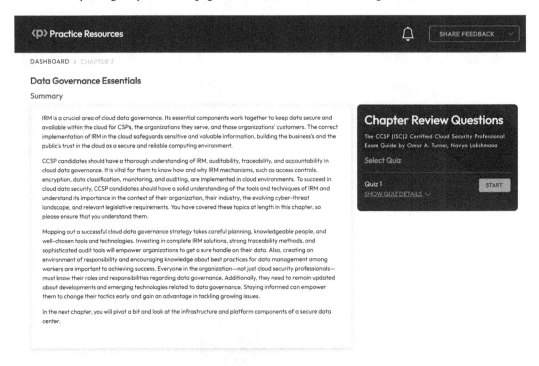

Figure 7.2 – Chapter Review Questions for Chapter 7

3. Once ready, start the following practice drills, re-attempting the quiz multiple times.

Exam Readiness Drill

For the first three attempts, don't worry about the time limit.

ATTEMPT 1

The first time, aim for at least **40%**. Look at the answers you got wrong and read the relevant sections in the chapter again to fix your learning gaps.

ATTEMPT 2

The second time, aim for at least **60%**. Look at the answers you got wrong and read the relevant sections in the chapter again to fix any remaining learning gaps.

ATTEMPT 3

The third time, aim for at least **75%**. Once you score 75% or more, you start working on your timing.

> **Tip**
>
> You may take more than **three** attempts to reach 75%. That's okay. Just review the relevant sections in the chapter till you get there.

Working On Timing

Target: Your aim is to keep the score the same while trying to answer these questions as quickly as possible. Here's an example of how your next attempts should look like:

Attempt	Score	Time Taken
Attempt 5	77%	21 mins 30 seconds
Attempt 6	78%	18 mins 34 seconds
Attempt 7	76%	14 mins 44 seconds

Table 7.1 – Sample timing practice drills on the online platform

> **Note**
>
> The time limits shown in the above table are just examples. Set your own time limits with each attempt based on the time limit of the quiz on the website.

With each new attempt, your score should stay above **75%** while your "time taken" to complete should "decrease". Repeat as many attempts as you want till you feel confident dealing with the time pressure.

8

Essential Infrastructure and Platform Components for a Secure Data Center

The shift to the cloud is essential for organizations to remain competitive in today's business landscape; it represents a complete transformation of how you interact and conduct business. This transformation is driven by the backbone of cloud computing—an intricate web of services and architectures that enable scalable, flexible, and cost-effective solutions. Embracing the cloud allows businesses to innovate rapidly, enhance collaboration, and access powerful computing resources on demand. In this chapter, you will learn what it really takes to run a cloud service, who is responsible for maintaining its security, and how all these components interconnect to create a resilient and efficient cloud environment.

In this chapter, you will review the following:

- Cloud infrastructure and platform components
- Physical design
- Environmental design
- Logical design

This chapter covers important areas for the CCSP exam, including real-world best practices that will make you a better and more security-aware cloud practitioner. Let's dive right in!

Cloud Infrastructure and Platform Components

In previous sections of this book, you explored the main pillars of cloud computing: **Infrastructure as a Service (IaaS)**, **Platform as a Service (PaaS)**, and **Software as a Service (SaaS)**. Each model represents a layer in the cloud computing stack, offering various levels of abstraction and control to meet diverse business and technological needs. The full potential and operational excellence of these models are fundamentally anchored in the underlying **cloud infrastructure** and **platform components**. Most of these components are physically located with the CSP, but many are accessible via the network.

Cloud infrastructure is the collection of hardware and software resources that make up the cloud. These components include the physical environment, storage, computing, networking, management, and virtualization. Let's look at these components and learn how they function. In the shared responsibility model, the customer and CSP share security responsibilities, so you will also review responsibilities and security controls in each area.

Physical Environment

To end users, it appears as if the cloud conjures resources out of thin air. However, that couldn't be further from the truth. There is a whole lot of hardware that makes these resources possible. A single data center could store hundreds of thousands of physical components, ranging from servers, storage devices, processors, routers, switches, load balancers, and **Power Distribution Units (PDUs)**. Typically, a CSP would have several data centers across various locations. The CSP owns all aspects of physical security within its data center, including facilities, equipment, and personnel.

> **Note**
> For the exam, make sure you know who owns which roles from the shared responsibility model.

The main challenge with physical security in a cloud data center is unauthorized access to the physical infrastructure. Standard measures for physical security thus include the following:

- Locks
- Security personnel
- Lights
- Fences
- Cameras
- Visitor check-in procedures

Data centers must adhere to the highest standards of physical security and controlled access. This involves implementing a multi-layered perimeter defense strategy, including fences, gates, and walls, all under constant surveillance. The positioning of a data center significantly influences the kind and extent of physical security measures that can be employed.

When having a data center facility situated in a spacious open setting, it's possible to establish several layers of protective barriers around the perimeter, along with the necessary monitoring systems to oversee the space between these layers. In densely populated urban environments, the challenge of keeping vehicles and pedestrians at a safe distance increases, exposing the facility to potential risks. One risk in an urban setting is the likelihood of external fires spreading to the data center.

The NIST 800-53 framework provides excellent guidance and a list of 20 controls for physical and environmental protection. Among those that are the most relevant for cloud security are the following:

- **PE-2: Physical Access Authorizations**, **PE-3: Physical Access Control**, and **PE-6: Monitoring Physical Access** deal with preventative and detective measures to ensure that only authorized personnel have access to the cloud data center and physical infrastructure and that any authorized attempts are quickly dealt with.

- **PE-11: Emergency Power** and **PE-12: Emergency Lighting** ensure that the cloud remains operational in emergency situations, such as power outages. Minimizing downtime is critical for client satisfaction.

- **PE-18: Location of Information System Components** involves the strategic placement of cloud hardware in a way that minimizes environmental influence and the possibility of physical intrusions.

When auditing the physical environment of a CSP, these are some of the measures that the independent third-party auditor will look at to measure physical security.

Toward the end of this chapter, you will learn physical design, and how to plan an efficient secure layout for a data center. This includes selecting a location for the data center while considering potential risks.

Network and Communications

Networking comprises the technologies that connect all hardware, software, and computing capabilities together. It provides connectivity between the CSP and the cloud customer. Regardless of how resilient and robust the physical cloud infrastructure is, it will not function without working network connectivity, which means that none of the services and systems will be available for the end user.

Some elements of managing the network components lie in the hands of the cloud service provider. A slight exception to that rule is in IaaS, where the customer is responsible for configuring the **Virtual Machines** (**VMs**) and virtual network. Even in that scenario, the CSP will provide the tooling to secure the VMs, but the customer must configure those tools.

So, considering the criticality of network connectivity, how do CSPs ensure that they meet the uptime standard, also known as the five 9s (99.999%)?

Typically, they deploy at least two unique network connections at each data center location. Each connection comes with independent pathways, hardware, and **Internet Service Providers (ISPs)** to eliminate the risk of a single point of failure and maximize service availability for customers.

While many people reading this book will never have to physically deploy a cloud network, a CCSP candidate should understand some key cloud network functionalities, such as the following:

- **Routing**: The process of path selection, directing network traffic between two or more nodes. These nodes can be part of the same network or two separate networks.

- **Filtering**: A range of processes and technologies designed to screen and manage data traffic moving to and from cloud-based services.

- **Rate limiting**: Prevents resource overuse by controlling the amount of incoming and outgoing traffic to or from a network or application.

- **Address allocation**: Assigns IP addresses to cloud hosts such as VMs and containers, which can be managed by the CSP (PaaS, SaaS) or the customer (IaaS).

- **Bandwidth allocation**: Standard practice for fairly distributing resources in cloud computing, where many tenants share the same infrastructure.

Software-Defined Networking

When speaking about cloud networking, you have to mention **software-defined networking (SDN)**. SDN is an integral component of modern cloud networking, fundamentally changing how data centers and networks operate. By separating the control layer from the physical hardware, SDN introduces a level of flexibility and programmability previously unattainable with traditional network architectures.

This separation allows network administrators to manage, configure, and optimize network resources via software applications, making the network more adaptable to the ever-changing needs of cloud services. More specifically, the processes involved in directing traffic and the physical routing of data in SDN are distinctly separate, which allows cloud network admins to easily modify network paths and resource distribution in response to the immediate requirements and preferences of their users.

Since SDN enables easy management of cloud resources, it's imperative to be very selective about who is granted control and management privileges within the network. Access should be tightly controlled and monitored with security measures such as **Two-Factor Authentication (2FA)** and **role-based access control (RBAC)**.

Compute

These are the infrastructure components that deliver computing resources, such as the VMs, disk, processor, memory, and network resources. Compute resources take data, process it, and show you the output of the processing. Just like the physical environment and network infrastructure, the compute infrastructure is also the sole responsibility of the CSP.

Ultimately, cloud-based computing comes down to the number of CPUs and the amount of memory. This is no different from traditional data center models. What is unique in a cloud environment is that these computing resources must be managed and allocated among multiple customers. There are several mechanisms CSPs use to achieve this:

- **Reservation** offers the ability to share resources efficiently based on need. A reservation guarantees that a cloud customer can access a minimum amount of cloud computing resources. These capabilities protect customers from denial of service due to other resource-intensive tenants.

- **Limits** can be set either for a specific VM or a user's computing allocation as a whole. Limits help customers ensure they don't use more resources than they can afford and that a single cloud tenant doesn't use a tremendous amount of resources.

- **Shares** is a prioritization and waiting system within a cloud environment that prioritizes specific applications or customers to receive additional resources when requested or when there is resource contention. Resource contention occurs when there are too many requests for resources and not enough resources to handle the requests.

Before you move on, let's re-familiarize ourselves with two common computing assets: VMs and containers.

VMs

VMs are essential components of a cloud environment as they allow users to simulate physical systems and get all the benefits of running an **operating system (OS)**. Since the VMs are directly controlled by customers, they must pay close attention to how they are configured and allocated. The CSP may or may not provide resources for VM management, but it's mainly the customer's responsibility.

Containers

Containers are a technology that's being increasingly used, particularly in multi-cloud or hybrid environments. Containers are lightweight packages that package applications together with libraries and other dependencies and run them in isolated environments—outside of their original environment. It's a very efficient way of deploying, running, and managing applications across different computing environments without the need to make any changes to the code.

Storage

If you were to walk into a cloud storage room, you wouldn't see much difference between that environment and the one found in traditional data centers. There will be **hard disk drives (HDDs)** and **solid-state drives (SSDs)** grouped together and used according to the needs and requirements of customers. The drives are typically arranged as **RAID** (which stands for **Redundant Array of Inexpensive Disks**) within a **Storage Area Network (SAN)**.

By grouping several drives into a single logical unit, RAID can provide data redundancy, increase storage capacity, and improve performance. Depending on the specific RAID level implemented, the system can tolerate one or more drive failures without losing data, thereby ensuring data integrity and availability.

The SAN connects to various storage devices, including HDDs and SSDs, and makes them accessible to host computers via a high-speed network.

The storage units have no filesystem of their own. Instead, a filesystem is created inside the guest OS whenever a tenant deploys a VM. This is also known as volume storage.

Volume Storage

Volume storage allocates disk space directly to a VM, configuring it similarly to a conventional hard drive and filesystem on a server. This configuration allows the storage, although centrally located and network-connected, to appear as a dedicated resource to the server, similar to how other computing and infrastructure services are presented in a virtualized OS.

This model provides a reserved slice of storage for the VM, with configuration, formatting, usage, and filesystem-level security managed by the host VM's OS and administrators. Volume storage is particularly effective when a specific amount of storage with a filesystem is needed for runtime operations. It allows for the installation and execution of programs or the creation of a filesystem for data storage, presenting itself as an attached drive to the user's VM, much like a physical drive on a tangible device.

However, volume storage is not without its vulnerabilities. It's affected by traditional data storage threats such as malware and accidental data deletion. Additionally, the cloud-based nature of volume storage introduces the risk of intermediary threats, such as man-in-the-middle attacks, as data is transferred to and from the cloud.

Object Storage

Object storage stores data as objects rather than files or blocks. These objects comprise not only the data itself but also metadata and a unique identifier, which enhances retrieval and management capabilities. End users can access their objects via APIs, network requests, or a web interface, which is more efficient and secure than hosting the data in a filesystem. Unlike conventional storage, which organizes data in a hierarchical structure, object storage utilizes a flat architecture, assigning a unique key value to each object for access. These values are then used to retrieve and manage the data directly.

While not suitable in all scenarios, cloud providers use object storage to offer highly scalable, durable, and secure storage solutions for unstructured data. It is best suited for unstructured data such as images, documents, and web pages, providing scalability, redundancy, and detailed metadata for better indexing and classification.

Storage Security

Since data spends most of its time at rest, it's crucial to distinguish who is responsible for securing cloud storage. The importance of the shared responsibility model is emphasized throughout this chapter, and for good reason. As a cloud security practitioner, you must know these different responsibilities and look at the relationship between a customer and a CSP as a partnership.

When it comes to storage, the following apply:

- CSPs are responsible for the physical protection of data centers and the storage infrastructure they contain. They're also responsible for the maintenance and timely security patching of the underlying data storage technologies and services they provide.

- Customers are responsible for properly configuring and using the storage tools provided by the CSP. Customers must also ensure the logical security and privacy of the actual data they store in the cloud environment.

To expand on the customer's role in ensuring storage security, it's essential to recognize that most cloud service providers offer a variety of security controls and tools designed to protect data. However, the effectiveness of these controls largely depends on how they are configured by the customer.

Ultimately, it's up to the customer to decide the extent to which they implement these measures. While CSPs can provide the means for security, they do not dictate the level of security that a customer chooses to enforce on their data and applications. Lack of implementation is sometimes caused by a lack of awareness that these controls exist in the first place.

Virtualization

Virtualization is what makes cloud computing so efficient that it can allow a single piece of hardware to be shared by multiple customers. Cloud benefits such as resource pooling, on-demand service, and scalability are all made possible through virtualization. Come to think about it, you probably wouldn't be reading this book if it wasn't for virtualization, as the cloud as you know it today wouldn't exist.

Here are some of the benefits of virtualization:

- **Increases scalability**: Virtualized environments are perfect for scaling operations based on need. Simply spinning up a new VM will always be cheaper and easier than buying new hardware. It's also a lot quicker.

- **Reduces downtime**: Physical components are prone to damage, failure, and disruptions. Recovering them can be costly and time-consuming, especially in large environments.

- **Saves time**: Virtualized resources are much easier to manage and maintain. System administrators can quickly deploy, clone, modify, and decommission VMs using management software without the need for physical intervention.

Hypervisors

A hypervisor, also known as a **Virtual Machine Monitor** (**VMM**), is software that creates and manages VMs by abstracting the underlying physical hardware. Hypervisors allow multiple OSs to run concurrently on a single physical machine by allocating and isolating hardware resources such as CPU, memory, storage, and network interfaces to each VM. This enables efficient utilization of hardware resources, improved scalability, and flexibility in deploying and managing applications.

There are two types of hypervisors:

- **Type I:** A Type I hypervisor, also known as a bare-metal hypervisor, is virtualization software that runs directly on the host's physical hardware, rather than within an OS. This allows it to efficiently manage hardware resources and provide a more secure and performant environment for running VMs. Type I hypervisors are typically used in enterprise data centers and cloud environments due to their high efficiency and direct access to hardware resources. Examples of Type I hypervisors include VMware ESXi, Microsoft Hyper-V, and Xen. By running directly on the hardware, Type I hypervisors can offer better performance, scalability, and security compared to Type II hypervisors, which run on top of a host OS.

- **Type II:** A Type II hypervisor, also known as a hosted hypervisor, is virtualization software that runs on top of a host OS, which in turn runs on the underlying physical hardware. This type of hypervisor relies on the host OS for device support and management, and it provides a layer of abstraction that allows multiple guest OSs to run concurrently in isolated VMs. Because it operates within an existing OS environment, a Type II hypervisor can be easier to set up and manage, making it suitable for development, testing, and desktop virtualization scenarios. However, this reliance on the host OS can introduce additional overhead, potentially reducing performance and security compared to Type I hypervisors. Examples of Type II hypervisors include VMware Workstation, Oracle VirtualBox, and Microsoft Virtual PC.

Management Plane

With all of the previous components, the CSP has most, if not all, of the control. When it comes to the **management plane**, however, the end user assumes a more prominent role, gaining the capability to oversee, configure, and optimize their own cloud resources within the defined parameters set by the provider. In other words, the management plane provides the tools (web interfaces and APIs) necessary to control a cloud environment. The management plane gives administrators the capability to control resources, such as turning VMs on and off, provisioning VM resources, or migrating a VM.

As a CCSP-certified cloud professional, you will likely spend most of your time in the management plane, ensuring that cloud configuration is in line with company policies and industry standards.

How you interact with the management plane will vary based on your chosen CSP. Almost all CSPs offer a **graphical user interface (GUI)** that is accessible through a web interface for managing cloud resources. Some providers also support **command-line interfaces (CLIs)**, such as PowerShell/SSH, or provide access to the management plane via a set of **application programming interfaces (APIs)**, enabling more automated or script-based management capabilities.

Apart from the cloud portal (the main web interface of the cloud platform), here are the other key interfaces of the management plane:

- **Scheduling**: Provides the ability to start or stop resources at a scheduled time

- **Orchestration**: Automates processes to manage resources, services, and workloads

- **Maintenance**: Updates, upgrades, installs security patches and fixes, and so on

To secure these interfaces, you will typically deploy measures such as **Multi-Factor Authentication (MFA)**, role management, and access control.

Designing a Secure Data Center

You can't protect something you don't truly understand. And what better way to understand cloud infrastructure than to learn about all the details of its design, from the layout and construction of the facility to choices regarding hardware placement and environmental controls? The CCSP exam expects candidates to understand the principles of secure cloud architecture and how different design decisions can impact the cloud's security posture.

In this section, you will learn how to design a secure data center, which includes physical, logical, and environmental designs.

Physical Design

A solid physical design is the backbone of any efficient and secure data center. Since physical design is among the first things to do when designing a data center, getting this part of the process right will be instrumental for success. Making any impactful changes after the physical design phase will be very challenging as all of the buildings, equipment, and contracts are already in place.

The physical design incorporates the actual data center and its internal components, along with the established policies and procedures to prevent physical damage or compromise.

Choosing a Location

The first step in physical design is finding a suitable location. Building a data center is a hefty investment, so exposing it to any kind of unnecessary risk is, well, unnecessary. While the costs of building in a safer geographical location are generally higher, the long-term benefits often outweigh these initial expenses.

Locations less prone to natural disasters such as floods, earthquakes, or hurricanes can significantly reduce the risk of catastrophic data loss and associated downtime. Moreover, choosing a region with political stability and a low risk of conflict can provide a more secure environment for both the physical infrastructure and the personnel operating the data center.

Some other considerations when selecting a location for a data center include the following:

- The availability of stable, affordable, and resilient electricity and internet connectivity
- Cost and availability of skilled labor
- Geographic location relative to customer base

It's also important to recognize that finding an optimal location with minimal risk can be challenging. Often, organizational goals and client demands necessitate building data centers in less-than-ideal locations from a risk perspective. In such cases, organizations should conduct a thorough risk analysis and cost-benefit analysis to inform their decision-making process. When a higher-risk location is unavoidable, the focus should shift to designing and constructing data centers with appropriate countermeasures to manage and mitigate location-specific risks effectively.

Buying versus Building

Organizations interested in owning a data center have two options:

- Build a facility from the ground up
- Buy or lease an existing facility

For simplicity's sake, you will refer to buying and leasing as the same thing. Both approaches are viable and have advantages and disadvantages, as shown in *Table 8.1*:

Feature	Build	Buy
Investment	Requires a significant investment	Generally cheaper, especially in a shared scenario
Control	Requires a significant investment	Less flexibility in design, limited to what the provider offers

Feature	Build	Buy
Skill	Requires knowledge and skill to match the quality of a pre-existing data center that you would get with the "buy" option	No specific requirement
Speed	Slower than buy	Will allow you to get up and running a lot quicker

Table 8.1 – Difference between build and buy

CSPs are the most common choice for organizations of all sizes because they provide "build"-level quality and flexibility for a "buy"-level price tag.

However, some organizations may still want to go the "own" route. If the said organization has the time and money (which isn't a small feat), building would generally be a better option. Building a data center from scratch will give the organization full control of the facility's design, layout, and operations, allowing for customization that aligns perfectly with business needs.

Buying or leasing is the much quicker, easier, and, in the case of leasing, financially safer option. Going this route will eliminate the enormous initial investment that comes with building a data center. The organization will also save a ton of time and get up and running much quicker, which is crucial in the IT world.

When it comes to compliance, building a data center will give the organization the opportunity to comply with industry standards and regulations from the start. Buying or leasing necessitates **Service-Level Agreement (SLA)** enforcement and may involve hiring third-party auditors or contractors to ensure these requirements are met.

Environmental Design

Cloud security is way more than technical measures. It involves a comprehensive approach that includes policies, procedures, and personnel to ensure robust security practices are maintained. Another important consideration is the environment in which the cloud systems exist.

Heating, Ventilation, and Air Conditioning (HVAC)

Data centers consume a ridiculous amount of power, generating an enormous amount of heat. A well-designed HVAC system is indispensable for maintaining optimal temperature and humidity levels, ensuring the longevity of hardware and the reliability of cloud services.

An HVAC failure can reduce the availability of computing resources. A lot of customers will audit the quality of HVAC design before moving forward with a CSP, so it's critical for providers to ensure robust and reliable environmental controls to maintain service continuity.

Per the **American Society of Heating, Refrigeration, and Air Conditioning Engineers (ASHRAE)** (`https://www.ashrae.org/`), the optimal temperature and humidity levels in a data center are as follows:

- Temperature: 64.4–80.6 degrees F (18–27 degrees C)

- Humidity: 40–60 percent relative humidity

The primary ASHRAE thermal guidelines for data centers are as follows:

- **Class A1**: Suitable for all types of IT equipment:

  - **Temperature Range**: 59°F to 89.6°F (15°C to 32°C)

  - **Recommended Range**: 64.4°F to 80.6°F (18°C to 27°C)

- **Class A2**: Suitable for less critical IT equipment:

  - **Temperature Range**: 50°F to 95°F (10°C to 35°C)

  - **Recommended Range**: 64.4°F to 80.6°F (18°C to 27°C)

- **Class A3**: Suitable for IT equipment designed for broader environmental ranges:

  - **Temperature Range**: 41°F to 104°F (5°C to 40°C)

  - **Recommended Range**: 64.4°F to 80.6°F (18°C to 27°C)

- **Class A4**: Suitable for IT equipment with the highest tolerance to temperature variations:

  - **Temperature Range**: 41°F to 113°F (5°C to 45°C)

  - **Recommended Range**: 64.4°F to 80.6°F (18°C to 27°C)

The difficulty of achieving these conditions will depend on several factors. The main one is the number of physical systems present in the data center. More systems generate more heat, requiring a higher amount of energy and cooling. Another factor is the layout of the systems. Physical racks should be designed in a way that maximizes airflow and facilitates efficient cooling.

Alternating hot and cold aisles, where cold air is directed toward equipment intakes and hot air is channeled away from equipment, can significantly improve cooling efficiency. Properly spaced and organized racks prevent hot spots and ensure that cooling efforts are not wasted, leading to a more energy-efficient operation overall.

Multi-Vendor Pathway Connectivity

Multi-vendor pathway connectivity is the connectivity to data locations from more than one ISP. This is another key element of environmental design, as it improves redundancy and allows CSPs to mitigate the risk of losing network connectivity. The best practice for multi-vendor pathway connectivity is dual-entry, dual-provider. This means that two providers enter the building from two separate locations.

Aside from CSPs, customers should also consider communicating with their cloud vendor from multiple paths. Network connectivity issues can originate from both sides of the client-vendor relationship.

Logical Design

The **logical design** is where cloud infrastructure begins to differentiate from a traditional data center. Because of that, you can expect the CCSP exam to place particular emphasis on this part of the design process. Key components of secure logical design include the following:

- Tenant partitioning or isolation
- Limited and secure remote access
- Cloud monitoring
- Patching and updating systems

To help us understand the logical components of a secure data center, it's important to familiarize ourselves with the concept of the Zero-Trust architecture.

NIST 800-207 defines **zero trust** as a collection of concepts, ideas, and component relationships (architectures) designed to eliminate the uncertainty in and enforce accurate access decisions in information systems and services.

In simpler terms, zero trust is a security model built on the idea that no entity within or outside an organization should be trusted by default. Understanding the zero trust architecture and why it's used will help you digest the measures discussed and give you a stronger foundation for implementing strong security controls and protecting an organization's IT environment.

With that out of the way, let's now shift our focus to the main tenets of secure logical design. The CCSP exam heavily focuses on tenant partitioning and access control within logical design, so I advise you to read the following sections carefully.

Tenant Partitioning

In a cloud environment, multiple tenants, sometimes thousands, are utilizing the same physical resources. If isolation measures between tenants are breached, it could significantly jeopardize the confidentiality and integrity of customer data. Both the CSP and the tenant share responsibility in ensuring the implementation and enforcement of controls that address multitenant risks in the public cloud.

Since physical partitioning isn't possible in the cloud, the CSP and tenants rely on logical network separation of systems and applications. This is achieved through a variety of mechanisms, including the use of **Virtual Private Clouds** (VPCs), subnetting, and network **Access Control Lists** (ACLs), which allow for the creation of segmented network spaces that are isolated from one another.

Access Control

You already covered access control when discussing the physical design of a data center. There, you were mostly concerned with protecting the physical infrastructure from unauthorized access, ensuring that only authorized personnel could enter sensitive areas and interact with critical systems.

When talking about logical access control, our focus shifts to restricting access to the organizational data and resources stored in the cloud. This includes the VMs as well as the management plane and hypervisor tiers.

One way to simplify access control is to use a single point of access, but that also means having a single point of failure.

Large organizations with hybrid cloud environments typically deploy hybrid identity measures—a single login for on-premises and the cloud. This facilitates a smoother user experience while satisfying **Identity and Access Management (IAM)** requirements.

There are multiple remote and local access mechanisms available, including the following:

- **Remote Desktop Protocol** (RDP): The native remote access protocol in Windows systems.
- **Secure Shell** (SSH): A commonly used protocol for remote access to networking devices. It's also the native remote access protocol for Linux machines.
- **Secure Terminal**: Also known as console access, it is used for secure local access.

All of these protocols support access control encryption of credentials and MFA, making them ideal for securing remote connections, and thus promoting a resilient logical design of the cloud environment.

Summary

So you made it to the end of this chapter! You reviewed how modern cloud infrastructure is heavily reliant on important elements, each critical to its functionality and efficiency. When selecting the location for a data center, organizations must decide whether to build a new facility or purchase an existing one, taking into account factors such as cost, scalability, and proximity to users. Optimizing data center operations involves managing environmental influences such as heat and humidity to ensure optimal performance and energy efficiency. Additionally, secure logical mechanisms are essential for CSPs to segregate resources among multiple tenants, ensuring proper data flow and maintaining data security and privacy across shared environments.

For the exam, make sure you know the cloud platform components: physical environment, network, storage, compute, virtualization, and management plane. More importantly, you should understand their roles and how they interconnect. Focus on specific technologies and mechanisms for each component, such as SDN for networking, reservations, limits, and shares for computing, and hypervisors for virtualization.

From a data center design perspective, you must fully grasp the logical architecture, as it focuses on core cloud mechanisms such as tenant partitioning, which are key for maintaining secure and efficient cloud operations.

The physical and environmental designs are also important. Understand the risk assessment and strategic decision-making involved in choosing a data center location, considering the susceptibility to natural disasters, proximity to threat vectors, and the balance of accessibility versus security.

Environmental design measures, such as aisle configuration to enhance cooling efficiency and multi-vendor connectivity to maximize redundancy, are also subjects you should expect to encounter on the exam.

In the next chapter, you will look at the top risks to physical, logical, and virtual environments as a cloud consumer and provider. You will discuss how to analyze, assess, and address the risk with safeguards and countermeasures—a very important topic, not only for the exam but for cloud security professionals as well.

Exam Readiness Drill – Chapter Review Questions

Apart from a solid understanding of key concepts, being able to think quickly under time pressure is a skill that will help you ace your certification exam. That is why working on these skills early on in your learning journey is key.

Chapter review questions are designed to improve your test-taking skills progressively with each chapter you learn and review your understanding of key concepts in the chapter at the same time. You'll find these at the end of each chapter.

> **How to Access These Materials**
>
> To learn how to access these resources, head over to the chapter titled *Chapter 25, Accessing the Online Resources.*

To open the Chapter Review Questions for this chapter, perform the following steps:

1. Click the link – `https://packt.link/CCSPE1_CH08`.

    Alternatively, you can scan the following **QR code** (*Figure 8.1*):

Figure 8.1 – QR code that opens Chapter Review Questions for logged-in users

2. Once you log in, you'll see a page similar to the one shown in *Figure 8.2*:

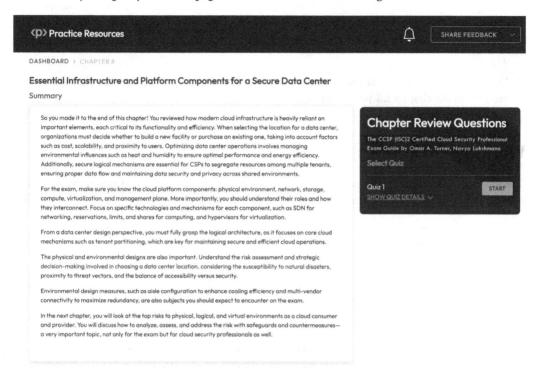

Figure 8.2 – Chapter Review Questions for Chapter 8

3. Once ready, start the following practice drills, re-attempting the quiz multiple times.

Exam Readiness Drill

For the first three attempts, don't worry about the time limit.

ATTEMPT 1

The first time, aim for at least **40%**. Look at the answers you got wrong and read the relevant sections in the chapter again to fix your learning gaps.

ATTEMPT 2

The second time, aim for at least **60%**. Look at the answers you got wrong and read the relevant sections in the chapter again to fix any remaining learning gaps.

ATTEMPT 3

The third time, aim for at least **75%**. Once you score 75% or more, you start working on your timing.

> **Tip**
>
> You may take more than **three** attempts to reach 75%. That's okay. Just review the relevant sections in the chapter till you get there.

Working On Timing

Target: Your aim is to keep the score the same while trying to answer these questions as quickly as possible. Here's an example of how your next attempts should look like:

Attempt	Score	Time Taken
Attempt 5	77%	21 mins 30 seconds
Attempt 6	78%	18 mins 34 seconds
Attempt 7	76%	14 mins 44 seconds

Table 8.2 – Sample timing practice drills on the online platform

> **Note**
>
> The time limits shown in the above table are just examples. Set your own time limits with each attempt based on the time limit of the quiz on the website.

With each new attempt, your score should stay above **75%** while your "time taken" to complete should "decrease". Repeat as many attempts as you want till you feel confident dealing with the time pressure.

9
Analyzing Risks

In this chapter, you will be going over cloud security risk management. This will include some definitions, how the risk management process differs from the on-site model, how to approach, assess, and mitigate risk, and the roles of both **cloud service providers (CSPs)** and the customer. In today's rapidly evolving digital landscape, where cloud computing has become the backbone of global business operations, understanding the intricate dynamics of cloud security is imperative. For CCSP candidates, mastering the art of identifying and mitigating risks across physical, logical, and virtual environments is not just a skill, it's a necessity.

This chapter dives deep into the definitions and methodologies that cloud consumers and providers must understand to safeguard their infrastructures. By exploring how to analyze, assess, and address potential vulnerabilities with effective countermeasures, you will be equipped with some of the knowledge needed to fortify your cloud environment against ever-present and emerging threats and you will have a better idea of how to answer the risk-centric questions in the exam. By the end of this chapter, you will be able to answer questions on risk management definitions, risk identification, and assessment steps, as well as understand the cloud attack surface area and risk response strategies.

By the end of this chapter, you'll be able to answer questions on:

- Risk management
- Risk identification and analysis
- Analyzing and assessing cloud security risks
- Cloud attack surface area, vulnerabilities, threats, and attack vectors
- Addressing cloud security risks – safeguards and countermeasures
- Implementing cloud security best practices, controls, and countermeasures

Let's begin.

Overview of Risk Management

The primary goals of risk management in the cloud are to identify, assess, and mitigate potential security threats and vulnerabilities specific to cloud environments, ensuring the protection of data, applications, and infrastructure. Let's now look at the important definitions of risk management.

Key Concepts in Risk Management

Let's now look at some key definitions and concepts that will help you on exam day. They are as follows:

- **Risk**: This can be defined as the potential for loss, damage, or destruction of assets, resulting from a threat exploiting a vulnerability. It represents the likelihood and impact of adverse events that could undermine an organization's ability to achieve its objectives and execute its strategies effectively.

- **Risk management**: In cloud security, this is the ability to manage cybersecurity risks in cloud environments to the business, operations, data, infrastructure, and customers.

- **Risk appetite**: This is an organization's capacity for risk and different risk types, based on stakeholders, regulations, affiliations, and business goals. It's the amount of risk that the organization is willing to accept. A high-risk appetite would mean riskier decision-making.

- **Risk tolerance**: Within its established risk appetite, this is the specific level of risk an organization is willing to tolerate.

- **Threat**: This is a security event that can negatively impact a business.

- **Vulnerability**: This is an existing system flaw that can be exploited by a threat.

- **Likelihood**: This is an expression of how likely the event is to happen: once a year, highly unlikely, and so on.

- **Impact**: This is the amount of damage that would be caused in the event of a particular security incident. This may be across several areas such as fines, reputation damage, and contract breaks.

- **Risk assessment**: This is a comprehensive set of activities designed to evaluate the risks existing in a particular system, often giving guidance on the next steps.

- **Risk response**: Whether a risk is accepted, transferred, mitigated, or avoided. The risk response is closely tied to risk appetite.

For the CCSP exam, understanding risk equations is crucial as they form the basis for evaluating and managing security risks in cloud environments. Here are some fundamental risk equations typically relevant to the CCSP curriculum:

- **Risk equation**:

  Risk = threat × vulnerability × impact

- This equation helps in assessing the level of risk by considering the probability of a threat exploiting a vulnerability and the potential impact it would have on the organization.

- **Annual loss expectancy (ALE)**: ALE is used to quantify the expected loss each year due to a specific risk, helping organizations in making decisions about implementing controls.

  ALE = single loss expectancy (SLE) × annual rate of occurrence (ARO)

- **SLE**: The monetary loss or impact of each occurrence of a threat.

- **ARO**: The expected frequency of a threat occurring within a year.

- **Cost-benefit analysis for countermeasures**:

  Cost of countermeasure ≤ ALE before countermeasure − ALE after countermeasure

- This equation helps to determine whether a security control is cost effective. The cost of implementing the control should not exceed the reduction in the ALE it provides.

Understanding these equations allows cloud security professionals to quantitatively assess risks and make informed decisions about where to apply resources and security controls to mitigate those risks effectively. These risk management concepts are essential to know and are likely to be included in the CCSP exam based on experience. Questions based on risk assessment calculations such as "Determine the ALE" or "Calculate the risk exposure" could appear, so familiarize yourself with these equations and definitions.

Risk Management in Cloud Environments

Now that you have understood the key concepts in risk management, let's look at risk from the perspective of cloud environments and in particular cloud service model types.

CSPs offer the outsourcing of similar IT services to the types that organizations run on-premises, data storage, processing, networking, platforms and applications, and so on. These services are offered under different cloud service model types: **infrastructure as a service (IaaS)**, **platform as a service (PaaS)**, and **software as a service (SaaS)**.

The risks involved in using CSP services remain the same as on-premises but with the added risks of outsourcing and cloud-based architecture and controls. In leveraging a CSP, you give up some visibility and management capabilities of owned on-premises services and data center hardware and software, in exchange for an easier, more portable, and abstracted service-based model. Each cloud computing service model also has different risks attached, which you will cover in a minute.

Verifying the confidentiality, integrity, and availability of your data in the cloud is difficult. CSPs provide service-level agreements and compliance-adherent products, but this is still a level of residual risk that you accept; they are doing their due diligence to ensure services remain compliant with their promises.

IaaS

IaaS refers to virtualized hardware resources available for rent in the cloud. The CSP owns, runs, and maintains the underlying hardware in a physical data center. You, as the customer, can provision, deploy, and run IT services from the CSP's catalog on top of the CSP's infrastructure, choosing from a pre-defined list of **operating systems (OSs)**, regions, applications, storage, and access controls.

IaaS offers the most customer control out of the three cloud computing service models, which generates high maintenance responsibility for the customer. The major risks involved in the IaaS model are in how the customer's team configures and runs the services. It is up to them to do tasks such as patch OSs and ensure correct data access.

Virtualized infrastructure carries with it many of the same risks as physical infrastructure; customer-created virtual networks must be protected by access controls such as AWS security groups, Azure network security groups, **Google Cloud Platform (GCP)** firewall rules, and virtual load balancers.

Typically, IaaS services are self-service and configurable. Managed services have a higher degree of abstraction and less visibility than you would have when running the same virtualized components on your own infrastructure.

PaaS

PaaS is an entire cloud architecture designed for a specific application development and deployment purpose. PaaS is a service that includes an environment, helpful services, and a collection of tools that are pre-assembled to support software development and deployment activities. Unlike IaaS, PaaS customers do not architect, configure, and run an underlying infrastructure. Instead, PaaS services typically run on a proprietary cloud infrastructure configuration. There may or may not be the option for customers to add controls on the underlying infrastructure or export the data to a different format.

PaaS environments may have limited customization options, restricting the ability to tailor the platform to specific business needs. Integration challenges can arise when connecting PaaS solutions with existing on-premise systems or other cloud services, potentially leading to interoperability issues.

SaaS

SaaS is a third-party software service or application running in the cloud, accessed by the customer through the internet via a dedicated application or browser. SaaS differs from PaaS in that it is a single solution, rather than a sprawling, all-encompassing environment and toolset.

Customers, however, have limited control over the underlying infrastructure and must rely heavily on the provider for security measures and updates. This reliance extends to compliance and regulatory requirements, which can be challenging to meet when dependent on a provider's adherence to standards. Vendor lock-in is another significant risk, where migrating data and applications from one SaaS provider to another can be complex and costly, leading to dependence on a single provider.

The availability and reliability of the service are also crucial, as any outages or downtime can disrupt business operations. Additionally, data residency and sovereignty issues arise when data centers are located in regions with differing data protection laws. Integrating SaaS applications with existing on-premises systems or other cloud services can present interoperability challenges.

SaaS poses inherent risks, including data security and privacy concerns, limited control over infrastructure, and dependency on the provider for compliance and reliability.

Distinction between Cloud Service Models and Deployment Models (Public, Private, Hybrid, and Community)

Cloud deployment models differ from cloud service models in how and where these services run. Let's look at the risks that one should be aware of for these models.

Public Cloud Deployments

Public cloud refers to services running on physical infrastructure owned and managed by a CSP on their premises. Cloud services are provisioned in a shared tenancy model (multi-tenancy), meaning that customers could potentially have their own services running on the same underlying infrastructure as multiple other customers.

There is risk in trusting that the public cloud's multi-tenancy virtualization and hypervisor are secure; if other tenants have security issues, they won't cross over to your own tenancy.

Private Cloud Deployments

Private cloud refers to customer-provisioned services running on an underlying physical machine that is dedicated to a single customer, known as the single-tenancy model. The physical infrastructure itself may be owned and run by a CSP on their premises, housed in a data center, or even housed on-premises. When housed on CSP or data center premises, there is always a risk to the physical security of infrastructure that can't be fully managed in-house.

Hybrid Cloud Deployments

Hybrid cloud is when customers use a combination of public cloud and private cloud deployments. Hybrid cloud may also refer to when public or private cloud services are mixed with on-site hardware and software (virtualized or not). Hybrid cloud can be tricky to configure, especially when services are from different companies, which can increase risk.

Community

Community cloud is a type of multi-tenancy where multiple customers agree to share underlying infrastructure. Community cloud can exist when organizations share regulatory and data governance requirements that are managed in the same way. An example is multiple government organizations sharing servers in a data center to run their own **virtual machines** (**VMs**). *Figure 9.1* depicts the shared responsibility model:

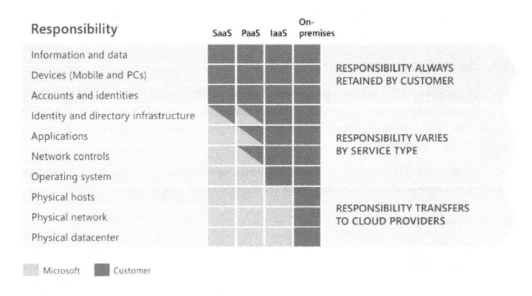

Figure 9.1 – Understanding the shared responsibility model in cloud security

Here's an example of the shared responsibility model in practice.

The US **Health Insurance Portability and Accountability Act** (**HIPAA**) of 1996 requires healthcare-associated organizations to securely store, process, and transfer protected health information. While CSPs offer HIPAA-compliant services, organizations must still ensure they operate the running services in a HIPAA-compliant way. This might include controls such as user- and group-level access controls.

Cybersecurity

Cybersecurity is everyone's responsibility, and cloud computing is no different. CSPs and customers must both work to ensure that the security and compliance of cloud services meet industry and legal standards.

Depending on the cloud service model and CSP, various cybersecurity responsibilities are either shared by the customer and CSP, managed wholly by the CSP, or managed wholly by the customer. The security of identity management, data, and on-premises and owned devices is always the responsibility of the customer in the relationship.

Many CSPs offer managed security, other solutions, and controls built into the service itself, such as an IaaS on-demand cloud instance (VM) that comes with encrypted data at rest as a standard feature. The CSP hardware server the VM runs on is in a secure facility, with hardware and the OS updated to the current secure version, but you must trust that the CSP is actually running in this way.

CSPs offer additional services and settings to help customers comply with additional standards, regulations, and security frameworks; however, this doesn't absolve the customer from all responsibility. For example, customers may need to upgrade and patch the OSs on their VMs running in the cloud. Some cloud services offer a greater degree of managed security (performed automatically by the CSP) on certain services than others.

Importance of Risk Management in Cloud Computing

Cloud security risks are the chance of a threat succeeding, and the impact that it would have on the organization. Threats succeed when there are vulnerabilities in a cloud configuration that are not addressed, and while common risks include financial losses, reputation damage, downtime, contract violations, fines, and legal penalties, the types and impacts will be unique to each organization.

The Outsourcing Model

In cloud computing, there is an inherent risk: the outsourcing model. By outsourcing business functionality, organizations relinquish control of a great deal of infrastructure and data management, entrusting that to a third party. Outsourcing is inherently risky.

Vendor-Lock In

Some cloud services and ecosystems make it more difficult to migrate to other systems or combine with another CSP's services. They may have proprietary data types and limited export capabilities. Different cloud services and deployment models may come with their own set of SLAs, regardless of any other agreements in place with a CSP.

Regulatory and Legal Requirements

Legal and regulatory requirements can be complex, requiring a careful legal and expert eye over SLAs and cloud contracts to ensure both the CSP and the customer will be fulfilling their compliance objectives.

Data Residency – Processing, Storage, and Transfer Considerations

Particularly when it comes to compliance and legal obligations, data residency requirements can be complex and absolute in their rulings. For strict data residency obligations, cloud services and products must either comply by design or be configured to comply with these data residency requirements.

Access to Resources

As you are aware, part of the **CIA (Confidentiality, Integrity, Availability)** triad is availability. In cloud computing, resources must be accessible and available to authorized and authenticated users as needed. In delivering cloud services over public networks and the CSP's own infrastructure, availability must be guaranteed.

Resources that are not persistent (ephemeral)

The ephemeral nature of cloud resources adds complexity to ensuring the persistence and historical recording of business operations and data. This is both in the sense of catastrophic events and in eDiscovery processes for legal obligations.

Risk Identification and Analysis

You will now explore the methodologies that can be used to uncover and scrutinize risks within cloud computing environments. This segment will cover essential topics such as risk assessments and the utilization of various risk frameworks that guide cloud security professionals in systematically identifying, categorizing, and prioritizing risks. You will discuss how these frameworks provide a blueprint for assessing risks based on their likelihood and impact.

Risk Frameworks

Risk frameworks provide a systematic methodology for identifying, assessing, and managing risks. This structured approach ensures that all potential threats and vulnerabilities are considered systematically, reducing the chances of oversight, and enhancing the effectiveness of the risk management process.

Some of the popular cloud cybersecurity risk management frameworks are the NIST **Risk Management Framework (RMF)**, ISO 31000:2018, ENISA's cloud computing risk assessment tool, and NIST 800-146 (cloud computing synopsis and recommendation) and 800-37 (risk management framework for information systems).

It will be essential to understand and determine the internal risk profile and appetite based on the company, regulations, stakeholders, and so on. Each organization is different in how much risk they are willing to accept in their cloud security. For organizations with low-value assets, minimal cloud service use or data processing and storage, and/or no applicable regulations, risk management activities may be minimal. This would mean the company has a high-risk appetite for its cloud cybersecurity program.

Assessing the CSP's Risk

Entrusting IP and operations to a CSP means a level of trust in the CSP itself. The Cloud Security Alliance's **Security, Trust, and Assurance Registry (STAR)** program (`https://cloudsecurityalliance.org/star/registry`) maintains a registry of **STAR Level 2** (third-party audited) CSPs, which are the most trusted in the world, to help guide cloud vendor selection. **Level 1**, self-assessed, isn't recommended for secure operations. The **EuroCloud Star Audit (ECSA)** program is a similar European model. **ISO 27001** is a security standard that should be critical in vendor selection, too, as it provides a baseline standard for information security management systems, plus 27002 for security controls. If you can get your hands on a **Service Organization Control (SOC)** 2 Type 2 report (controls implementation) from the CSP, they can validate security, although they are not made public. The SOC 2 Type 1 report only relates to control design, and SOC 1 reports are for financial auditing. A SOC 3 report (audit) says the controls have been audited.

SLAs

SLAs, the binding contracts outlining the delivery of service from the CSP to the customer, vary between CSPs, services, and levels of service delivery, and some are customized to the customer. SLAs are an important area of risk assessment and shouldn't be agreed to before evaluation by a customer's legal team. Beyond SLAs, contracts will have other items, such as compliance with regulations, financial restitution in the case of CSP failure, and the full obligations of the customer.

Identifying Cloud Security Risks

Cloud threat modeling is a set of processes designed to identify, analyze, and mitigate security risks, based on threats, within an existing cloud system. This starts by modeling the system, its boundaries, including entry and exit points and connected systems, each of the assets and their criticality, data flows, and mapped design. From here, threats are identified, assessed, and prioritized, and then plans are created to remediate risks. In contrast to non-cloud threat modeling, this set of processes considers the common cloud security risks in relation to threats. There are cloud threat modeling frameworks, outlining common threats, and tools dedicated to helping organizations with this process. The following are cloud threat modeling frameworks (you will go into these later in the book):

- **STRIDE**: The STRIDE model covers the main categories of spoofing, tampering, repudiation, information disclosure, denial of service, and elevation of privilege.

- **DREAD**: The DREAD model covers damage, reproducibility, exploitability, affected users, and discoverability.

- **CSA Top Threats**: The Cloud Security Alliance's Top Threats (`https://cloudsecurityalliance.org/artifacts/cloud-threat-modeling`) guidance covers core threat modeling activities, and a range of different cards to outline threats, assets, vulnerabilities, impacts, and controls.

Other frameworks include MITRE ATT&CK, **Process for Attack Simulation and Threat Analysis** (**PASTA**), and OWASP Threat Modeling.

Tools and Practices for Identifying Risks in Cloud Environments

Identifying risks in cloud environments is crucial for maintaining security and compliance, and it requires robust tools and practices. Tools such as cloud threat modeling, risk assessments, and continuous monitoring systems are essential for detecting vulnerabilities and potential threats. Employing these tools and practices ensures a proactive approach to cloud security, protecting sensitive data and maintaining business continuity.

Risk Assessments and Tools

Risk assessments and tools are used both by CSPs and customers to identify risks in cloud environments. The tools used by CSP will be outlined in contracts and there will also be security surrounding access to the reports generated.

Penetration testing involves white hat hacking to attempt to gain access to systems, or otherwise exploit cloud vulnerabilities.

Static application security testing (**SAST**) is used to identify vulnerabilities in non-running application code, whereas **dynamic application security testing** (**DAST**) is used for testing applications and systems while they are in use. SAST is a form of white box testing, where everything is known, whereas DAST is a form of black box testing, where the internals are unknown.

External Third-Party Assessments

CSPs will have risk and vulnerability assessments with trusted external assessor partners available for customers who need to meet specific compliance objectives.

Analyzing and Assessing Cloud Security Risks

Once risks have been identified, they must be analyzed, assessed, classified, and prioritized, to determine the correct response to each risk.

In this phase, you determine the following:

- The likelihood of a threat succeeding
- The threat's anticipated impact on both critical and non-critical assets
- A **business impact analysis** (**BIA**) to evaluate the consequences of a threat succeeding (including financial losses, reputation damage, downtime, and time to full recovery)
- Which threats should be prioritized for remediation
- The cost and time to remediate

Qualitative versus Quantitative Risk Analysis Methods

Qualitative analysis methods are subjective, usually manually performed, and are the first step in analysis and assessment (e.g., matrices, categorization).

Quantitative analysis methods are objective, typically involve numerical data collection, are often automated, and are the next step in analysis and assessment (e.g., simulations, **Cloud Security Posture Management (CSPM)**).

Tools and Frameworks for Cloud Risk Assessments

Selecting the right tools and frameworks is crucial for conducting thorough cloud risk assessments. These tools and frameworks provide a structured approach to identify, analyze, and prioritize potential security vulnerabilities within your cloud environment. This helps organizations make informed decisions about their cloud security posture.

Here's a look at some prominent options:

- **CSA Cloud Controls Matrix (CCM)**: The CSA CCM is used by both CSPs and customers and covers 17 categories of controls, including **Audit and Assurance (A&A)**, **Application and Interface Security (AIS)**, and **Business Continuity Management and Operational Resilience (BCR)**.

- **NIST Frameworks and Others**: NIST publishes many guidance documents and frameworks to help organizations with cybersecurity governance. These include the NIST RMF (`https://csrc.nist.gov/Projects/risk-management`), with seven activities: prepare, categorize, select, implement, assess, authorize, and monitor, plus guidance such as NIST SP 800-37 (the guide for implementing the RMF). Other RMFs include ISO 31000:2018 (Risk Management), ISO27005, and the ENISA **Operationally Critical Threat, Asset, and Vulnerability Evaluation (OCTAVE)**. These frameworks are publications from international and respected government organizations.

- **Cloud Security Posture Management**: CPSM tools provide quantitative security scores and recommended remediations across a number of different security verticals. These tools are often vendor-supplied products.

- **Security Information and Event Management (SIEM)**: SIEM tools are log and event management tools that aggregate and correlate data from disparate services and systems. Within the tool, features include security alerting and reporting and may have suggested remediation or escalation functions. These tools are typically cloud native and designed to integrate into many different vendor ecosystems.

- **Cloud Workload Protection Platforms**: Cloud workload protection platforms examine the security of the workloads running on cloud infrastructure. These platforms are particularly useful when assessing the risk of running containerized applications in a distributed, remote cloud architecture.

- **Endpoint Detection and Response (EDR)**: EDR tools can identify suspicious activity at endpoints and use historical data to learn anomalous events and behaviors.

- **Cloud or Service-Specific Monitoring Tools**: CSPs or SaaS vendors may have additional tools that help with risk management within their proprietary systems or offer a view to the customer of their own SIEM and other monitoring solutions.

Cloud Attack Surface Area, Vulnerabilities, Threats, and Attack Vectors

Cloud computing offers a multitude of benefits for organizations, but it also introduces a unique set of security challenges. Unlike traditional on-premises environments, cloud infrastructure presents an expanded attack surface area for malicious actors to exploit. This section delves into the key concepts related to cloud security risks: attack surface, vulnerabilities, threats, and attack vectors.

Cloud Attack Surface and Vulnerabilities

The same cyber threats exist in the cloud as on-premises infrastructure and services but with additional new points of attack.

Hypervisor

The hypervisor is a common target, as attacks on the hypervisor can affect the host machine, the OS running on the host machine (in the case of a software-based Type 2 hypervisor), the hypervisor itself, and its VM tenants. In the case of a Type 1 hypervisor (a bootable hypervisor that runs directly on the machine), there is a smaller attack surface, as there is no underlying OS.

Hypervisor compromise can infiltrate multiple virtualized instances at once. In the case of multi-tenancy, this can mean attacks on multiple organizations at once.

In some cases where hypervisors have been misconfigured, it is possible for attacks, data, or process information in a VM to escape the VM. Guest escape is when users can access other instances running on the hypervisor. Information bleed is when users can either see other tenants' information or learn about their activities. There may also be a conflict of interest if a tenancy is shared with competitors.

Direct Connections from Remote Devices

Rather than accepting direct connections to resources from remote devices, services can be delivered from a secure server, so the data itself is not stored on the device.

Threats, Attack Vectors, and Incident Response (IR) in Cloud

Threats actors come in the form of malicious external actors (e.g., criminal hackers, hacktivists, competitors, and gray hat hackers), non-malicious external actors (e.g., script kiddies, gray hat hackers), disgruntled insiders (e.g., employees or contractors stealing IP to give to the competition, taking down systems), and negligent insiders (e.g., employee being phished, accidentally losing a device). Each of these threat actors must be considered in mitigation strategies.

CCSPs must consider these threat actors and the cloud attack surface and corresponding vectors must be considered from a remote access, ephemeral perspective.

IR Planning for Cloud Environments

Business Continuity and Disaster Recovery (BCDR) is a significant part of cybersecurity planning activities, and within this field, IR is key to recovery after an incident. Cloud providers and customers should have their own complementary BCDR and IR policies and plans in place, which are shared with each other and part of the SLA contract. The goal of business continuity is to maintain operations even in the case of disruption while disaster recovery is about recovery to full operational health, incident investigation, and so on. Communication between the CSP and the customer during IR is critical.

The goals of IR include minimizing losses, ensuring the threat has been eradicated, and meeting business continuity goals such as time to recover business-critical services. The phases of IR are prepare, detect, respond, recover, and post-incident. In the post-incident phase, root-cause analysis should be conducted.

IR plans or playbooks should be developed in line with each likely risk scenario so you know which steps to follow should an incident occur. IR teams should be prepared to respond to an incident at any time, which is why many organizations run a 24/7 SOC. Both CSP and customer internal logs and monitoring will typically be how an incident is discovered, although sometimes the alert may be from a customer or the public. Incidents are classified by impact and urgency.

Non-cloud-specific incident management standards include the Carnegie Mellon University **Software Engineering Institute (SEI)** – Incident Management Capability Assessment, NIST SP 800-61, Computer Security Incident Handling Guide, and ISO 27035 (Incident Management).

Risk Response Strategies

A risk response strategy is crucial for cloud security as it directly influences how organizations prepare for, react to, and recover from potential threats. The importance of having a well-defined risk management strategy or policy in the cloud involves several key components, including risk avoidance, mitigation, transfer, and acceptance. Here's how each plays a vital role in shaping an organization's overall cloud security posture.

- **Avoidance**: Avoidance refers to the proactive approach of eliminating activities, operations, or assets that introduce risk, thereby preventing potential threats or vulnerabilities from materializing. This can involve not adopting certain cloud services, redesigning systems to exclude risky components, or completely abstaining from specific actions that pose significant security or compliance risks.

- **Mitigation**: Mitigating risks is when countermeasures are put in place to help stop threats. Mitigation strategies were mentioned before and include making use of cloud security services such as virtual intrusion detection systems (IDSs) and stricter SLA contracts.

- **Transfer**: Transferal of risk is offloading risk to another entity. For example, cyber security insurance transfers the risk of monetary losses, or outsourcing cloud cybersecurity management with strict SLAs to a third party can help transfer overall risks.

- **Acceptance**: If a risk's likelihood and impact are low, the risk may be accepted without mitigation, transfer, or avoidance.

Addressing Cloud Security Risks – Safeguards and Countermeasures

Safeguards and countermeasures are critical for addressing cloud security risks, such as unauthorized access and data breaches. Implementing strong access controls, such as multi-factor authentication and **role-based access control** (**RBAC**), ensures that only authorized users can access sensitive data and systems. Encryption, both in transit and at rest, protects data from being intercepted or compromised. Regular security audits, continuous monitoring, and IR plans enable organizations to detect and respond to potential security incidents swiftly. These measures collectively enhance the security posture of cloud environments, protecting against malicious activity and ensuring data integrity and confidentiality.

Now, let's look at how to address certain risks from a safeguard and countermeasures perspective.

Data Breaches and Data Loss

Data breaches and data loss can arise from misconfigurations, non-authorized access, malware, black hat and grey hat hacking, and other reasons. While CSPs can provide baseline security here, customers still have a significant amount of responsibility. **Data Loss Prevention** (**DLP**) tools and processes can help mitigate data loss incidents. For plaintext **Personally Identifiable Information** (**PII**), use techniques such as tokenization (replacing sensitive data with unique tokens), masking (modifying data so that it looks similar), and obfuscation (replacing data with other data). Regular patching, audits, staff training, and all manner of tools and devices can be deployed to mitigate data loss.

In cloud storage clusters, there are two additional concepts that can be used for resiliency, which prove useful in guarding against data loss: replication and erasure coding. Replication involves creating multiple copies of data across different nodes or regions, ensuring availability even if one or more nodes fail. Erasure coding, on the other hand, splits data into fragments and distributes them across multiple storage nodes along with redundant data pieces, enabling data reconstruction even if some fragments are lost or corrupted. These architectures are essential for ensuring high availability, durability, and fault tolerance in cloud storage systems.

Non-Authorized Access

Since access control and permissions lie with the customer and not the CSP, and cloud services and networks are remote and not owned and fully managed by the customer, misconfiguration or rule-based access that doesn't follow best practices can be used to breach systems. To mitigate this, deploy **Identity and Access Management** (**IAM**) solutions (potentially via **Cloud Access Security Brokers** (**CASBs**) and policies such as **Multifactor Authentication** (**MFA**) and IP checking, as mentioned previously.

Administrative Concerns

Administrative concerns include non-intended elevated privileges, lack of correct privilege revocation policies, and insider threats. Mitigation requires privilege audits for software and infrastructure, automation of policies (potentially through IAM), and separation of duties.

Virtualization Risks

Virtualization means that resources are ephemeral and can be "lost" or destroyed at any time, which can impact operations or regulatory compliance. Clusters with failover, backups, versioning of VM images and states, data dispersion, and monitoring and alerting can prevent loss.

Containers add another layer of complexity and resilience in cloud environments. They encapsulate applications and their dependencies into isolated units, ensuring consistency across different deployment environments. The ephemeral nature of containers means that they can be quickly replaced or scaled, enhancing operational flexibility. However, this also necessitates robust orchestration tools such as Kubernetes to manage container life cycles, ensure high availability, and automate failover processes.

Regulatory Non-Compliance

In industries and organizations where regulatory compliance is essential, it can be difficult to ensure this is being met, due to the nature of cloud services, data flows, and operations. Work with CSPs and third-party assessors to ensure compliance and keep up to date with laws.

Distributed Denial of Service (DDoS) Attacks

DDoS network attacks are a risk to network-accessible cloud services. While **intrusion prevention systems (IPSs)** and web application firewalls are important, many CSPs already have provisions in place to guard against DDoS attacks. That being said, DDoS protection services are not enough on their own to ensure resilience against DDoS attacks. To truly safeguard your cloud workloads, it is imperative to architect your systems with built-in redundancy and scalability.

Man-in-the-Middle Attacks

Man-in-the-middle attacks are where threat actors insert themselves at a point in the network to spy on (unencrypted) data flows or impersonate the sender or receiver. Using secure transmission protocols such as **Transport Layer Security (TLS)**, **virtual private networks (VPNs)**, and tunneling helps prevent these attacks.

Vendor Issues

If vendors change the terms of their contracts, do not honor contracts go out of business, have diminishing **quality of service (QoS)**, or other vendor-specific issues, customers may find it difficult to ensure the confidentiality, integrity, and availability of their cloud-based operations. There may be assessment and audit risks, depending on how the vendor offers and runs these solutions. Going over contracts and SLAs with a legal team, selecting a vendor from the CSA's STAR program, and conducting regular CSP assessments can help mitigate these risks.

Shadow IT

Shadow IT is when employees deploy cloud services that do not follow organizational policy and are not known to administrators, which can introduce unseen threats. This can be avoided with access permissions and a strict set of services that are permissible by user and group.

Natural Disasters

Unlikely infrastructure changes due to natural disasters can have security consequences, depending on the cloud configuration. To mitigate the chance of a natural disaster causing an impact, organizations can have secondary regional zones ready to spin up, for built-in redundancy and resiliency in addition to proper resiliency architecture design and planning.

Insider Threats

Insider threats can be mitigated by implementing strict hiring policies and background checks, running internal training, two-person integrity for highly sensitive operations and access, and separation of duties. Regular auditing and continuous monitoring are critical for detecting and mitigating insider threats, as they help identify anomalies, policy violations, and suspicious activities in real time. Implementing robust auditing processes ensures compliance with security policies and the effectiveness of controls, providing an additional layer of protection against potential insider threats.

Insecure APIs

APIs that are vulnerable to injection attacks, expose PII, or have other flawed securities need remediation. This includes full testing, input validation, authentication practices, and encryption of data in transfer.

Misconfigurations

Avoiding misconfiguration vulnerabilities can be achieved by following vendor playbooks, prescriptive frameworks such as the **OWASP Top 10**, and running vulnerability scanners.

Forensic Challenges in Cloud Environments and Solutions

For IR, especially in the case of legal obligations in conducting eDiscovery in the cloud, there are notable challenges. These include non-permanent resources and virtualization, multi-tenancy (whereby machine operations might disrupt other tenants), and data residency concerns. Single-tenancy, data dispersion, backups, and versioning help here.

Implementing Cloud Security Best Practices, Controls and Countermeasures

Effectively securing your cloud environment requires a multi-layered approach that combines best practices, controls, and countermeasures. These strategies work together to mitigate potential security risks and ensure the integrity of your data and systems.

Best Practices

The following subsections outline some key best practices for cloud security:

- **Follow Frameworks:** Organizations should follow the industry-standard frameworks mentioned and remain up to date with current and emerging threats.

- **Follow CSP Best Practice Guides:** CSPs have documented best practice guides and tools that are specific to their own services, configurations, and infrastructure. Integrating these into organizational processes and documentation, and keeping it up to date, must be a priority.

- **CASBs:** CASBs are products that can be integrated with cloud services to help with tasks such as policy enforcement, traffic management, and endpoint detection. At their core, the focus is on IAM functions and monitoring.

- **Continuous Monitoring via SOC:** Continuous monitoring, alerting, and triage systems for cloud infrastructure, access, data flows, and so on, help organizations respond to threats and potential threats in a timely manner. Typically, continuous monitoring is performed within a security operations center , where security staff have access to monitoring tools such as SIEM and IDS.

- **Implementing Effective Cloud Security Policies and Governance:** Cloud security policies and governance frameworks, such as the NIST Cybersecurity Framework, can also help organizations decrease risk by having clear standards to follow. Maturity models, such as the **Cloud Security Maturity Model**, can help gauge how well your organization is doing among the wider community.

- **Role of Cloud Security Certifications and Standards in Risk Management:** Attaining and refreshing vendor-specific certifications helps admins to understand and manage these environments more effectively. Attaining other third-party certifications, such as PCI-DSS certification to ensure safe financial data processing, may be essential to operating legally. Remaining up to date with the latest standards in cloud security is essential and provides peace of mind both internally and for clients.

- **Importance of Continuous Risk Management and Security Monitoring in Cloud Environments:** Cloud environments must be monitored at all times to identify and triage anomalies, alert on new vulnerabilities, and deflect attacks. These systems must incorporate up-to-date information such as threat feeds, new patches, and zero-day exploits. Risk management as an activity must also be continuous: as new functions, systems, and cloud services come online, they must first be risk assessed, deployed securely, and added into monitoring systems.

- It's your duty as a cloud security professional to ensure that continuous risk management and security monitoring are a pillar of your work environment or work towards implementing these practices.

- **Encouraging a Culture of Security Awareness and Responsibility Across the Organization:** A cloud security professional must help increase security awareness around the organization and ensure that people know security is everyone's responsibility. While you may not be tasked specifically with training and awareness, if there is a successful security culture within your organization, cybersecurity risks decrease, and your job is made easier.

Controls

Cloud security relies heavily on a robust set of controls to safeguard sensitive information. These controls encompass various techniques and measures designed to protect data at rest, in transit, and during processing.

Data Encryption and Protection Techniques in the Cloud

From a cloud security perspective, several data encryption and protection controls are essential for safeguarding sensitive information. These controls include the following:

Encryption at rest: This control ensures that data stored on cloud storage solutions is encrypted, typically using AES-256 encryption algorithms. Encryption at rest mitigates the risk of unauthorized access to data if physical storage devices are stolen or accessed without proper authorization.

Encryption in transit: Data encryption during transit ensures that data being transferred between user devices and cloud services is encrypted using protocols such as TLS or **Secure Sockets Layer** (**SSL**). This mitigates the risk of interception or eavesdropping during data transmission, protecting data from man-in-the-middle attacks.

Encryption in use: Also known as confidential computing, this control protects data while it is being processed. Utilizing technologies such as hardware-based secure enclaves, encryption in use ensures that data remains protected even during computation, mitigating risks associated with unauthorized access or breaches during processing.

Key management services (**KMS**): Effective key management involves generating, storing, rotating, and revoking cryptographic keys securely. Cloud providers often offer KMS solutions to handle encryption keys' life cycles. This control mitigates the risk of key compromise, which can lead to unauthorized decryption and access to sensitive data.

Data masking and tokenization: These techniques replace sensitive data with non-sensitive equivalents, making real data unusable to unauthorized users while maintaining its usability for applications. Data masking and tokenization mitigate the risk of exposing sensitive information during testing, analytics, or when accessed by internal or external parties.

Access control and identity management: Implementing strict access controls and identity management practices, such as RBAC and MFA, ensure that only authorized users can access encrypted data. This mitigates the risk of unauthorized access and potential data breaches.

Encryption control plays a crucial role in mitigating the risk of unauthorized access and data breaches and ensuring data privacy and compliance with regulatory requirements. By encrypting data at rest, in transit, and in use, and managing encryption keys securely, organizations can protect sensitive information against various security threats, maintaining the integrity and confidentiality of their data in cloud environments.

In addition to encryption, several other data protection controls are essential for securing cloud environments. Access controls such as RBAC and **Attribute-Based Access Control** (**ABAC**) ensure that only authorized personnel can access sensitive data, thereby reducing the risk of data breaches and unauthorized access. DLP systems monitor and control data transfers to prevent unauthorized sharing or leakage of sensitive information. Regular data backups are critical for maintaining data availability and integrity, allowing data restoration in the event of accidental deletion, corruption, or a cyberattack. MFA enhances security by requiring multiple forms of verification before granting access to critical systems and data. Audit logging and monitoring provide continuous oversight of data access and usage, enabling the detection and investigation of suspicious activities and compliance with regulatory requirements. Secure data disposal practices ensure that data is irretrievably deleted when it is no longer needed, preventing unauthorized recovery of sensitive information. Together, these controls provide a comprehensive approach to data protection in cloud environments, ensuring data confidentiality, integrity, and availability.

IAM in Cloud Environments

IAM must be a core security activity. Digital identity management can be guided by publications such as **NIST Special Publication 800-63 Digital Identity Guidelines**, as well as the CSP's own set guidance.

Cloud security practitioners must consider group and user access and controls. Access to networks, resources, apps, files, and so on should be set at the user and group level, with processes in place for when roles change, people resign, or unusual activity is flagged.

There are also remote access considerations. Identity management providers are typically tasked with identity management, authenticating remote access users, crypto key management, MFA, remote kill switches, and more.

Information Rights Management (IRM) / Digital Rights Management (DRM)

IRM or DRM solutions can be used at the file level to control access permissions, usually for guests (e.g., clients or customers), such as read-but-not-copy permissions on a PDF, and may have time-expiry periods. IRM solutions are important when sharing sensitive IP more widely, and DRM particularly when it comes to copyright. IRM solutions are complementary to IAM solutions.

Virtualization

Virtualization can help mitigate some of the risks associated with cloud security. By utilizing VMs, the risk of an exploit escaping a VM and affecting other VMs is somewhat reduced due to robust isolation technologies. However, that is not a guarantee, so effective malicious monitoring methods are essential. Additionally, virtualization allows for centralized patching and policy roll-outs, enabling updates and security measures to be deployed consistently across all VMs, regardless of their physical location. This centralized management enhances security and compliance by ensuring that all virtual environments adhere to the same security policies and updates.

Network Segmentation

Logical network segmentation in cloud networks allows organizations to subnet zones based on restricted data, critical infrastructure, general use, vendor access, a public access **demilitarized zone** (**DMZ**), and more. By implementing network segmentation, there is a decreased risk of threats crossed zone borders.

Firewalls and Other Devices

Virtual security devices operate similarly to their physical counterparts, using techniques such as state-based inspection and rule-based filtering and alerting. Historically, organizations would include stateful inspection firewalls for traffic filtering, routers for traffic management within the network, VPN gateways for secure remote access, IDSs, and IPS. Now, CSPs will deploy these on their own systems, but customers can also introduce their own into their environments.

DLP Tools / Egress Monitoring

DLP tools and processes can help mitigate data loss incidents through data inventorying, classification, and policy enforcement. DLP tools are used at the boundaries of the production environment before data leaves systems. Here, the data can be checked against known classifications, policies, access permissions, keywords, templates, and so on. This can trigger alerts, halt data transfer, flag users, or any number of actions set by the security team.

Summary

In this chapter, you covered risk management across CSPs, various cloud service and deployment models, and the methods to identify, assess, and address risks, threats, and vulnerabilities. A CCSP candidate must be able to explain and implement risk management concepts, including understanding the risks associated with different cloud service models (IaaS, PaaS, and SaaS) and deployment models (public, private, hybrid, and community). They should be familiar with the shared responsibility model, common cloud risks, and industry-standard risk frameworks such as NIST RMF and ISO 31000:2018.

Additionally, candidates should know how to evaluate CSPs using tools such as CSA STAR and ISO 27001, differentiate between threats, vulnerabilities, and risks, and outline risk identification and assessment techniques. Best practices in risk countermeasures, understanding common cloud attack vectors and mitigation strategies, identifying threat actors, and IR in cloud computing are also crucial areas. Finally, they must be able to explain risk response strategies and name cloud security governance frameworks. This domain is essential, especially for those not regularly working with risk concepts, and further resources are provided for deeper exploration. The next chapter will focus on selecting, planning, and implementing security controls in cloud environments.

Exam Readiness Drill – Chapter Review Questions

Apart from a solid understanding of key concepts, being able to think quickly under time pressure is a skill that will help you ace your certification exam. That is why working on these skills early on in your learning journey is key.

Chapter review questions are designed to improve your test-taking skills progressively with each chapter you learn and review your understanding of key concepts in the chapter at the same time. You'll find these at the end of each chapter.

> **How to Access These Materials**
>
> To learn how to access these resources, head over to the chapter titled *Chapter 25, Accessing the Online Resources.*

To open the Chapter Review Questions for this chapter, perform the following steps:

1. Click the link – `https://packt.link/CCSPE1_CH09`.

    Alternatively, you can scan the following **QR code** (*Figure 9.2*):

Figure 9.2 – QR code that opens Chapter Review Questions for logged-in users

2. Once you log in, you'll see a page similar to the one shown in *Figure 9.3*:

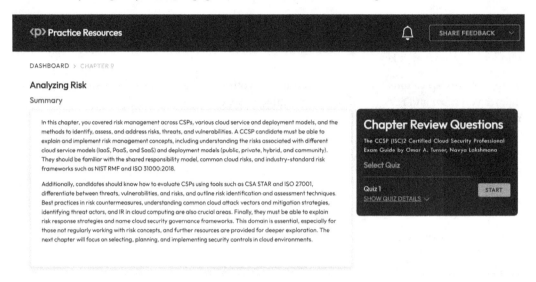

Figure 9.3 – Chapter Review Questions for Chapter 9

3. Once ready, start the following practice drills, re-attempting the quiz multiple times.

Exam Readiness Drill

For the first three attempts, don't worry about the time limit.

ATTEMPT 1

The first time, aim for at least **40%**. Look at the answers you got wrong and read the relevant sections in the chapter again to fix your learning gaps.

ATTEMPT 2

The second time, aim for at least **60%**. Look at the answers you got wrong and read the relevant sections in the chapter again to fix any remaining learning gaps.

ATTEMPT 3

The third time, aim for at least **75%**. Once you score 75% or more, you start working on your timing.

Tip

You may take more than **three** attempts to reach 75%. That's okay. Just review the relevant sections in the chapter till you get there.

Working On Timing

Target: Your aim is to keep the score the same while trying to answer these questions as quickly as possible. Here's an example of how your next attempts should look like:

Attempt	Score	Time Taken
Attempt 5	77%	21 mins 30 seconds
Attempt 6	78%	18 mins 34 seconds
Attempt 7	76%	14 mins 44 seconds

Table 9.1 – Sample timing practice drills on the online platform

---

Note

The time limits shown in the above table are just examples. Set your own time limits with each attempt based on the time limit of the quiz on the website.

With each new attempt, your score should stay above **75%** while your "time taken" to complete should "decrease". Repeat as many attempts as you want till you feel confident dealing with the time pressure.

10
Security Control Implementation

In the previous chapter, you learned about the top risks that exist today in the physical, logical, and virtual environments as a cloud consumer and provider. You also reviewed how to analyze, assess, and address the risks with safeguards and countermeasures.

This chapter delves deeper into the vital security controls necessary for the robust protection of cloud environments. We'll explore how these controls are strategically implemented to safeguard the confidentiality, integrity, and availability of data and services, essential for maintaining secure and compliant cloud infrastructures.

The topics covered in this chapter include the following:

- Physical and environmental security controls
- System, storage, and communication security controls
- **Identity and Access Management** (**IAM**) solutions for identification, authentication, and authorization
- Cloud security audit mechanisms to ensure the proper implementation of these controls

Each of those topics plays a critical role in a comprehensive cloud security strategy. Having a solid understanding of these security controls will be very important for the CCSP exam and the rest of your career as a cloud professional. Let's dive right in!

Physical and Environmental Protection Controls

Physical and environmental protection covers all of the physical devices and infrastructure components of a data center. This includes devices and components inside the actual data center, such as servers and server racks, networking gear, cabling, cooling units, and **Power Distribution Units** (**PDUs**).

Security extends beyond the internal mechanisms to include the physical premises, where barriers, surveillance cameras, and controlled access points protect against unauthorized entry.

In addition to the data center and its physical assets, physical protection must also include the end user devices used to connect to the cloud environment. These devices can include desktops, laptops, mobile devices, and any other endpoint devices used by the customer. As a CCSP, you must remember that even physical protections have a shared responsibility.

The main priorities regarding physical and environmental protection are as follows:

- Restricting physical access at all entry points
- Ensuring a secure and uninterrupted power supply
- The availability of essential utilities such as water and power
- The availability of an adequate workforce

In a public cloud environment, all of these are the responsibility of the **Cloud Service Provider** (**CSP**). With that said, this is considered basic knowledge for a certified cloud security professional, so it is important for candidates to have a general understanding of how these protections are implemented and managed by a CSP.

The guiding principle for physical and environmental protection is a **defense-in-layers** (also known as defense-in-depth) approach. These layers would look something like this:

- A strong perimeter defense outside of the actual data center building. There, you would have typical security measures such as fences, lights, and cameras.
- Access to the building would be tightly controlled with vehicle and personnel access measures such as ramps and ID cards.
- Inside the data center, sensitive systems are protected with further defense mechanisms such as cages, monitoring, and access control.
- Power and cooling must have multiple independent supplies to ensure continuous operation.

The necessary protections should be tailored according to a thorough risk assessment specific to the CSP's environment. This includes evaluating compliance risks, which are crucial for maintaining physical and environmental security.

With cloud providers increasingly recognized as critical infrastructure, regulatory oversight is intensifying. In the United States, for instance, CSPs fall under stringent scrutiny, such as that enforced by the **North American Electric Reliability Corporation Critical Infrastructure Protection (NERC CIP)** standards.

Site Selection and Facility Design

The primary consideration around physical security is the cloud location, as it will impact both physical and environmental protections. The key elements for **site selection** include visibility, area composition and accessibility, and the potential effects of natural disasters. For example, you wouldn't want to build a data center in an area that isn't easily accessible by car or in a region prone to frequent flooding or earthquakes.

Remember that all these decisions are the responsibility of the CSP. As a cloud security professional, you will likely only have to make decisions about selecting the CSP's target regions that meet your organization's specific security, regulatory, and operational requirements.

System, Storage, and Communication Protection Controls

From a hardware and systems perspective, the underlying makeup of cloud infrastructure is somewhat similar to a traditional data center. The main challenge in the cloud is the sheer number of assets and the scale at which they need to be managed and secured.

The best approach to securing complex systems is breaking them down into smaller components, making it more manageable to implement appropriate security controls at each step. The level of exposure the customer has to the underlying infrastructure and the necessity to protect it depends on the cloud model (IaaS, PaaS, or SaaS).

There are three security mechanisms in play regarding controls at this stage:

- **Policy and procedures**: These are the established requirements for system protection. These requirements should be well defined, including the purpose, scope, role, and responsibilities needed to achieve them. These policies and procedures must be regularly reviewed and updated to adapt to new security challenges and changes in regulatory requirements.

- **Separation of duties**: A core security principle that ensures that no single person can control all elements of a critical function or system. Responsibilities are divided among multiple individuals to prevent fraud and errors. Separating user and admin access can prevent users from altering processes or misconfiguring systems, whether intentionally or not.

- **Security function isolation**: Another useful approach is isolating security roles from non-security roles, also called security function isolation.

Let's now shift our focus to the specific controls you must implement to protect the data and systems within a cloud environment.

Protecting Data

Following the flow of data and aligning security controls based on the data state ensures that security measures are optimized, addressing the specific vulnerabilities and threats at each stage of the data life cycle. Here are the three main data states and their corresponding security controls:

- **Data at rest**: This data is stored in any type of persistent storage (object storage, block, file, or any type of persistent database) and is not actively moved or processed. The main security controls here are encryption and access permissions, which ensure that the data is unreadable to unauthorized identities (human or application, service, and computer accounts), and tokenization – the process of hiding the contents of a dataset by replacing sensitive or private elements with a series of non-sensitive, randomly generated elements. Strong access control mechanisms, such as **Role-Based Access Control** (**RBAC**), should be implemented to ensure only authorized identities can access and modify the data.

- **Data in transit**: This state involves data that is transmitted across networks, whether within a data center, between data centers, or over the internet. Once again, encryption plays a key part here, this time using protocols such as **Transport Layer Security** (**TLS**) or HTTPS to secure data as it travels.

- **Data in use**: Protecting data in use involves several key practices to ensure it remains secure during processing and operations. Utilizing encryption at the application level allows data to stay encrypted as it is processed, mitigating the risk of exposure even in memory. Implementing robust access controls and authentication measures ensures that only authorized users and systems can access the data when it's being used. Employing **Trusted Execution Environments** (**TEEs**) provides a secure area within processors where sensitive data can be processed away from the main operating system, reducing the risk of unauthorized access.

Security measures such as encryption and access control are applied across all data stages. For the exam, you should understand how these measures are implemented and managed differently, depending on whether data is at rest, in transit, or in use.

Cryptographic Key Establishment and Management

Cryptography provides several security functions to secure data and communications, including confidentiality, integrity, and nonrepudiation. Encryption tools and protocols such as TLS and **Virtual Private Networks** (**VPNs**) are used to provide confidentiality by encrypting data during transmission.

Hashing is another data protection measure you should be familiar with. It involves transforming plaintext data into an unreadable string of characters that are difficult to decipher. Hashing is typically used for storing sensitive data such as passwords and provides integrity.

Additional security measures such as digital signatures and **Hash-Based Message Authentication Code (HMAC)** can be used to detect intentional tampering. HMAC can simultaneously verify data integrity and message authenticity.

Managing a Network to Protect Systems and Services

Apart from protecting the data itself, you must also protect the cloud systems and services that

handle the data. A cloud security professional must be a master of secure network management. This includes the implementation and configuration of firewalls, IDSs/IPSs, and other security controls that prevent and detect anomalous or malicious network activity.

In the cloud, you will deal with virtual firewalls and IDSs/IPSs, so the main concern will be configuring rules and policies in these systems. Large CSPs have a policy engine that allows you to configure centralized policies and apply them to your virtual appliances.

Disruptive attacks at scale, such as **Denial-of-Service/Distributed-Denial-of-Service (DoS/DDoS)** attacks, can cripple cloud systems, causing significant downtime and potentially long-lasting damage to a business's reputation and operations. Most CSPs offer DoS/DDoS mitigation as a service (Azure DDoS, AWS Shield, and Google Cloud Armor). A basic version of these services is typically available free of charge, while advanced features may require an expensive pricing tier. There are also specialized third-party providers such as Cloudflare. It is important to note that to effectively withstand a major DDoS attack, your architecture must incorporate redundant services across all layers, include auto-scaling capabilities, and utilize load balancers or potentially CDN services. These elements are essential for maintaining resilience and operational continuity during such incidents.

Landing Zones

A cloud landing zone is a pre-configured, well-architected environment that establishes the foundation for secure and scalable cloud application development and service deployment. It incorporates best practices, security policies, and governance controls from the outset of cloud adoption, which is crucial for maintaining network security. Landing zones facilitate network segmentation and isolation, creating defined boundaries between different environments (such as development, testing, and production) or application tiers (such as web, application, and database), thus limiting breach scope and lateral movement within the network. Centralized security controls within the landing zone ensure consistent deployment of threat management systems, firewalls, and intrusion detection systems, simplifying security management and reducing misconfiguration risks. Additionally, landing zones are designed to integrate compliance with standards such as GDPR, HIPAA, or PCI DSS directly into the architecture, ensuring all deployments remain compliant through automated checks and continuous monitoring.

Robust IAM features such as RBAC, **Multi-Factor Authentication** (**MFA**), and least-privilege policies are foundational to landing zones, which manage access to resources effectively. Comprehensive logging and monitoring are also integral, with systems set up to capture and analyze all network traffic and resource usage logs, providing real-time alerts to potential security threats. Furthermore, landing zones support the automation of security deployments and configurations, which ensures that as a cloud environment expands, security measures are consistently applied to new instances and services, thereby reducing human error and keeping security measures in step with cloud development. This strategic approach not only bolsters security but also streamlines the deployment and management of cloud resources, making cloud landing zones an essential element in secure cloud computing.

Virtualization Systems

The significance of virtualization in the cloud and how it enables many of the great features associated with cloud computing has already been stressed. Considering the vital role virtualization plays, attackers see it as the ideal target to disrupt cloud operations and security.

There are two critical areas you need to know about to protect virtualization systems – the **hypervisor** and the **management plane**. The management plane has full control over the environment and exposed APIs to allow administrative tasks. As such, it's the most visible and important aspect to fully protect. The management plane is made up of a series of APIs, exposed function calls, and services, as well as web portals or other client access to allow its use. Any of these points are potential vulnerabilities. Cloud security professionals must develop a holistic picture of the threats and vulnerabilities at each level and component of the management plane. Not doing so could expose an entire system to a missed weakness that an attacker can exploit.

Like all systems, the management plane and virtualization infrastructure significantly benefit from RBAC. It's essential to maintain strict regulation of all administrative access, which should also be consistently and thoroughly reviewed through audits.

IAM Solutions for Identification, Authentication, and Authorization

Users with access to systems must be held accountable for following policies and procedures. This is typically done by logging and monitoring system activity. Enforcing accountability in the cloud can be challenging. For example, users may access SaaS apps as they travel, which makes identifying anomalous behavior more difficult. In simple terms, identification and authentication validate a person's identity, while authorization ensures the person in question should have access to the system or resource. Users should be granted the minimum necessary privileges to do their job, which in the security world is known as the principle of least privilege.

For the exam, you should know about cloud-based identity services such as OpenID and OAuth. These services offer a centralized and scalable solution for identification, authentication, and authorization across your cloud applications and APIs, saving you from the complexities of building these functionalities from scratch and the challenges of relying on non-cloud-based services for user management.

OpenID

OpenID is an authentication protocol that allows users to be authenticated across cooperating sites using a third-party service. It eliminates the need for users to create new passwords for every site, instead enabling them to use one set of credentials across multiple sites. This is particularly beneficial in cloud environments where users may need to interact with numerous applications and services. OpenID operates under the principle of a trusted **Identity Provider (IdP)** that manages the user's identity.

The OpenID authentication process follows a well-defined sequence:

1. **The user chooses to log in**: A user visits a website (known as the "relying party") and chooses to log in using their OpenID provider (e.g., Google or Facebook).

2. **Discovery**: The website discovers the user's OpenID provider from the provided OpenID identifier.

3. **Authentication request**: The website redirects the user to their OpenID provider with an authentication request.

4. **User authentication**: The user logs in to the OpenID provider (if not already logged in) and grants the website permission to use their identity.

5. **Assertion**: After successful authentication, the OpenID provider sends an assertion back to the website, confirming the user's identity.

6. **Access granted**: The website validates the assertion and logs the user in.

This process eliminates the need for multiple usernames and passwords, reducing password fatigue and simplifying the user experience while maintaining security through centralized authentication.

OAuth

OAuth is a more comprehensive framework compared to OpenID and is used for authorization. It allows third-party services to exchange web resources on behalf of a user. The typical scenario involves a user giving a third-party application access to their information, stored with another service provider, without exposing their access credentials.

It's important to note that OAuth is about authorization, not authentication (although it is often used in conjunction with authentication protocols such as OpenID Connect). Here are the steps OAuth follows:

1. **Authorization request**: A user logs in to a service (client), which needs to access resources hosted by another service (resource server) on behalf of the user. The client directs the user to the authorization server (often combined with the IdP in implementations such as OpenID Connect) with a request for specific access rights.

2. **The user grants permission**: The user authenticates with the authorization server (if not already authenticated) and explicitly grants the client permission to access the resources on their behalf.

3. **Access token**: The authorization server issues an access token to the client after successful authorization.

4. **Resource access**: The client uses the access token to request resources from the resource server.

5. **Token validation**: The resource server validates the access token with the authorization server and, if valid, serves the resources to the client.

OAuth provides robust security controls by ensuring that user credentials are not shared with the client. Instead, access is granted via tokens, which can be restricted to specific types of data, have a limited lifespan, and be revoked by users at any time.

Let's now explore the concepts of identification, authentication, and authorization in more detail. You will also learn about federation, which is another important aspect of cloud computing, especially in public clouds with a large number of users.

Identification

To maintain security, every entity inside a cloud environment must have a unique identifier. This includes users, servers, applications, and devices. These identifiers allow for precise tracking and control of access to resources, ensuring that only authorized entities can interact with sensitive data and system functionalities.

The most common identifier for users is a username. For devices and servers, it's typically an IP address or a unique device ID. These identifiers are crucial in the initial step of securing access, as they help to map out which entities interact with the system and how they are related to each other within the network. Most larger organizations use Microsoft Active Directory for IAM. Conversely, small businesses and nonprofits tend to use open source identity systems such as FreeIPA or OpenLDAP.

Authentication

While identification determines the presence of an entity or person within a system, authentication deals with ensuring that the entity or person is indeed who they claim to be. Authentication is a very broad topic, so let's focus on the essentials that you must know for the CCSP exam. For a very long time, passwords were considered the primary and, in many cases, sole factor for authentication. However, as the risk of credential exposure has become very high, it's a security best practice to use additional forms of authentication on top of a password. This is called MFA.

Understanding MFA is crucial for the CCSP exam. You will likely encounter one or more questions about this, so make sure you understand the following concepts.

MFA requires at least two out of the three possible authentication factors. These factors are as follows:

- **Something you know**, such as a password or PIN

- **Something you have**, such as a smartphone or security token

- **Something you are**, such as biometric data, including fingerprints and facial recognition

The most common second factors are an authentication app, voice call, SMS message, and OAuth hardware token (a **time-based one-time password**, or **TOTP**). In many cases, implementing MFA is not just a security best practice but also a requirement. For example, requirement 8.3 of PCI DSS explicitly mandates the use of MFA for any personnel with non-console administrative access and all remote access to the cardholder data environment. This makes MFA a compulsory requirement under PCI DSS in specific scenarios involving access to sensitive cardholder data environments.

Authentication Methods

Perhaps the most popular authentication method these days (aside from a password) is **authentication apps**. This is a software-based authenticator that implements two-step verification with the use of a TOTP. The user is typically given a code that expires every 30 seconds. These one-time passcodes are generated using open standards developed by the **Initiative for Open Authentication** (**OATH**). Popular authenticator apps include Google Authenticator and Microsoft Authenticator.

Push notifications are another popular method. Here, a login attempt prompts the server to push a notification to your mobile device, which is used to confirm or deny the access attempt directly. The identity platform typically uses an associated authentication or general-purpose app to deliver the notification. For example, Google uses the Gmail app to send login confirmations.

Conditional Authentication Policies

This capability is increasingly common in **Identity-as-a-Service** (**IDaaS**) cloud platforms, including Microsoft Azure Active Directory and Google Cloud Identity. Conditional authentication policies look at the signals relating to the authentication attempts, which include the user's location, their device, their application, and the real-time risk rating of the user (determined with machine learning and AI based on past behavior). The signals are then processed together, and the platform determines whether to block access, allow access, or request MFA.

Authorization

Once an identity has been established and authenticated, the next step is to determine the level of access it should get to internal resources. Authorization is the act of granting an authenticated party permission to do something. This is where the principle of least privilege comes into play. Organizations should assign the minimum level of access necessary for users to perform their job functions. The goal is to minimize the risk of accidental or malicious data exposure and enhance overall security.

Authorization begins during the authentication phase, where the IdP sends predetermined attributes about a user to the relying party or service. These attributes, which can include names, roles, group memberships, and locations, are used to make decisions about what resources the user is allowed to access and what actions they are permitted to perform. Permissions, rights, and privileges are granted to users based on their proven identity. If a user has been assigned rights to a resource, they are granted authorization.

When deciding to permit access and its level, an application can utilize various attributes provided by the IdP about a user. Decisions might be based on a simple single attribute or a complex set of conditions and multiple attributes. For instance, consider a healthcare portal where hospital staff accesses patient records. Typically, the primary condition for access under the hospital's policy is verified employment with the hospital. The critical information needed from the IdP is confirmation that the individual is a current hospital employee. Once confirmed that the access control criteria have been met, the system allows entry.

Federation

Federation is a collection of domains that have established trust and share identity information. Trust is established via policies and guidelines that each member organization must adhere to, and it often involves the use of shared security protocols and encryption standards. The level of trust can vary, typically requiring authentication and almost always requiring authorization.

Federation allows users from one domain to access resources and services in another without needing to re-authenticate. This capability streamlines user access across different systems and organizations, enhancing collaboration and efficiency while maintaining security controls. When a user or system within a federation needs to access a federated application, they authenticate locally to obtain tokens. These tokens are then provided to the application, known as the relying party, for access. Each member of the federation operates their own IdP to manage this process.

For example, let's say you have a website hosted on Microsoft Azure (using Active Directory for IDaaS). To enable a user without an Active Directory account to authenticate on your site, you would establish federation with another service the user might use, such as Google or Facebook. What you're essentially doing is trusting Google and Facebook as IdPs. This will allow users to log in using their existing Google or Facebook credentials, leveraging OAuth for secure token-based authentication.

Key Cloud Control Audit Mechanisms

The final key to the security control puzzle is ensuring that appropriate audit mechanisms are in place to monitor and record activities within a cloud environment. This includes the implementation of logging and tracking systems that can detect, alert, and report on security incidents or anomalies. Regular audits help to ensure compliance with regulatory standards and internal policies, provide insights into operational effectiveness, assist in incident response processes, and offer a means to verify that security controls function as intended.

Auditing is the process of ensuring compliance with policies, guidelines, and regulations. Auditing is an integral part of most business operations. With cloud environments, things can get a bit more complicated, as audits have to consider two entities – the CSP and the customer. Additionally, data centers typically span across multiple countries and jurisdictions. The scope of an audit in cloud environments – whether IaaS, PaaS, or SaaS – will vary depending on the level of access granted to the customer and the specifics agreed upon in the contract with the cloud provider.

The agreement and **Service-Level Agreements (SLAs)** should explicitly define the audit requirements, the division of responsibilities, and the frequency of audit and penetration testing activities. In a multi-tenant cloud setup, it is crucial that any penetration or audit testing is coordinated between the cloud provider and the customer to prevent interference or disruption to the services used by other customers within the same environment.

There are three main audit mechanisms for cloud environments – log collection, correlation, and packet capture.

Log Collection

The cloud management plane makes it very easy for customers to centrally aggregate their logs. All CSPs, including Azure, offer dedicated services that will collect and store logs across various cloud sources. The problem with log collection in the cloud is that it is way more limited in terms of the type of logs you have available compared to having your own data center.

Cloud services will offer different controls over what information is logged, but they all collect a minimum level of security-related events. Log aggregators are commonly used to ingest logs from all on-premises and cloud-based log sources for review. The most common log aggregator type is a **Security Information and Event Management System (SIEM)**. This is similar to what you would see in a traditional data center. The challenge with cloud log aggregation comes with determining which logs a customer has access to. This will depend on the type of cloud deployment and the contract with the CSP.

In IaaS, the customer has the most access to logs, ranging from operating systems and virtual devices, such as virtual firewalls, to platform and application logs. With PaaS, customers have comprehensive logs on the platform and application levels, but not on the operating system and network levels.

Understanding the different levels of log access and what your organization gains and loses regarding log collection will be very important in your role as a cloud security professional.

Collecting audit records in cloud environments is crucial for monitoring, compliance, security analysis, and forensics. Various frameworks and standards provide guidelines and best practices for this purpose. Here are some prominent ones that are especially relevant to cloud computing:

- **The Cloud Security Alliance (CSA) Cloud Controls Matrix (CCM)**: The CSA CCM is a cybersecurity control framework for cloud computing that comprises a comprehensive set of security controls, categorized across different domains. The audit and assurance domain specifically addresses requirements related to audit logging and monitoring, offering structured guidance on how to collect, manage, and analyze audit records in the cloud.

- **NIST Special Publication 800-53**: Developed by the **National Institute of Standards and Technology (NIST)**, SP 800-53 provides a catalog of security and privacy controls for federal information systems and organizations. It includes detailed controls for audit and accountability (the AU family), which cover aspects such as audit event logging, monitoring, analysis, and the protection of audit information. NIST's guidelines are widely respected and can be adapted for private sector organizations as well.

- **ISO/IEC 27017:2015**: This is an international standard that provides guidelines for information security controls applicable to the provision and use of cloud services. It extends the ISO/IEC 27002 standard to cloud environments, detailing controls that include mechanisms for the collection and preservation of audit logs.

- **Payment Card Industry Data Security Standard (PCI DSS)**: While specifically focused on organizations that handle credit card information, PCI DSS requirements include detailed mandates to track and monitor all access to network resources and cardholder data. The standard outlines specific expectations for audit logging, including what events to log, how to secure logs, and how long to retain them.

- **Health Insurance Portability and Accountability Act (HIPAA)**: For healthcare applications, HIPAA requires the safeguarding of **Protected Health Information (PHI)**. This includes implementing hardware, software, and procedural mechanisms that record and examine activity in information systems containing or using PHI. It doesn't provide a framework as detailed as others for non-healthcare-specific environments but is crucial for compliance in healthcare sectors.

- **Service Organization Control 2 (SOC 2)**: SOC 2 is a framework for service organizations, which includes cloud providers, to ensure their information security measures are in line with specific criteria. The Trust Services Criteria for Security, Availability, Processing Integrity, Confidentiality, and Privacy include aspects of logging and monitoring that are essential for audit purposes.

- **General Data Protection Regulation (GDPR)**: Although GDPR is a legal framework rather than a technical one, it requires detailed record-keeping of processing activities, including the need for logging and audit trails to demonstrate compliance with the regulation, particularly regarding the processing of personal data.

Correlation

Collecting many different logs is great, but organizations need a way to make these logs useful. Correlation is the ability to identify a connection between two or more events across different log sources. The correlation of logs in cloud environments is vital for effective auditing, enhancing security, ensuring compliance, and improving operational efficiency. By analyzing and synthesizing logs from various sources, such as servers, applications, and network devices, organizations can detect patterns and anomalies that indicate security threats, such as unauthorized access or malware activity.

This process is crucial for accurate anomaly detection, as it helps filter out noise and focus on significant events, enhancing the detection of subtle security or operational issues. For instance, correlating an unusual login attempt with a subsequent large data download can flag a potential data exfiltration attempt. Furthermore, correlated logs streamline an incident response by providing a comprehensive view of security incidents, facilitating quicker and more coordinated troubleshooting and mitigation efforts.

Compliance with regulatory requirements often mandates maintaining and reviewing logs to protect sensitive data. Correlation simplifies compiling detailed reports that demonstrate compliance and supports forensic analysis by tracing an attacker's steps if there is a security breach. Beyond security and compliance, log correlation also boosts operational efficiency by monitoring system performance and aiding in troubleshooting, which is particularly valuable in cloud environments where dependencies between services can complicate root cause analysis.

Additionally, continuous monitoring and correlation of logs enable proactive risk management by identifying potential vulnerabilities and configuration errors, helping you maintain the integrity and security of cloud-based services. Overall, log correlation is an indispensable tool for auditing cloud architectures, as it enhances the visibility and analysis of interconnected activities across the cloud. This capability is fundamental to maintaining a secure, compliant, and efficient cloud environment, providing organizations with the intelligence necessary to manage risks effectively.

Packet Capturing

Packet capturing is crucial in auditing cloud environments because it provides detailed insights into network traffic, essential for several key reasons. First, it enhances security monitoring and threat detection, allowing teams to detect, investigate, and respond to potential threats by identifying suspicious activities such as unusual data transfers or unauthorized access. It also helps in verifying compliance with regulations that mandate monitoring network traffic and controlling access to sensitive information, which is pivotal during audits to prove adherence to compliance standards. Moreover, packet capturing aids in identifying network performance issues, assisting auditors in analyzing traffic to assess network efficiency, and recommending improvements for optimal performance.

In the event of a security incident, packet captures are invaluable for forensic analysis, helping to reconstruct events, understand their impact, and identify the source of the compromise. This detailed record supports accountability and transparency, ensuring that all transactions within the cloud are recorded, thus upholding internal policies and external regulations. Finally, regular packet capturing verifies that security policies, such as firewall rules and intrusion detection settings, are correctly enforced, making it a foundational tool for the effective governance and management of cloud resources.

Packet capture tools, also called protocol analyzers, are used to analyze and log network traffic that passes over a network or part of one. Wireshark is the industry standard for packet capturing and protocol analysis. It is a free, open source tool that provides detailed insights into network traffic, allowing users to dissect and examine packets for a variety of purposes, including troubleshooting, network performance evaluation, and security investigations. Wireshark has both CLI and GUI versions and is available for Windows and Linux. Many cloud environments don't provide any facility to capture packets, especially in SaaS scenarios where the customer isn't responsible for anything related to the environment.

However, within an IaaS environment, CSPs will offer some facilities for packet capture. This typically comes in the form of full support for industry-standard packet analyzers such as Wireshark. Some CSPs have built-in, specialized services to perform packet capture on virtual networks. For example, Microsoft Azure uses a specialized package called Network Watcher, while AWS supports Wireshark directly.

It's worth noting that most of these analyzers save the data as a PCAP file, so it can be easily shared and analyzed using various tools across different platforms.

It's important to understand some of the inherent challenges in initiating packet captures in a modern cloud architecture. Packet capturing in cloud environments poses several challenges that differ from traditional on-premises networking. One of the primary challenges is limited access and visibility, as users do not have physical access to the underlying network hardware, restricting the visibility of network traffic. Additionally, the dynamic and scalable nature of cloud services, where resources are frequently scaled and reconfigured across different regions, demands a robust and flexible monitoring setup. Legal and compliance issues are particularly challenging, especially in jurisdictions with strict data protection laws such as GDPR, necessitating adherence to all relevant laws and regulations when capturing packets.

Finally, the storage and analysis of the large volumes of data generated by packet capturing can be both technically challenging and costly, requiring significant resources for efficient data management and analytical tools.

Summary

Good job on getting to the end of this chapter! You learned about some very important concepts that will prepare you for the CCSP exam. A lot was covered, so let's recap some of the main points to ensure that they're well ingrained in your memory:

- **Physical and environmental controls**: The critical need to secure the physical infrastructure of data centers, including protective measures such as surveillance, access restrictions, and environmental controls to maintain the integrity and availability of services, was discussed.

- **System and data protection**: Emphasizing the importance of safeguarding data across all states – whether at rest, in transit, or in use – encryption techniques, secure API implementations, and the necessity for strict access controls were covered. Here, you also learned how to secure networks within cloud environments through the use of virtual firewalls, IDSs/IPSs, and the strategic segmentation of network trust zones to enhance security postures.

- **IAM**: You explored the roles of OpenID and OAuth in managing identities and accesses effectively across multiple platforms, thereby reducing complexity and improving security in cloud-based systems.

- **Comprehensive audit mechanisms**: You learned about the importance of audit mechanisms such as log collection, correlation analysis, and packet capture to detect, prevent, and respond to potential security threats efficiently.

Some of the key areas you should remember from this chapter are related to distinguishing between identification, authentication, and authorization. Identification involves verifying a user's identity, authentication confirms it, and authorization determines the user's access rights to resources. Get familiar with the key mechanisms and technologies used in implementing these processes, such as MFA, RBAC, and OAuth protocols.

Understanding comprehensive audit mechanisms in cloud environments is crucial for the CCSP exam. Auditing encompasses log collection, correlation, and packet capture to ensure compliance and monitor security. You should have a solid grasp of how audits can be scoped based on the cloud model (IaaS, PaaS, and SaaS) and contractual agreements. Additionally, make sure you understand the specific security controls to protect data, systems, and communication. This includes protecting data at all stages (at rest, transit, and in use) through appropriate encryption protocols, access controls, and monitoring practices.

In the next chapter, you will dive into how organizations can prepare to withstand disasters and business disruptions so they can continue the delivery of products and services within acceptable time frames. The use of cloud services can be seen as a key element in supporting critical business functions if there is a major disruption, so understanding business continuity and disaster recovery as it relates to the use of the cloud is very important.

Exam Readiness Drill – Chapter Review Questions

Apart from a solid understanding of key concepts, being able to think quickly under time pressure is a skill that will help you ace your certification exam. That is why working on these skills early on in your learning journey is key.

Chapter review questions are designed to improve your test-taking skills progressively with each chapter you learn and review your understanding of key concepts in the chapter at the same time. You'll find these at the end of each chapter.

> **How to Access These Materials**
>
> To learn how to access these resources, head over to the chapter titled *Chapter 25, Accessing the Online Resources*.

To open the Chapter Review Questions for this chapter, perform the following steps:

1. Click the link – https://packt.link/CCSPE1_CH10.

    Alternatively, you can scan the following **QR code** (*Figure 10.1*):

Figure 10.1 – QR code that opens Chapter Review Questions for logged-in users

2. Once you log in, you'll see a page similar to the one shown in *Figure 10.2*:

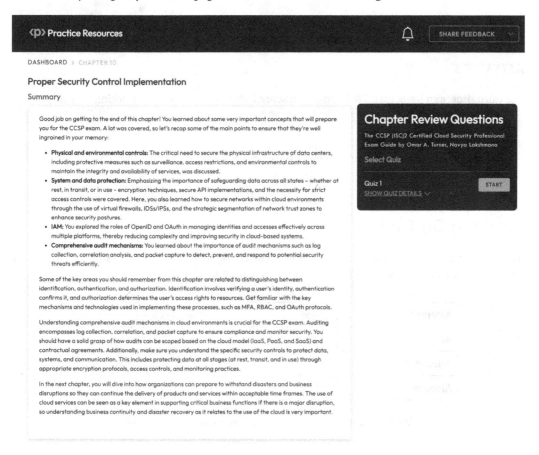

Figure 10.2 – Chapter Review Questions for Chapter 10

3. Once ready, start the following practice drills, re-attempting the quiz multiple times.

Exam Readiness Drill

For the first three attempts, don't worry about the time limit.

ATTEMPT 1

The first time, aim for at least **40%**. Look at the answers you got wrong and read the relevant sections in the chapter again to fix your learning gaps.

ATTEMPT 2

The second time, aim for at least **60%**. Look at the answers you got wrong and read the relevant sections in the chapter again to fix any remaining learning gaps.

ATTEMPT 3

The third time, aim for at least **75%**. Once you score 75% or more, you start working on your timing.

Tip

You may take more than **three** attempts to reach 75%. That's okay. Just review the relevant sections in the chapter till you get there.

Working On Timing

Target: Your aim is to keep the score the same while trying to answer these questions as quickly as possible. Here's an example of how your next attempts should look like:

Attempt	Score	Time Taken
Attempt 5	77%	21 mins 30 seconds
Attempt 6	78%	18 mins 34 seconds
Attempt 7	76%	14 mins 44 seconds

Table 10.1 – Sample timing practice drills on the online platform

---

Note

The time limits shown in the above table are just examples. Set your own time limits with each attempt based on the time limit of the quiz on the website.

With each new attempt, your score should stay above **75%** while your "time taken" to complete should "decrease". Repeat as many attempts as you want till you feel confident dealing with the time pressure.

11

Planning for the Worst-Case Scenario – Business Continuity and Disaster Recovery

Cloud environments are generally very reliable as they depend on **CSPs** with vast amounts of resources and infrastructure designed to ensure high availability, scalability, and fault tolerance. However, unforeseen events can still cause major disruptions, so having a **business continuity and disaster recovery (BCDR)** plan is essential.

In this chapter, you will review the following:

- BCDR – definitions, importance, and key concepts
- BCDR strategies
- Understanding an organization's business requirements
- Creation, implementation, and testing of BCDR plans

Ensuring **business continuity (BC)** and providing **disaster recovery (DR)** solutions are tasks you will likely encounter on the job as a cloud professional as well as in questions in the CCSP exam.

BC and DR are essential topics for CCSP exam candidates because they are fundamental to safeguarding cloud environments against disruptions. This knowledge is crucial for negotiating robust **Service-Level Agreements (SLAs)** that align with organizational needs for uptime and data recovery.

Additionally, understanding BCDR is integral to designing resilient cloud architectures that support quick recovery and minimal service interruption, covering technological, procedural, and human resource dimensions. Ultimately, a good knowledge of BCDR strategies enhances a cloud security professional's ability to maintain business stability and stakeholder trust, underpinning strategic operations and contributing significantly to a comprehensive security posture in dynamic cloud environments.

This chapter will look to equip you with the insight needed to do well with questions related to this domain. Let's get started!

BCDR Definitions

BCDR is a set of practices that support an organization's ability to remain operational and recover from adverse events such as cyberattacks, system failure or outages, natural disasters, and other disruptions.

Cloud computing, by its nature, offers significant advantages in terms of redundancy and data distribution across multiple locations, which can inherently bolster an organization's resilience to disruptions. However, the dependence on cloud-based services also introduces unique vulnerabilities, such as data loss due to cyberattacks, service interruptions from technical failures, or downtime resulting from natural disasters affecting data center regions.

For example, consider a retail company that uses a cloud service for storing customer data and managing inventory. If the cloud provider experiences a data center outage due to a severe weather event, the company could face significant operational disruptions, from an inability to process transactions to challenges in inventory management. This scenario underscores the need for a well-defined BCDR plan that ensures the quick restoration of services and data access, minimizing downtime and financial loss.

Thus, integrating effective BCDR strategies in cloud computing is not just about safeguarding data – it's about ensuring operational continuity and maintaining service delivery in the face of adverse events. This approach not only protects the business but also upholds customer trust and compliance with regulatory requirements.

It merges three major concepts: **resiliency**, **system recovery**, and DR:

- **Resiliency**: Systems and processes should be designed to handle disruptions smoothly by including backups and extra capacity across all critical areas – from power and cooling to system availability and staffing. This ensures that users experience minimal or no interruption in service, even when issues arise.

- **System recovery**: System recovery often occurs when the hardware (like servers hosted by the cloud provider) is functioning correctly, but data has been lost or corrupted due to software or system errors. In such instances, backups are utilized to return the system to its state before the disruption. The management team decides on the acceptable amount of data loss for these scenarios. Generally, in cloud environments, services can be quickly restored to their previous condition using the same resources.

- **DR**: DR comes into play when catastrophic events such as natural disasters, terrorist attacks, or major fires cause widespread damage. In these scenarios, quick recovery within the affected environment is often impossible. Thus, there needs to be a way for the business to shift to an alternate hosting arrangement where data and services can be restored and made operational within an acceptable timeframe. Planning for such a move should be considered a last resort, reserved for severe outages because of the significant time, cost, and resources involved in making the transition and eventually returning to normal operations.

BCDR consists of two parts:

- **BC**, whose prime concern is ensuring that critical business functions continue to operate during and after a disaster. The goal is not just to keep the business running but to do so with minimal disruption and to ensure that essential operations such as customer service, IT services, and other vital activities can continue even under adverse conditions.

> **Note**
>
> There is a slight chance you will encounter a BC plan called a **Continuity of Operations Plan (COOP)** on the exam. For the CCSP, these two terms are one and the same.

- **DR** deals with restoring the cloud infrastructure and its data to a working state after a disaster (such as a breach or a network outage) has occurred. DR plans typically outline detailed steps for reinstating critical technological resources, applications, and data that are essential for business operations to resume at an acceptable level of performance.

Merging these two parts allows organizations to quickly respond to and recover from expected and unexpected disruptions.

While you will probably see BCDR discussed together in the exam, it's important to understand the distinction between BC and DR. BC focuses on maintaining business functions or quickly resuming them in the event of a major disruption. On the other hand, DR is more technically focused and deals specifically with the recovery and restoration of IT infrastructure and data after a disaster.

The Importance of BCDR

To understand BCDR's value in cloud environments, let's discuss the core goals of BCDR planning.

The main goal of BCDR is minimizing the effects of a disaster, which is done by improving **employee responsiveness** in different situations. A BCDR plan will erase confusion by providing written procedures that employees can follow in the event of a disaster. Ultimately, BCDR improves **logical decision-making** during a crisis.

BCDR is critical in cloud environments because even short disruptions can lead to significant operational challenges. For instance, take a healthcare provider that uses cloud services to store patient records and manage appointments. If this cloud system experiences downtime, not only could the provider be unable to access essential patient information, potentially delaying or compromising care, but it could also disrupt scheduling, causing confusion and dissatisfaction among patients. This example underscores the importance of robust BCDR strategies in maintaining uninterrupted access to vital services and protecting sensitive data in the cloud.

Here are some of the main benefits of BCDR in the cloud:

- **Minimizes downtime**: Downtime can cost businesses a lot of money and even damage their reputation. With BCDR in place, you can quickly recover from outages and minimize downtime, ensuring that you can continue to serve your customers without any interruptions.

- **Protects data**: Data is critical to any business, and losing it can be catastrophic. BCDR solutions help you ensure the protection and availability of critical data at all points in time, even in the event of an unforeseen mishap or disaster.

- **Ensures compliance**: Many industries have compliance regulations that businesses must follow. BCDR measures help businesses meet these standards by incorporating regulatory requirements into their resilience and recovery strategies. You will touch on some of the most popular compliance regulations a bit later in the chapter.

Key Concepts and Terminology

Before delving deep into the main topics, you must understand the main terms and definitions you will encounter around this topic in the CCSP exam. A great way to learn them is to remember what each acronym stands for, which will typically give you the answer to what they represent in the context of BCDR.

In the context of BCDR planning, especially within cloud computing environments, several critical metrics help organizations prepare for, respond to, and recover from disruptive incidents. These metrics are **Recovery Time Objective (RTO)**, **Recovery Point Objective (RPO)**, **recovery service level (RSL)**, **maximum tolerable downtime (MTD)**, **mean time between failures (MTBF)**, and **mean time to recovery (MTTR)**. Understanding how these metrics interrelate can greatly enhance an organization's resilience and operational continuity.

Here are the main terms and concepts you need to memorize:

- **RTO**: This is the maximum time a system, service, or network can be down after a disruption or disaster before it starts negatively affecting the organization. It essentially defines the window of time in which recovery must be completed to avoid severe consequences.

- **RPO**: This term refers to the maximum acceptable amount of data that can be lost in the event of a disruption, measured in time. It helps define how frequently data should be backed up. An RPO of one hour, for example, means that in the event of a disaster, the organization could lose up to one hour's worth of data without significant harm.

- **RSL**: This is a lesser-known but essential metric that measures the percentage of the total, typical production service level that needs to be restored to meet BCDR objectives in case of failure. For example, if an organization determines that operating at 70% of normal service levels is sufficient to maintain critical operations during a disaster, the RSL would be set at 70%.

- **MTD**: This is the amount of time you can have without an unavailable asset before you must declare an emergency and initiate our DR plan. The tolerable downtime mainly depends on the criticality of the asset and the impact its unavailability has on business operations.

- **MTBF**: MTBF measures the average time between failures of a system or component. A higher MTBF indicates a more stable system with fewer disruptions. In the cloud, there are often redundancies and failover mechanisms in place to ensure service continuity even when individual components fail.

- **MTTR**: MTTR refers to the average time it takes to repair or recover a system after a failure. This includes the time it takes to identify and troubleshoot the problem, as well as implementing and verifying a fix to restore operations. The lower the MTTR, the better.

BCDR Strategies

BCDR is a lengthy process that requires a great deal of analysis, planning, and ongoing refinement. The first step in this process is laying the groundwork for the BCDR plan with a clear strategy that will guide all subsequent actions.

Let's look at the main strategic elements of a robust and responsive BCDR plan.

Defining the Scope of the BCDR Plan

The first strategic step toward creating a BCDR plan is defining its scope. Engage senior leadership from IT, information security, and other key business areas early in the process. Involving these stakeholders in defining roles, responsibilities, and key areas of concern helps craft a comprehensive plan that integrates security concerns intrinsically rather than trying to retrofit them later after the plan has already been developed.

Gathering Requirements and Generating Objectives

The next step is the most extensive one. It includes gathering the requirements for the BCDR plan, including the RTO and RPO objectives you covered previously. It's these requirements that will give us directions about the types of solutions and recovery strategies you need to implement via the BCDR plan.

During this phase, remember that while it's important to understand the financial aspect of your requirements and objectives, you also need to consider regulatory and contractual obligations. You will discuss identifying business requirements based on critical business functions, risks, and regulatory compliance in depth later in this chapter. For now, keep in mind that effectively balancing these factors is essential for developing a robust BCDR strategy.

Integrating Requirements

The final part of this step is integrating the previously defined requirements and scope to establish the objectives and create a roadmap for designing the actual BCDR plan. It involves an in-depth analysis of the current hosting location for systems or applications, identifying which components must be replicated for DR, and assessing the associated risks.

Assessing Risk

Transitioning to a new environment introduces new risks due to differences in configurations, support models, and the unfamiliarity of new hosting staff with the systems. For example, one of the key concerns is ensuring that a secondary hosting provider selected for BCDR purposes can manage the expected load and performance requirements as effectively as the primary host.

Risk assessments are a continuous element of any BCDR strategy. When preparing for the CCSP exam, you might face scenarios that require you to evaluate risks. It's important to remember that your perspective in these scenarios is from the customer's side. Therefore, your primary focus should be on evaluating the risks related to potential BCDR providers. However, this evaluation must also take into account the criticality of your own assets and the capability of potential CSPs to ensure BCDR operations for those assets. The decision-making process should always weigh the value proposition and cost-benefit analysis, specifically the cost of the selected controls against the value of the assets being protected.

There are two types of risks associated with BCDR development – those that require the execution of the plan and those that emerge as a result of the plan itself.

Risks that require the execution of the BCDR plan include the following:

- Natural disasters such as earthquakes, tornadoes, and floods
- Technical failures such as system crashes or software bugs
- Operational and utility disruptions such as power outages or critical infrastructure failures
- Acts of terrorism, war, or deliberate cyberattacks such as ransomware

Once the BCDR plan is executed, there are additional risks that arise from the plan's execution itself. These are outlined in the following subsections.

Performance Issues Due to Location Change

When an outage in one cloud data center necessitates a failover to another geographic region, significant latency and performance issues can arise. The change in hosting location may result in slower system or application performance, impacting both the user and customer experience. These latency issues are particularly problematic for security and encryption systems that depend on precise time synchronization.

To mitigate these risks, it is advisable to choose a cloud provider that offers multiple geographically suitable regions to ensure that data can be accessed swiftly and reliably, even during a failover scenario.

Cost and Effort to Maintain Redundancy

Maintaining redundancy in a BCDR plan involves significant effort and cost, especially if the plan operates in a hot-hot configuration where both the primary and secondary sites must be equally staffed, configured, and maintained. Cloud infrastructures facilitate the syncing of primary and secondary locations, allowing near real-time updates and data maintenance across multiple regions.

While this level of redundancy ensures compatibility and readiness in case of an unforeseen emergency, it also significantly increases operational costs due to increased usage fees.

Failover Issues

Risks can arise if the primary site fails, and the customer has to fully failover to a secondary site. The main risk is the amount of time this would take and the resulting impact on business operations. Luckily, most cloud environments provide high levels of elasticity that ease such transitions. Your job as a security professional will be to ensure that the CSP has the necessary bandwidth to deliver a smooth transition between sites.

Another option that helps with redundancy is having two separate CSPs – one for the main site and one for BCDR. To address risks in failover scenarios, it is essential to have mechanisms in place such as networking and **Domain Name System (DNS)** changes, along with global load balancers.

Legal and Compliance Issues

The CSP and cloud customer share the responsibilities related to compliance with regulations and contractual agreements. However, since the customer is the data owner, it's ultimately up to them to ensure that their CSP complies with relevant legal, regulatory, and contractual requirements.

The customer must conduct due diligence to ensure that the BCDR site meets all the same standards as the primary site. For example, if your operations fall under GDPR compliance due to handling personal data from EU citizens, it's critical that your BCDR provider also adheres to GDPR standards. Failure to ensure this can result in severe penalties and reputational damage.

The following section will detail how to gather business requirements, including conducting a business impact assessment. After that, you will have all the information you need to finally create and implement the BCDR plan.

Cloud Environment Options for BCDR Planning

Traditionally, on-premises BCDR involves maintaining physical hardware and backup facilities directly managed by the organization. For the CCSP exam, you will focus on the various cloud options organizations have for selecting a BCDR provider.

The first option is to maintain a **private IT architecture**, with a CSP used for BCDR. This setup allows organizations to retain full control over their primary infrastructure while leveraging the cloud for backup and recovery purposes.

The other options are as follows:

- **Primary CSP as BCDR**: In this model, the organization uses a single CSP as the primary means for both ongoing operations and DR. This involves hosting all critical systems and data on the CSP's infrastructure, relying on the provider's built-in resilience and redundancy features.

- **Secondary CSP as BCDR**: This approach involves using a different CSP for DR than the one used for primary operations. The primary infrastructure might be either on-premises or with a primary CSP, with a secondary CSP engaged specifically for backup and recovery purposes.

One CSP for primary use and BCDR purposes is usually cheaper and more scalable than using two CSPs. It's also easier to maintain, with a single point of contact for support and management. However, a dependency on a single provider can be risky if there are widespread outages. Additionally, a secondary CSP may offer better DR features than your main provider.

Understanding an Organization's Business Requirements

One of the key elements of building an effective BCDR plan is understanding an organization's business requirements, which will serve as the basis for developing tailored recovery strategies. Recovery strategies include identifying critical functions that must continue during a disruption, determining the technological and human resources needed to support these functions, and assessing the potential financial and operational impacts of downtime.

This is where key metrics such as RTO and RPO are established. Management determines these calculations strictly from a risk tolerance perspective. It's then up to the IT and security team to start forming a BCDR strategy to implement the necessary processes and measures to ensure these objectives are met.

Identification of Critical Business Functions

During a disaster, the most important consideration is ensuring that critical business functions continue to operate with minimal disruption. So, an effective BCDR plan must recognize these critical business functions and build adequate measures to ensure their resilience and continuity.

One way to identify critical business functions is through risk assessment and impact analysis.

Risk Assessment and Impact Analysis

Business Impact Analysis (BIA) is used to determine which processes are critical and which are not.

BIA measures the impact a disruption to each process could have on the business. Processes that are deemed critical to the organization's functioning must be prioritized in emergency situations.

A BIA also contains these two calculations:

- **Cost-benefit analysis (CBA)**: For businesses, investments in resilience must make financial sense to justify spending. CBA helps organizations decide whether the financial investment in resilience measures aligns with the expected gains in stability and reduced impact from potential threats.

- **Return on investment (ROI)**: ROI calculates the financial return on investments made to protect and sustain critical business processes. It compares the costs of implementing BCDR functions with the benefits of reduced downtime and loss mitigation.

Legal and Regulatory Compliance

Aside from their own requirements, most organizations must adhere to external legal and regulatory compliance standards that dictate how they must manage and protect data, maintain privacy, and ensure continuity of service. This can include industry-specific regulations, such as the HIPAA for healthcare, GDPR for data protection in the EU, and SOX for financial reporting in the US.

Let's look at an example.

The HIPAA mandates strict controls on how patient information is handled in the healthcare industry. Healthcare providers, insurance companies, and any third-party service providers must ensure that all personal health information (PHI) is protected both in transit and at rest. This means implementing security measures such as encryption, access controls, and regular audits. Non-compliance with HIPAA can result in hefty fines and severe reputational damage.

If you're working for a hospital or another organization dealing with PHI, when building a BCDR plan, you must ensure that the BCDR provider complies with HIPAA regulations as well.

SLAs and Vendor Management

In the cloud, much of the BCDR capabilities will be in the hands of the CSP. Therefore, learning how to manage relationships with CSPs, including SLAs, is one of the most essential skills for a cloud security professional.

> **Note**
>
> An SLA is a contract between a service provider and its customers that documents what services the provider will furnish and defines the service standards the provider is obligated to meet.

Organizations must meticulously assess and align their BCDR requirements with the capabilities that their CSP offers. This involves reviewing the CSP's DR plans and ensuring they meet the organization's RTOs and RPOs. Cloud security professionals should also evaluate the CSP's ability to handle different disaster scenarios and their readiness to quickly restore services to minimize downtime.

Negotiating the right SLAs is vital in this process. The organization and the CSP must clearly define and agree on terms concerning availability, uptime, data protection, incident response times, and the specific responsibilities of the CSP in the event of a disruption. Regular audits and reviews should be conducted to ensure that the CSP complies with the agreed terms. Additionally, the SLA should include provisions for regular updates to the BCDR strategies in response to evolving threats and changes in the business environment, ensuring that the SLA remains relevant over time.

Creation, Implementation, and Testing of a BCDR Plan

Once you've identified your requirements, defined the scope of the BCDR plan, and assessed its risks, it's time to create and implement your plan. This part of the process relies heavily on the research, planning, and analysis you've done already, and it's all about translating these findings into a practical, executable, and comprehensive BCDR plan.

Creating Your Plan

The process begins with a thorough creation (or design) phase where the organization evaluates potential BCDR solutions against previously established technical and business requirements. This is where the organization identifies the best-fit solution that meets its needs.

During the design phase, technical requirements for the BCDR solution, including required services, features, and capacity, must be defined. Support requirements should also be included in SLAs and contracts with the BCDR provider. The plan should clearly determine the events, situations, and scenarios that will trigger the BCDR plan and outline the necessary procedures for activating the BCDR solution during a disruption.

A great way to make all policies and procedures easier to follow is to include a **responsible, accountable, consulted**, and **informed (RACI)** matrix. A RACI matrix identifies the roles and responsibilities of all relevant teams and personnel involved in the BCDR process.

Implementation

Once the plan is fully designed, before doing anything, make sure that your plan is fully vetted and approved by leadership, as this is the final step before deployment. It moves to the implementation stage, which may require modifications from both technical and policy standpoints to align with the actual operational environment.

The implementation of the BCDR plan involves several action steps, both within your primary site and the BCDR site. For the plan to work, both environments need to be well synchronized with the necessary data and configuration replication services.

> **Note**
>
> Non-technical leadership may push back against incurring costs by modifying the existing infrastructure, insisting on a solution that can be dropped into place without prior modification. It is up to the cloud security professional to perform a cost-benefit analysis to demonstrate the necessity of these modifications for ensuring resilience and compliance.

Some modifications that may be required on the primary infrastructure include the following:

- Configuring the connection between the primary site and the BCDR site. This typically includes programming API calls, but it sometimes may require physical connections as well.

- Implementing systems that monitor events and have the capability to trigger the BCDR plan.

The BCDR site should have all the capabilities, services, and products needed to ensure a smooth transition during failover.

Testing and Maintenance of the BCDR Plan

Since business requirements and circumstances are constantly changing, it is essential to regularly test and maintain the BCDR plan to ensure its effectiveness and relevance. Regular testing helps identify any gaps or weaknesses in the plan, allowing timely modifications and updates.

Each test aims to verify that the designed and implemented plans can meet the established RPOs and RTOs. The testing plan should specify the scope of the tests, which applications are involved, and the methods used to emulate a genuine BCDR scenario. It is essential to conduct these tests at least annually, more often if dictated by relevant certifications, laws, regulations, or internal policies.

There are several types of tests organizations can conduct to ensure the effectiveness of their BCDR plan.

Tabletop Exercises

These are scenario-driven discussions that involve key personnel reviewing and discussing the specific actions they would take in response to a simulated disaster scenario. This is a purely conversational, non-technical method, which means that it's relatively easy and cheap to conduct. Tabletop exercises help clarify roles and responsibilities during DR and identify gaps in the plan without impacting actual operations.

Aside from organizational roles, a tabletop exercise also includes a facilitator who drives the discussion and ensures that the scenario progresses in a meaningful and structured way.

Here is how a tabletop exercise typically plays out:

- A meeting is scheduled involving all key personnel. At this stage, a relevant and plausible disaster scenario that impacts the organization is created.

- During the exercise, participants discuss their roles and the steps they would take in response to the scenario. Facilitators guide the discussion, ensuring all aspects of the BCDR plan are covered.

- Document the outcomes of the exercise, noting any issues or gaps identified. Discuss potential improvements and update the BCDR plan accordingly.

Full Interruption

A complete contrast to tabletop exercises, **full interruptions** are the most rigorous form of testing, where the organization simulates a real disaster that causes a complete shutdown of their operations. Despite being very thorough, in the real world, very few organizations opt for this type of testing. The main reason is that it's expensive, and there's a real chance for the test to cause a major disruption of operations if it fails.

In full interruption testing, planning has to be extensive. All stakeholders must be fully briefed on the process, and contingency measures should be in place to mitigate any negative consequences that may arise during the test.

After the planning is complete, the test is executed by shutting down operations at the primary site and shifting to the recovery site. The shift tests the organization's ability to operate from an alternative location under emergency conditions. This involves activating all necessary systems and resources at the recovery site and ensuring that they function as intended to handle the company's operational load.

Walk-Through Tests

Walk-through tests expose team members to a specific type of disruptive event. Each team member then walks through the actions and steps they would take in the given scenario, surfacing any gaps or deficiencies. Unlike full interruption tests, simulation tests allow the organization to evaluate the response to specific elements of a disaster, such as system recovery procedures or communication channels, without disrupting all operations. There is a particular set of steps to conduct a walk-through test:

1. A walk-through exercise starts with a detailed script or scenario document, outlining the nature of the disruptive event and its impact on the organization. Then, the exercise officially begins with a briefing session where the facilitator introduces the scenario and sets the stage for the simulation.

2. Participants then sequentially articulate and walk through the specific actions they would take in response to the scenario. This can include contacting emergency services, activating recovery sites, and retrieving data backups.

3. Throughout the exercise, the facilitator interjects with new challenges and developments to the scenario to test the team's dynamic response capabilities.

4. The test ends with a debriefing session where participants gather to assess the execution of the plan.

Functional Drills

A **functional drill** is a focused test against a specific function or component of the BCDR plan. It will enable the organization to assess whether adhering to the plan effectively restores essential systems at an alternative site.

The first step in a functional drill is identifying a critical function that's vital for the continuity of operations. Then, specific scenarios are developed that disrupt these critical functions. Since a functional drill fully tests the BCDR plan, all employees participate in the process. The drill should be planned and executed as realistically as possible, which involves activating the specific component or function under test and then following the planned response procedures.

Each test, regardless of the method used, should be followed by a comprehensive report that management can review to analyze what worked, what didn't, and what improvements or adjustments need to be made to the plan. It's important to maintain regular testing schedules to ensure that all aspects of the BCDR plan are always ready to be activated.

As threats evolve and new risks emerge, the BCDR plan must also be updated to reflect these changes. Regular reviews should be conducted at least annually or more frequently if significant changes in the business environment occur.

Example Scenarios

So, to help you grasp how the testing or implementation of a BCDR plan would work in the real world, let's look at two fictional organizations and their approach toward a BCDR strategy.

Scenario 1 – a Multi-Region Deployment and BCDR Strategy for a Sample Retail Company

Background: A world-renowned retail corporation, Global Retail Inc., has recently undertaken a major initiative to migrate its critical business operations to the cloud. This transition aimed to enhance operational efficiency, increase scalability, and improve data accessibility across its global branches. Global Retail Inc. partnered with a leading CSP to leverage its expansive network of global data centers.

Pre-disaster preparation: As part of their cloud migration strategy, Global Retail Inc. implemented a robust BCDR plan that emphasized resilience and rapid recovery. The key components of their strategy included the following:

1. **Multi-region deployment**:

    - The corporation deployed its critical applications and data across multiple geographic regions. This strategy was designed to mitigate the risk of a localized natural disaster impacting all operational capabilities.

    - Each region operated independently but was closely synchronized to ensure consistency and real-time data availability.

2. **Synchronized data replication**:

    - Global Retail Inc. set up continuous data replication between primary and secondary data centers. This ensured that all transactional data, customer information, inventory data, and other critical datasets were consistently mirrored across multiple regions.

    - The replication process was configured to be near-real-time to minimize data loss (low RPO) in the event of a disaster.

Automated failover protocols:

    - Automated failover mechanisms were put in place to ensure that if the primary data center in any region went offline, the system would automatically switch to a secondary data center without manual intervention.

    - These protocols were regularly tested to ensure they functioned correctly under various failure scenarios.

Disaster event: A significant earthquake struck near one of the primary data centers, causing immediate power outages and physical damage to the infrastructure. This event triggered the following response:

1. **Immediate activation of the BCDR plan**:

   - The automated monitoring systems detected the outage and triggered DR protocols.

   - Notifications were sent to the IT DR team and key stakeholders about the incident and the initiation of the DR plan.

2. **Traffic redirection and load balancing**:

   - Network traffic and user requests were automatically redirected to the nearest operational data center. This redirection was facilitated by the cloud provider's global traffic management system.

   - Load balancers adjusted to the increase in traffic to the secondary sites, ensuring performance stability and response times remained within acceptable limits.

3. **Data integrity checks and synchronization**:

   - Once the failover was complete, automated systems performed data integrity checks to confirm that all replicated data was up-to-date and consistent with the primary site before the disaster.

   - Any transactions in process at the time of the disaster were reconciled and verified to prevent data corruption or loss.

4. **Stakeholder communication**:

   - Continuous communication was maintained with all stakeholders, including updates on the recovery status and expected resolution times. This communication helped manage expectations and reduce potential business impacts.

5. **Post-disaster review and recovery**:

   - After the primary data center was restored and deemed safe for operations, a controlled failback process was initiated.

   - A comprehensive review of the disaster response was conducted to identify lessons learned and opportunities for improvement. This review led to further refinement in the BCDR strategy.

Outcome: Thanks to Global Retail Inc.'s proactive BCDR planning and investment in cloud technology, the impact of the natural disaster on their operations was minimal. Customer transactions continued seamlessly, and the company maintained its reputation for reliability and customer service, even in the face of unforeseen challenges.

Scenario 2 – a Multi-Region Deployment and BCDR Strategy for a Financial Services Firm

Background: FinanceSecure Ltd., a prominent global financial services firm, recognized the critical need to enhance its operational resilience in light of increasing cyber threats and the potential for natural disasters. As part of its strategic initiatives, FinanceSecure Ltd. embarked on a substantial cloud migration project to bolster its infrastructure's reliability and scalability. This transition was facilitated through a partnership with a top-tier CSP known for its robust global network of data centers.

Pre-disaster preparation: FinanceSecure Ltd. carefully crafted a BCDR strategy that was tailored to the specific needs and regulatory requirements of the financial services industry. This strategy incorporated several critical elements:

1. **Multi-region deployment**:

   - Critical financial systems, including transaction processing platforms, customer databases, and trading platforms, were deployed across multiple geographical regions. This distributed approach aimed to safeguard against regional disruptions.

   - Each deployment was equipped with independent operational capabilities yet synchronized to ensure data consistency and real-time access across the globe.

2. **Synchronized data replication**:

   - FinanceSecure Ltd. implemented real-time data replication between primary and secondary data centers to ensure the high availability of financial data. This replication was crucial for maintaining up-to-the-minute accuracy of financial transactions and customer data.

   - Replication processes were compliant with financial industry regulations concerning data security and integrity.

3. **Automated failover protocols**:

   - Automated failover mechanisms were established to seamlessly switch operations from a primary data center to a secondary one in the event of a disruption.

   - These failover procedures were integrated with the firm's risk management systems and were subjected to regular drills to ensure their efficacy under actual disaster conditions.

Disaster event: A severe hurricane impacted one of the regions hosting FinanceSecure Ltd.'s primary data centers, resulting in power failures and significant infrastructural damage. The firm's BCDR strategy kicked into action:

1. **Immediate activation of the BCDR plan**:

   - Automated systems detected the disruption and instantly initiated the DR protocol.

   - The firm's crisis management team was alerted, and they began coordinating the recovery operations in line with the predefined DR plan.

2. **Traffic redirection and load balancing**:

   - Incoming requests and data traffic were automatically rerouted to the nearest operational data center without manual intervention.

   - Load balancing adjustments were made to handle the increased load on secondary sites, ensuring that transactional performance and customer access were not compromised.

3. **Data integrity checks and synchronization**:

   - Post-failover, automated systems conducted integrity checks to confirm the accuracy and completeness of the replicated data.

   - Any ongoing financial transactions at the time of the disaster were secured and reconciled to prevent any discrepancies.

4. **Stakeholder communication**:

   - Continuous updates were provided to customers, regulatory bodies, and internal teams about the status of operations and recovery efforts. This transparency helped maintain trust and compliance with financial communication regulations.

5. **Post-disaster review and recovery**:

   - After restoration efforts at the primary site were completed, a controlled failback procedure was implemented.

   - A thorough review of the disaster's handling was conducted to identify improvements for the BCDR strategy, focusing on enhancing resilience and reducing recovery times.

Outcome: Thanks to the robust BCDR planning and the strategic use of cloud technologies, FinanceSecure Ltd. was able to maintain critical financial operations during a severe hurricane. This resilience minimized potential financial losses and upheld the firm's reputation for reliability and regulatory compliance amidst challenging conditions.

The detailed scenarios of Global Retail Inc. and FinanceSecure Ltd. provide critical insights into effective BCDR planning in cloud environments. Key takeaways include the necessity of multi-region deployment to mitigate risks of localized disasters and ensure uninterrupted operations. Real-time data replication is crucial for maintaining data integrity, while automated failover protocols enhance responsiveness and operational continuity during disruptions. Compliance with industry-specific regulations, particularly in sectors such as finance, is vital to avoid legal repercussions. Effective communication with stakeholders during disasters builds trust and ensures transparency. Regular testing and reviews of BCDR plans are essential for identifying potential improvements and adapting to new threats or technological changes.

These practices highlight that integrating BCDR strategies into the core business operations and considering them during initial system design is crucial for building resilient systems capable of withstanding and quickly recovering from unexpected disruptions.

Summary

Good job on making it to the end of this chapter! The information you covered will be highly actionable both in the CCSP exam and in your career as a cloud professional. You covered a lot, so let's recap some of the main ideas:

- BCDR is a set of practices that support an organization's ability to remain operational and recover from adverse events.

- BCDR in the cloud is important because it provides clear guidelines for employees to follow in the event of a disaster. It also supports operational stability and ensures that critical business functions can continue with minimal disruption.

- To develop a BCDR plan, you first need to understand and analyze the organization's specific business requirements and risks, along with relevant compliance standards.

- BCDR implementation includes steps to synchronize the primary and secondary systems to ensure data consistency and quick recovery in the event of a disaster.

- Ongoing testing and refinement are necessary to ensure that objectives (including RTO and RPO) are being met.

In our next chapter, you will look at the critically important topic of application security. You will review the development basics, the challenges organizations face, and the common cloud vulnerabilities of web applications.

Exam Readiness Drill – Chapter Review Questions

Apart from a solid understanding of key concepts, being able to think quickly under time pressure is a skill that will help you ace your certification exam. That is why working on these skills early on in your learning journey is key.

Chapter review questions are designed to improve your test-taking skills progressively with each chapter you learn and review your understanding of key concepts in the chapter at the same time. You'll find these at the end of each chapter.

> **How to Access These Materials**
>
> To learn how to access these resources, head over to the chapter titled *Chapter 25, Accessing the Online Resources*.

To open the Chapter Review Questions for this chapter, perform the following steps:

1. Click the link – `https://packt.link/CCSPE1_CH11`.

    Alternatively, you can scan the following **QR code** (*Figure 11.1*):

Figure 11.1 – QR code that opens Chapter Review Questions for logged-in users

2. Once you log in, you'll see a page similar to the one shown in *Figure 11.2*:

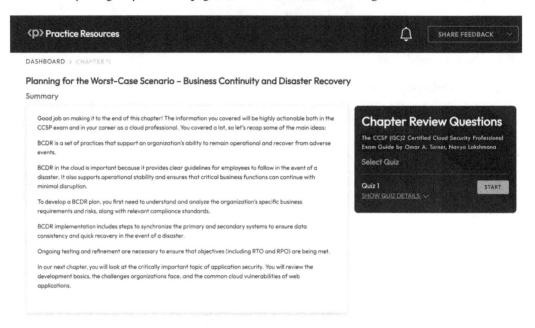

Figure 11.2 – Chapter Review Questions for Chapter 11

3. Once ready, start the following practice drills, re-attempting the quiz multiple times.

Exam Readiness Drill

For the first three attempts, don't worry about the time limit.

ATTEMPT 1

The first time, aim for at least **40%**. Look at the answers you got wrong and read the relevant sections in the chapter again to fix your learning gaps.

ATTEMPT 2

The second time, aim for at least **60%**. Look at the answers you got wrong and read the relevant sections in the chapter again to fix any remaining learning gaps.

ATTEMPT 3

The third time, aim for at least **75%**. Once you score 75% or more, you start working on your timing.

> **Tip**
>
> You may take more than **three** attempts to reach 75%. That's okay. Just review the relevant sections in the chapter till you get there.

Working On Timing

Target: Your aim is to keep the score the same while trying to answer these questions as quickly as possible. Here's an example of how your next attempts should look like:

Attempt	Score	Time Taken
Attempt 5	77%	21 mins 30 seconds
Attempt 6	78%	18 mins 34 seconds
Attempt 7	76%	14 mins 44 seconds

Table 11.1 – Sample timing practice drills on the online platform

> **Note**
>
> The time limits shown in the above table are just examples. Set your own time limits with each attempt based on the time limit of the quiz on the website.

With each new attempt, your score should stay above **75%** while your "time taken" to complete should "decrease". Repeat as many attempts as you want till you feel confident dealing with the time pressure.

12
Application Security

In the last chapter, you learned how organizations prepare to withstand disasters and business disruptions and can continue the delivery of products and services within acceptable time frames. The use of cloud services can be seen as a key element in supporting critical business functions in case of a major disruption. As you turn your attention to the next critical area of focus, you will dive into the realm of application security. Ensuring that applications are secure from vulnerabilities and threats is paramount in protecting sensitive data and maintaining the integrity of business operations.

Application security, especially in terms of modern software development in cloud environments, is extremely important considering the nature of breaches based on code vulnerabilities. This chapter will review development basics, challenges organizations face, and the common cloud vulnerabilities based on web applications.

You will also look at why application security is extremely important, not only because it is a critical domain for the CCSP exam, but also because of its importance to the security and digital transformation of many modern organizations.

By the end of this chapter, you will be able to confidently answer questions on the following topics:

- Criticality of application security
- Common challenges in securing web applications
- Vulnerabilities with **Open Web Application Security Project (OWASP)**
- Cloud application security tools and solutions

Why is Application Security Critical?

Software applications are everywhere—in computers, televisions, watches, cars, smartphones, and airplanes. Billions of people worldwide rely on millions of applications to do their jobs, collaborate with colleagues, communicate with friends, listen to music, play games, and more.

At the enterprise level, applications—both on-premises and cloud-based—can significantly impact operations, revenues, profitability, and competitiveness. They help to streamline processes, improve functional efficiencies, and increase workforce productivity. Cloud-based applications and collaborative cloud environments are also essential for today's hybrid workforces and agile enterprises.

Clearly, applications are vital to the modern-day business landscape. Unfortunately, many applications are riddled with security vulnerabilities that malicious adversaries exploit to author devastating attacks, steal enterprise data and intellectual property, and compromise sensitive resources. Organizations need to protect their applications and safeguard their business-critical assets from these adversaries. This is where application security comes in.

Application security is a discipline that aims to protect applications from cybersecurity threats and risks. Any organization that uses or develops applications needs robust application security processes, tools, and practices. However, application security is especially vital for enterprises that utilize the cloud infrastructure and depend on its agility, flexibility, and scalability. The following are some of its goals:

- One of its overarching purposes is to minimize the vulnerabilities in enterprise applications

- Another is to improve the organization's development and security practices to help create more secure applications

- A third aim is to find and fix security issues as early as possible in the **software development life cycle (SDLC)**

To meet the third goal, security testing must be a part of the SDLC from the very beginning, not added to the process as an afterthought. This philosophy of **shifting left** helps to minimize the costs of rework and the risk of delivering insecure applications to the customer or end user.

For best results, it's a good idea to incorporate application security practices into the entire application life cycle, including phases such as the following:

- Requirements analysis
- Design
- Coding
- Testing
- Implementation
- Maintenance

One way to promote application security throughout the SDLC is to introduce security standards and tools such as vulnerability scanners during design and development. It's also important to perform continuous security testing to protect applications in production environments.

Strong authentication and controls such as **web application firewalls (WAF)** and **intrusion prevention systems (IPS)** should also be part of the application security ecosystem. Securing applications in the cloud is an integral part of the broader field of application security. An effective security ecosystem for cloud-based applications may include various policies, processes, technologies, tools, and controls designed to protect these applications from vulnerabilities and threats.

Together, these elements enable organizations to protect their cloud-based applications, workloads, and data throughout the SDLC. Equally important, security tools provide visibility into cloud assets, allowing admins to limit access to authorized users, identify open vulnerabilities, and protect applications from cyberattacks and data breaches.

Common Challenges in Securing Web Applications

Security vulnerabilities are a common problem in many web applications. The number of vulnerabilities added to the **National Vulnerability Database (NVD)** kept increasing between 2017 and 2022.

Organizations must step up their efforts to secure their applications. They must pay more attention to application security, particularly in the cloud. As a CCSP candidate, you will be tasked with how to help organizations and their development teams tackle this massive problem of vulnerabilities.

Common Application Vulnerabilities

Threat actors try to exploit open vulnerabilities in applications. Most of these vulnerabilities—82% of them—are found in application code. The following vulnerabilities are listed in the OWASP Top 10:

- Injection
- Broken access control
- Cryptographic failures
- Security misconfigurations
- Identification and authentication failures
- Server-side request forgery
- Vulnerable or outdated components

These flaws present a significant security risk to organizations because they allow attackers to gain unauthorized access to enterprise assets, compromise credentials, steal data, and launch ransomware, SQL injections, and other kinds of devastating attacks. Furthermore, many of these applications are connected to or reside in the cloud. These vulnerabilities can affect any of these applications.

Cloud applications are also vulnerable to other security threats such as the following:

- **Credential exposure**: The exposure of credentials in the cloud can lead to account hijacking and long-term attacks.

- **Security misconfigurations**: Improperly configuring cloud service permissions and leaving cloud storage open can lead to an increase in the risk of data leakage.

- **Automated attacks**: Bots may launch automated attacks against cloud services or web-facing applications, leading to data breaches, service disruptions, and unauthorized access to sensitive information. These attacks can overwhelm systems with traffic, exploit vulnerabilities, steal user credentials, or perform malicious activities.

- **Insecure APIs**: Feature and data-rich APIs that enable data sharing in the cloud are particularly vulnerable; attackers can exploit vulnerabilities in APIs to inject malicious code or commands, leading to unauthorized data access or manipulation.

- **Denial-of-service (DoS) attacks**: The DoS attacks against a **cloud service provider (CSP)** can have a substantial downtime impact on all their enterprise customers.

- **Runtime threats**: Users may mistakenly assume that cloud workloads are fully protected by the CSPs, allowing adversaries to target the application to obtain access to enterprise cloud resources.

In addition to these threats, outsourcing development, using legacy applications, and using third-party or open-source components add to the complexity of cloud-based software supply chains. All these factors create serious security risks and increase the potential for attack. To minimize these risks, organizations need to be hyper-focused on application security. Application security is vital for companies operating in hybrid or multi-cloud environments. It is also crucial for teams using cloud-based applications for remote operations, collaboration, and communications.

Cloud environments provide shared resources that can be accessed from anywhere and from any device, expanding the number of access points for a potential attacker. These resources are ephemeral, reducing visibility into the environment and making it harder to secure all components. Furthermore, data is transmitted across the internet between users and the application, increasing its vulnerability to unauthorized access or compromise.

For all these reasons, enterprises need solutions that can protect them across all cloud platforms. They also need help to monitor user privileges and ensure that only authorized users can access the data in cloud applications.

It's also important to shift left in the cloud because almost all the infrastructure, environments, and resources are determined at the development stage, often using declarative configurations, also known as **infrastructure as code (IaC)**. Shifting left enables teams to catch security vulnerabilities early and fix them before adversaries have a chance to exploit them.

Undoubtedly, cloud applications add a lot of value to organizations, particularly now when there's a lot of focus on remote work, hybrid teams, and organizational flexibility. However, the cloud also increases the organization's attack surface. It presents adversaries with many new avenues to breach enterprise networks and unleash catastrophic damage. The only way to keep them out is to step up cloud application security with the right set of tools, processes, and controls. For example, cybersecurity researchers found that 42% of organizations are concerned about keeping up with the increasing number of vulnerabilities in their web apps.

> **Note**
> You can learn more about security vulnerabilities here - *Welcome to the OWASP Top 10 – 2021*: `https://owasp.org/Top10/`.

These concerns are valid since almost half (44%) have experienced data breaches recently and serious security attacks against their applications, such as the following:

- Malware attacks (31%)
- **Distributed denial-of-service (DDoS)** attacks (23%)
- Attacks due to application misconfigurations (21%)
- Stolen credentials (20%)

Vulnerabilities in applications open the door to many kinds of cyberattacks that enable adversaries to compromise business-critical resources, steal sensitive data, and even persist within the network to author more devastating attacks in the future. The next section will detail this.

Vulnerabilities with OWASP

OWASP is a non-profit organization dedicated to web application security. Through documentation, tools, and community forums, OWASP provides guidance to organizations looking to shore up their web application security defenses.

It is not only a widely recognized categorization of the ten most serious security threats to web applications, but also a guide for security-conscious web developers, security researchers, and business stakeholders to assist them in identifying and mitigating the most serious online application security issues.

There will be questions related to your knowledge of the OWASP Top 10 in the exam. A CCSP candidate is expected to not only know the differences between each of the vulnerabilities of the OWASP Top 10 but also know the potential examples and mitigation strategies. The following are a few reasons why the OWASP Top 10 is critical:

- **Common language**: The OWASP Top 10 provides a common language and knowledge of the most serious security risks that can affect web applications. This standardized vernacular allows developers, security experts, and other stakeholders to communicate more efficiently and minimize risks proactively.

- **Awareness**: The OWASP Top 10 increases awareness of online application security threats. By identifying the Top 10, development teams and their security partners can be aware of potential attacks and limit their vulnerability exposure. In addition, developers may include security best practices in their code, which can result in applications that are more secure.

- **Prioritization**: The OWASP Top 10 provides a framework for prioritizing security initiatives. By prioritizing the most significant risks, those who perform threat modeling may address the web application vulnerabilities that represent the greatest harm.

> **Note**
>
> You can learn more about threat modeling here - *OWASP Threat Modeling Process:*
> `https://owasp.org/www-community/Threat_Modeling_Process.`

- **Compliance**: Some regulatory groups provide guidance that enterprises comply with the OWASP Top 10. Organizations can show compliance and avoid potential penalties or fines by adopting actions to manage these risks.

The OWASP Top 10 is periodically updated to reflect the dynamic nature of online application security threats and is amended when new attack vectors emerge to ensure that it remains relevant and up to date.

> **Note**
>
> The OWASP Top 10 list, first published in 2003, is updated every few years. It was last updated in 2021 and before that, in 2017.

Next, you will learn about the 10 application security risks in the OWASP Top 10 2021 list.

Broken Access Control (A01:2021)

One of the Top 10 risks to the security of web applications that the OWASP has identified is A01:2021-Broken Access Control. It refers to flaws that make it possible for unauthorized users to access private information or carry out actions that ought to be prohibited. The confidentiality, integrity, and availability of an application and its data can be at risk due to broken access control weaknesses.

The following are the examples of broken access control flaws:

- **Inadequate or nonexistent access controls**: A program that does not enforce proper authentication and authorization mechanisms makes it possible for anyone to view or alter data without restriction

- **Insecure direct object reference**: A circumstance where an attacker can change the value of an object identifier or parameter to gain unauthorized access to confidential information or functionality

- **Privilege escalation**: A flaw that enables an unauthorized user to obtain higher privileges than they are supposed to have, giving them access to sensitive information or functionality they shouldn't have

- **Horizontal and vertical privilege escalation**: An attacker can gain access to data or functionality outside of their authorized scope by escalating their privileges either within the same user role (horizontal) or across different roles (vertically)

The following are the mitigation techniques for broken access control vulnerabilities:

- **Strong policies**: A strong password policy, **multifactor authentication (MFA)**, and **role-based access controls (RBAC)** should all be enforced by applications to guarantee that only authorized users have access to restricted data or functionality

- **Creating secure and random session IDs**: Session IDs ought to be encrypted, kept in a secure cookie, and defended against brute force attacks

> **Note**
> You can learn more about the Session Management Cheat Sheet at `https://cheatsheetseries.owasp.org/cheatsheets/Session_Management_Cheat_Sheet.html`.

- **Validating user input**: Applications should validate user input to guard against injection attacks and make sure that user requests do not circumvent access controls

> **Note**
> You can learn more about the Input Validation Cheat Sheet at `https://cheatsheetseries.owasp.org/cheatsheets/Input_Validation_Cheat_Sheet.html`.

- **Using a well-established framework to implement access control**: Applications should make use of well-known frameworks, such as RBAC, to guarantee uniform and thorough access controls throughout the entire application

> **Note**
>
> You can learn more about the Authorization Cheat Sheet at `https://cheatsheetseries.owasp.org/cheatsheets/Authorization_Cheat_Sheet.html`.

- **Keeping error messages minimum**: Error messages should be kept to a minimum so that attackers cannot use them to access data or functionality that is restricted

> **Note**
>
> You can find more about the Logging Cheat Sheet at `https://cheatsheetseries.owasp.org/cheatsheets/Logging_Cheat_Sheet.html`.

- **Conducting routine user access reviews**: Organizations should conduct routine user access reviews to ensure that users have access to only the information they require to carry out their job roles

Cryptographic Failures (A02:2021)

Cryptographic Failures is a section of the OWASP Top 10 list that highlights common cryptographic mistakes that can lead to security vulnerabilities in web applications.

> **Note**
>
> You can read more about the Cryptographic Storage Cheat Sheet at `https://cheatsheetseries.owasp.org/cheatsheets/Cryptographic_Storage_Cheat_Sheet.html`.

The following are some instances of the cryptographic mistakes mentioned in A02:

- **Weak cryptographic algorithms**: They can leave web applications open to collision and brute force attacks, for example, when weak encryption algorithms such as RC4 or SHA1 are used

- **Insecure key management**: Unsecure key management techniques, such as hardcoding keys into application code or storing them in plaintext, can result in the theft of sensitive data

- **Insufficient entropy**: Using too little randomness can make keys predictable and jeopardize system security when it comes to cryptographic operations

- **Inadequate protection of sensitive data**: If the right cryptographic safeguards, such as hashing or encryption, are not used, sensitive data may be exposed to attackers who can take advantage of these flaws

It's crucial to do the following in order to reduce cryptographic errors:

- Use robust, attack-resistant cryptographic algorithms

- Use secure key management procedures and routinely change your key inventory

- Ensure that the cryptographic functions are using adequate entropy

- Use appropriate cryptographic safeguards such as hashing or encryption to protect sensitive data

- To make sure cryptographic implementations are secure, regularly review and update them

Injection (A03:2021)

The vulnerability specifically entails inserting malicious code or SQL statements into a web application's input fields, which could allow attackers to access sensitive data or interfere with the system. Consider a website where users can access their accounts by entering their username and password on a login page. To verify the user's credentials, the website runs the following SQL query:

```
SELECT * FROM users WHERE username = 'username' AND password =
'password
```

The `username` and `password` values are typically provided by the user through a web form. However, say a malicious user enters the following value for the password input:

```
password' OR '1'='1
```

Then, the resulting SQL query will be as follows:

```
SELECT * FROM users WHERE username = 'username' AND password =
'password' OR '1'='1'
```

Since `'1'='1'` is always true, this query will always return all data from the `users` table, giving the malicious user access to all user accounts.

This is just one illustration of how vulnerable web applications can be taken advantage of using SQL injection.

> **Note**
>
> You can read more about the Injection Prevention Cheat Sheet at
> `https://cheatsheetseries.owasp.org/cheatsheets/Injection_`
> `Prevention_Cheat_Sheet.html`.

Mitigation Strategies for A03:2021-Injection

Mitigation strategies for A03:2021-Injection are crucial in protecting applications from injection attacks, which are among the most severe and common security vulnerabilities. Effective mitigation involves a combination of validating all input data, using parameterized queries, sanitizing user input, and leveraging security frameworks with built-in defenses.

Validate all input data to make sure application inputs follow the desired format and are free of malicious code and SQL statements:

- **Parameterized queries**: When creating parameterized queries, make sure to properly sanitize any user input that will be incorporated into SQL statements

- **User input sanitization**: Make sure to sanitize any user input used in file paths, command-line arguments, or other system-level APIs

- **Utilize security frameworks**: Struts, for example, has built-in defenses against injection attacks, so use security frameworks that have similar features

Insecure Design (A04:2021)

Security risks associated with A04:2021-Insecure Design occur when security considerations are either ignored or given low priority when designing a system or application. This could be because functionality may be given precedence over security requirements or developers may lack knowledge of secure programming techniques. Unintentional access control, improper error handling, insufficient data input validation, and poor encryption techniques are just a few of the security flaws that can arise from insecure design and affect the application.

> **Note**
>
> You can read more about the Threat Modeling Cheat Sheet at `https://cheatsheetseries.owasp.org/cheatsheets/Threat_Modeling_Cheat_Sheet.html`.

The following are some mitigation techniques for A04:2021-Insecure Design:

- **Security by design**: It is essential to include security in the design stage of the application development process. Prioritizing security requirements while creating the application will help developers avoid vulnerabilities.

- **Adopting secure programming practices**: Developers should possess the necessary knowledge of secure programming techniques to minimize security risks in the application. This includes employing reliable input validation methods, robust encryption algorithms, and appropriate error handling.

- **Consistent testing**: Regular security testing can find security flaws early in the development life cycle, including vulnerability scanning and penetration testing.

- **Adhering to secure coding standards**: Observing secure coding standards can help maintain a more secure application by preventing insecure coding. Access control procedures must be put in place, encryption must be used when necessary, and insecure coding techniques must be avoided.

- **Threat modeling**: Threat modeling can assist in identifying potential risks during the application design phase, allowing developers to make the necessary changes to the application's design to obviate potential vulnerabilities.

- **Audits**: Regular code audits and reviews can help to keep the application's security posture strong by spotting issues that went unresolved in earlier stages.

Security Misconfiguration (A05:2021)

Security misconfigurations are situations in which the security settings or configurations of a system, application, or network are not correctly configured or are left at the default settings, resulting in possible security vulnerabilities. It is one of the most prevalent concerns that puts systems and networks at risk of cyberattacks, data breaches, or cyber threats.

Insecure settings include the use of weak passwords, unpatched software, default login credentials, exposed sensitive data, unsecured ports or services, obsolete security protocols, and insecure configurations.

In order to reduce security misconfigurations, security professionals implement the following strategies:

- **Frequent vulnerability and risk assessments**: Regular vulnerability scans and penetration testing can assist in identifying security misconfigurations and system vulnerabilities.

- **Adopt secure configuration standards**: Organizations should adhere to industry-standard security configurations for operating systems, applications, and network devices. This involves setting up firewalls, access restrictions, anti-malware or **endpoint detection and response (EDR)**, and encryption standards.

- **Patch management**: Organizations must maintain systems and applications with the most recent security patches and software upgrades to avoid the exploitation of known vulnerabilities.

- **Strong password rules**: Implement strong password policies and require regular password rotations to prevent the use of weak passwords or default credentials.

- **Training and awareness**: Employees and staff must be educated on security best practices and the hazards presented by improper security setups.

- **Installing access controls**: Organizations should implement access controls to restrict unauthorized access to sensitive data and resources.

- **Monitor system logs**: Organizations must consistently monitor system logs and network traffic to discover any malicious actions or abnormalities that may signal a security misconfiguration or vulnerability.

Vulnerable and Outdated Components (A06:2021)

A06:2021 is a software security flaw caused by the usage of insecure or out-of-date components or their dependencies. To be developed fast and efficiently, software applications generally rely on third-party libraries and frameworks. Although this is a cost-effective method of software development, it increases the attack surface of the program, as outdated or insecure components can lead to severe security breaches.

Typically, A06:2021 mitigation measures center on keeping software components managed and updated. The following details the preventive measures:

- **Updating libraries and dependencies on a regular basis**: Continuously updating obsolete libraries is one of the most effective methods for mitigating risks. Using a package manager to automate the process of upgrading software packages is suggested.

- **Performing routine security audits**: Another technique for minimizing susceptible and out-of-date components is to conduct routine security audits. This involves reviewing the software's source code in order to uncover obsolete or vulnerable third-party libraries and frameworks.

- **Monitor software supply chain**: Another way is to ensure that the software build process's dependencies are secured and will help mitigate such threats. Frequent monitoring and verification of the origins of these dependencies can aid in identifying and mitigating supply chain vulnerabilities and assaults.

- **Teaching developers**: Educating developers on the significance of utilizing just the most recent versions of software components and conducting rigorous testing can go a long way toward limiting the risks connected with A06:2021.

Identification and Authentication Failures (A07:2021)

Identification and authentication failures is a vulnerability that happens when an application is unable to properly verify and identify users or secure sensitive data during identification and authentication operations. It can result in unwanted data access and compromised user accounts.

The following are some A07:2021 mitigating strategies:

- **Deploy robust authentication protocols**: MFA, passwords, and biometric verification can substantially lower the risk of unwanted access to sensitive data

- **Employ secure transmission protocols**: Secure transmission methods, such as HTTPS and SSL/TLS, ensure that data transferred during authentication procedures is encrypted and shielded from interception

- **Monitor and update authentication systems routinely**: Organizations must ensure that authentication methods are routinely monitored, tested, and updated with the most recent security fixes

- **Restrict access to sensitive information**: Only authorized people who need access to sensitive data and resources should be granted access

- **Establish access control policies**: Organizations must guarantee that only authorized people have access to specified resources or data by adopting access control policies

> **Note**
>
> You can read more about the Authentication Cheat Sheet at `https://cheatsheetseries.owasp.org/cheatsheets/Authentication_Cheat_Sheet.html`.

Software and Data Integrity Failures (A08:2021)

Software and data integrity failures are vulnerabilities resulting from a program or system's inability to preserve the integrity of data. They may be caused by code mistakes, software design flaws, insufficient testing, or unanticipated system failures. A08:2021 can have significant consequences, including data loss, data tampering, and illegal access.

> **Note**
>
> You can find more information on Test Integrity Checks at `https://owasp.org/www-project-web-security-testing-guide/latest/4-Web_Application_Security_Testing/10-Business_Logic_Testing/03-Test_Integrity_Checks`.

To reduce the risks associated with A08:2021, developers must incorporate the necessary safeguards to protect the data and program integrity. The following are the mitigating methods for A08:2021:

- **Implement data validation checks**: This requires implementing data validation checks on input data to guarantee that only valid and relevant data is utilized in a process

- **Encrypt data**: By encrypting data, developers may increase the security of critical information and limit the possibility of data tampering

- **Perform code reviews**: Doing frequent code reviews can assist in identifying and correcting coding problems and design issues that could lead to events described by A08:2021

- **Employ secure coding methods**: Using secure coding practices, such as input validation, output encoding, and error handling, helps avoid data manipulation and injection attacks

- **Updates to software systems on a regular basis**: Updating software systems on a regular basis can assist in addressing known vulnerabilities and rectify coding flaws that could lead to data integrity issues

Security Logging and Monitoring Failures (A09:2021)

A09:2021 refers to the OWASP Top 10 2021 Security Logging and Monitoring Failures category. It is the responsibility of the systems to produce logs and monitor events in order to detect and respond to security problems. This can enable attackers to conduct additional assaults, proliferate throughout the system, or exfiltrate sensitive data without being discovered.

> **Note**
>
> You can find more information about the Logging Cheat Sheet at `https://cheatsheetseries.owasp.org/cheatsheets/Logging_Cheat_Sheet.html`.

The following measures may be utilized to minimize A09:2021:

- **Security information and event management (SIEM) products**: SIEM solutions are meant to identify and respond in real time to security concerns. They gather, correlate, and analyze event data from several sources in order to detect abnormalities and respond to potential risks.

- **Frequent monitoring**: Regular monitoring and examination of system logs can assist in the detection of suspicious activities. This includes real-time alerts, frequent log reviews, and setting the necessary monitoring thresholds.

- **Invest in staff training**: Employee training can play a crucial role in the prevention of A09:2021. Training should emphasize recognizing security danger indicators, the need for prompt reporting, and the employee's responsibility in protecting corporate security.

- **Repeated penetration testing**: Regular penetration testing can assist in detecting security vulnerabilities and weaknesses, including in monitoring and logging systems. It can identify flaws and assist businesses in proactively addressing them.

Server-Side Request Forgery (A10:2021)

A10:2021 describes **Server-Side Request Forgery** (**SSRF**) as a vulnerability that allows an attacker to send requests to other internal or external systems by manipulating the server. By using SSRF, attackers can get unauthorized access to sensitive data or execute operations on behalf of the server.

Imagine a web application designed to let users obtain information from external websites by providing a URL. The program retrieves data from the supplied URL and displays it on the user's screen using a server-side script.

A malicious user may provide a URL pointing to an internal resource, such as `http://192.168.1.1/private-info`, which would then be fetched by the application's server. While the server is situated on the same network as the resource, it is able to successfully access the private resource and show its contents on the user's screen.

In this instance, the attacker will access the sensitive internal resources by exploiting the web application's SSRF vulnerability. On the internal server, the attacker might have possibly retrieved sensitive data, such as user passwords or financial information.

> **Note**
> You can find more information about the SSRF Prevention Cheat Sheet at `https://cheatsheetseries.owasp.org/cheatsheets/Server_Side_Request_Forgery_Prevention_Cheat_Sheet.html`.

The following are a few methods for defending against SSRF attacks:

- **Input validation**: As SSRF attacks weaknesses inside inputs that are processed, it is vital to validate the inputs. Programs should only accept known content and reject any malicious or unexpected input.

- **Deny by default**: Firewall rules should be configured to refuse all outbound server requests unless expressly allowed. This will aid in preventing attackers' harmful conduct and unlawful access.

- **Access control**: Restricting access to sensitive information on a need-to-know basis on the server is an effective technique to decrease the danger of SSRF attacks. Users must have the permissions necessary to do their jobs, and access should be provided based on the need to know.

> **Note**
> Take time to review the OWASP Top 10 at their official site as you should be able to answer questions about specific examples and mitigation strategies.

Now that you have reviewed the OWASP Top 10 application vulnerabilities, and mitigation strategies for each of them, turn your attention now to some of the specific tools to assist in the overall security of applications in cloud environments.

Cloud Application Security Tools and Solutions

Organizations can implement all kinds of tools and solutions to boost the security of their cloud application portfolios. These solutions can assess and address threats, modify auditing permissions and access rights, implement encryption, and also add security features to prevent vulnerabilities from cropping up in application code.

Some tools are suited to secure specific elements of an application, while others can protect entire applications. Moreover, specialized tools that are available for various kinds of applications are as follows:

- Internal applications
- External or third-party applications
- Mobile applications
- Network-based applications
- Web applications
- APIs

Some application security tools can also protect legacy applications and microservices (on-premises and cloud-based). Any of these applications may be used by the organization's internal teams or by external stakeholders, such as customers, vendors, and other third parties.

Ideally, application security tools should integrate into the application development environment and cover all the stages of an SDLC.

For example, integrating security tests as part of a **continuous integration/continuous deployment** (**CI/CD**) pipeline is essential for ensuring that modern, cloud-native applications are developed and deployed securely. The CI/CD pipeline automates the process of code integration, testing, and deployment, enabling rapid and frequent updates. By incorporating security tests into this pipeline, organizations can identify and address vulnerabilities early in the development life cycle, reducing the risk of security breaches and ensuring compliance with security standards.

To integrate security tests effectively, one would start by embedding **static application security testing** (**SAST**) tools into the CI/CD pipeline. SAST tools analyze the source code for vulnerabilities without executing the code, allowing developers to detect issues such as SQL injection, **cross-site scripting** (**XSS**), and insecure configurations. These tests should be automated and triggered with every code commit to ensure continuous security validation.

Next, implement **dynamic application security testing** (**DAST**) tools that test running applications for vulnerabilities. DAST tools simulate attacks on the application to identify security flaws in a real-world scenario. These tests can be scheduled to run automatically during the staging phase of the CI/CD pipeline, ensuring that security vulnerabilities are detected before the application is deployed to production.

Additionally, include **software composition analysis** (**SCA**) tools to scan for vulnerabilities in third-party libraries and dependencies. As modern applications often rely on open-source components, SCA tools help ensure that all dependencies are secure and up to date.

Finally, integrate security testing with continuous monitoring and feedback loops. Use security dashboards and reports to provide real-time insights into the security posture of the application. Automated alerts and notifications can help developers address security issues promptly, fostering a culture of security awareness and responsibility within the development team.

Doing the preceding would make it easier to find coding mistakes and security flaws early. One example of such a coding error is allowing unverified inputs. The error opens the door to SQL injection attacks and data leaks. Integrated tools simplify the testing process and streamline the fixing workflow so such issues can be caught and remediated early.

The best solutions can analyze the code in cloud applications and test them for exploitable vulnerabilities. They also enable stakeholder collaboration and streamline security testing, development, and compliance.

Types of Cloud Application Security Tools and Solutions

Traditional testing tools can be used for cloud-native applications. However, most companies benefit from cloud-native security tools and solutions because they are designed to mitigate security threats to cloud applications. These tools can automatically scan cloud artifacts, provide fast feedback, and report on security issues. Read on to learn about some of the popular cloud application security tools and solutions.

Cloud-Native Application Protection Platform (CNAPP)

A CNAPP is a must-have for cloud application security. CNAPP tools combine **Cloud Workload Protection Platform** (**CWPP**) and **Cloud Security Posture Management** (**CSPM**) into a single platform and user interface.

The platform provides centralized and detailed visibility into cloud environments, allowing organizations to see all their cloud-native applications and take steps to protect them, even at runtime. They can also shift left and integrate security into development workflows to identify and fix flaws early in the SDLC.

Cloud Access Security Brokers (CASBs)

A CASB is a piece of cloud-based software between an enterprise and its CSP. It thus acts as a **gatekeeper** that enables organizations to safely use cloud services and safeguard their data in the cloud.

The CASB enforces an organization's granular security policies for the following areas:

- Authentication
- Authorization
- **Single sign-on (SSO)**

- Encryption

- Logging

- Tokenization

- Malware detection

The tool identifies the risks and threat actors that could adversely impact cloud applications. Such visibility enables enterprise teams to proactively stop security violations, potentially prevent data theft, and stop malware and other threats from infiltrating enterprise systems.

Cloud Web Application and API Protection (WAAP)

A cloud WAAP is a holistic multi-cloud platform designed to protect vulnerable web applications and APIs. It combines the functionalities of a WAF, **runtime application self-protection** (**RASP**), and other point solutions. It thus provides more robust security for web applications, APIs, and cloud microservices that access a lot of sensitive data and are exposed to the public internet.

Advanced WAAP solutions leverage machine learning and artificial intelligence to identify and block threats to web applications and APIs. They can also detect and respond to a wider range of threats than traditional WAF and other tools, including zero-day attacks, malicious bots, malicious users, and DDoS attacks.

The proper use and implementation of application security tools are critical to organizations that are heavily reliant on cloud-based applications. These solutions are crucial because they enable organizations to proactively assess and address threats, ensuring that security vulnerabilities are identified and mitigated before they can be exploited. From a CCSP exam perspective, you should have a solid understanding of the major types used and for what purpose.

> **Note**
>
> You can read more about application security verification here - *OWASP Application Security Verification Standard*: `https://owasp.org/www-project-application-security-verification-standard/`

Summary

The cloud has transformed the way businesses operate. Moving to the cloud enables organizations to increase workforce flexibility, improve process efficiencies, and enhance their long-term scalability and resilience.

At the same time, the cloud also creates serious security risks. Cloud applications are vulnerable to malicious actors because they are often open to the internet and carry sensitive data. Organizations must secure their cloud-based applications to minimize the threat of serious cyberattacks and business-crippling data breaches. The consequences of not doing so can be fatal.

Since most of the web applications are internet-facing, they enable threat actors to slip past enterprise security defenses and engineer all kinds of dangerous cyberattacks. To protect their applications, organizations must take application security seriously.

The OWASP Top 10 list provides a good starting point to identify vulnerabilities in web applications. By following its guidelines, organizations can write more secure code, strengthen their application testing processes, and act early to fix security gaps. The payoff of these efforts is huge – more secure applications that protect the company's data, customers, and reputation from harm. A CCSP with this knowledge will be critical to these ongoing efforts.

In the next chapter, you will delve further into the Secure SDLC, its implementation in software development, and the essential task of threat modeling.

Exam Readiness Drill – Chapter Review Questions

Apart from a solid understanding of key concepts, being able to think quickly under time pressure is a skill that will help you ace your certification exam. That is why working on these skills early on in your learning journey is key.

Chapter review questions are designed to improve your test-taking skills progressively with each chapter you learn and review your understanding of key concepts in the chapter at the same time. You'll find these at the end of each chapter.

> **How to Access These Materials**
>
> To learn how to access these resources, head over to the chapter titled *Chapter 25, Accessing the Online Resources.*

To open the Chapter Review Questions for this chapter, perform the following steps:

1. Click the link – `https://packt.link/CCSPE1_CH12`.

    Alternatively, you can scan the following **QR code** (*Figure 12.1*):

Figure 12.1 – QR code that opens Chapter Review Questions for logged-in users

2. Once you log in, you'll see a page similar to the one shown in *Figure 12.2*:

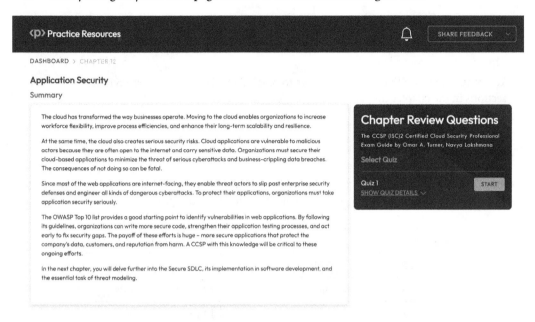

Figure 12.2 – Chapter Review Questions for Chapter 12

3. Once ready, start the following practice drills, re-attempting the quiz multiple times.

Exam Readiness Drill

For the first three attempts, don't worry about the time limit.

ATTEMPT 1

The first time, aim for at least **40%**. Look at the answers you got wrong and read the relevant sections in the chapter again to fix your learning gaps.

ATTEMPT 2

The second time, aim for at least **60%**. Look at the answers you got wrong and read the relevant sections in the chapter again to fix any remaining learning gaps.

ATTEMPT 3

The third time, aim for at least **75%**. Once you score 75% or more, you start working on your timing.

> **Tip**
>
> You may take more than **three** attempts to reach 75%. That's okay. Just review the relevant sections in the chapter till you get there.

Working On Timing

Target: Your aim is to keep the score the same while trying to answer these questions as quickly as possible. Here's an example of how your next attempts should look like:

Attempt	Score	Time Taken
Attempt 5	77%	21 mins 30 seconds
Attempt 6	78%	18 mins 34 seconds
Attempt 7	76%	14 mins 44 seconds

Table 12.1 – Sample timing practice drills on the online platform

> **Note**
>
> The time limits shown in the above table are just examples. Set your own time limits with each attempt based on the time limit of the quiz on the website.

With each new attempt, your score should stay above **75%** while your "time taken" to complete should "decrease". Repeat as many attempts as you want till you feel confident dealing with the time pressure.

13

Secure Software Development Life Cycle

Software vulnerability can be catastrophic for organizations, opening the door to all manner of adverse cyber events, such as cyber espionage or data leaks, just to name a few. That is why organizations need to take the security of their software more seriously.

Instead of considering security only in the later stages of the development process, you have to think about security at the inception stage itself or, at the very least, the early stages of a development project. This requires moving beyond the traditional **Software Development Life Cycle (SDLC)** and adopting the **Secure Software Development Life Cycle (SSDLC)**.

As a cloud security practitioner, you will be tasked with helping software development teams and their associated business stakeholders understand the importance of developing cloud-native applications with security in mind, leveraging not only the SSDLC but also other concepts such as threat modeling.

So, by the end of this chapter, you will know how the SSDLC is leveraged. You will also be able to explain how threat modeling applies to the SSDLC and learn ways to avoid writing vulnerable code. All this will be detailed in the following topics:

- Traditional SDLC to SSDLC
- SSDLC in cloud projects
- Threat modeling
- Code vulnerability

First, you will learn about the SSDLC. So, read on.

Traditional SDLC to SSDLC

Cloud-native applications are created to benefit from the scalability, flexibility, and cost-effectiveness of cloud computing. Cloud-native applications must be secure because a security compromise or data leak could have detrimental effects, such as monetary losses, reputational harm, and a decline in client and user trust. Therefore, it is crucial to put security first when developing these applications so as to safeguard against these risks and guarantee the privacy, availability, and integrity of the application and its data. This requires moving beyond the traditional SDLC and adopting the SSDLC.

Instead of considering security only later in the development process, organizations must think about it right from the beginning. The former (later in the development process) is known as **bolted-on** security and is typical of the traditional SDLC, while the latter (from the beginning) is about **baking in** security by conducting testing and following secure coding practices during all the SDLC phases. The result of the SSDLC is a more secure application that is better able to withstand threats and avoid breaches.

Benefits of the SDLC

The SDLC is the end-to-end structured process to plan, design, develop, test, and deploy a software application. It typically consists of activities across multiple phases—from planning and requirements gathering to coding, deployment, and post-deployment maintenance. It may also include activities to retire, replace, or upgrade the application.

The goal is to deliver a high-quality software product within budget and in the shortest possible time. Teams that adhere to its steps and best practices are more likely to meet their goals, minimize project risks, and deliver a successful project.

In addition, the SDLC remains an important aspect of software development because it does the following:

- Provides a standardized framework of activities, milestones, goalposts, and deliverables and a common **language** that everyone comprehends
- Helps with software project planning, scheduling, tracking, and control
- Provides visibility into statuses and challenges to increase transparency and accountability
- Enables teams to collaborate and communicate throughout the development cycle

Several SDLC models are in use today. One of the earliest is the **waterfall** model, a linear development model in which objectives are clearly defined from the beginning, and each stage is completed before the next can start. Over time, many other SDLC models have emerged, such as agile, scrum, spiral, lean, V-model, iterative, incremental, **Rapid Application Development (RAD)**, and **Development and Operations (DevOps)**. It is important to know these SDLC models from the exam's perspective:

- **Waterfall model**: This is a linear model in which each phase of the development process is completed before the next phase begins.

- **Agile model**: This is an iterative and incremental model in which the development process is divided into small increments, each of which is developed and tested in a short period of time. This model incorporates concepts such as DevOps. You will learn more about DevOps in the *Utilizing the SSDLC in Cloud Projects* section.

- **Scrum model**: This is an Agile model that is specifically designed for software development. It is based on the principles of transparency, inspection, and adaptation.

- **Spiral model**: This is a risk-driven model in which the development process is divided into a series of short cycles, each of which focuses on a specific aspect of the development process.

- **Lean model**: This model is based on the principles of the Toyota Production System and is designed to minimize waste and maximize value.

- **V-model**: This is a linear model that is similar to the waterfall model but includes more rigorous testing at each phase of the process.

- **Iterative model**: Software development progresses through repeated cycles, with each iteration focusing on a specific feature or functionality. This allows for early feedback and continuous refinement throughout the development process.

- **Incremental model**: This is a model in which the development process is divided into a series of small increments, each of which builds upon the previous increment.

- **RAD model**: This is a model that is designed to be flexible and allows rapid prototyping and development of software.

- **DevOps**: This model focuses on communication and collaboration between the development and operations teams in order to improve the pace and efficiency of software delivery while maintaining high-quality, high-security, and high-dependability standards.

> **Note**
>
> It is important to understand the major development models for the exam.

Drawbacks of the SDLC

Despite its many advantages, the traditional SDLC has a serious drawback: it doesn't focus on security. For several decades, the focus remained on completing development and bringing a software-based product to market.

Furthermore, all security activities were performed as part of the testing phase, which occurred toward the end when teams didn't have enough time to find and fix vulnerabilities. Consequently, many vulnerabilities remained unfixed, and the end product was both buggy and insecure.

Even when teams did discover these flaws pre-delivery, it was already too late. So, fixing the flaws incurred a lot of additional expense and often delayed release. The end result was the same: an insecure application.

Why the Traditional SDLC Is Obsolete

Today's world of software and application development is highly dynamic and fast-moving. Organizations are looking to develop applications and bring them to market before their competitors. To move faster and speed up time to market, teams follow agile methods and automate many activity steps in the SDLC. At the same time, they also adopt cloud-native tools, microservice architectures, and open source components. While all these changes benefit organizations, they can also introduce new vulnerabilities and risks into the software, which threat actors exploit to carry out devastating cyberattacks.

The traditional SDLC, where security testing happens mostly once and only toward the end, cannot catch these issues early, much less ensure they are removed to prevent cyberattacks or data breaches such as SolarWinds or any emerging zero-day vulnerability that surfaces unexpectedly, which can exploit previously unknown software flaws and pose immediate security risks to systems. That is why the SDLC has become somewhat obsolete and given rise to the idea of the secure software development life cycle, or SSDLC.

SDLC versus SSDLC

In the context of traditional SDLC models, the burden of ensuring application security typically falls disproportionately on a select group of security professionals, rather than being a collective responsibility shared by every member involved in the project—from developers to system architects. Testing happens toward the end of the SDLC when the application is already built and a lot of time and effort has already been expended. At this point, fixing vulnerabilities can be both time-consuming and costly. That's why teams often ignore the flaws or choose to implement the fixes after release. As a consequence, this often leads to numerous vulnerabilities slipping through the development process and becoming publicly exposed. These security gaps are then readily and frequently exploited by attackers, who take advantage of these oversights to launch cyberattacks.

The SSDLC approach aims to address these problems by taking a complete 180-degree turn from the **develop first, test later** approach of the SDLC. In the SSDLC, developers implement security-related practices and conduct security testing from the earliest phases. The goal is to find and remediate vulnerabilities early and deliver a more secure product to the end user.

Its primary purpose is to find vulnerabilities early in the development life cycle and fix them before they open the door to intrusions, attacks, and data breaches. Ultimately, the goal is to deliver a more secure product to the end user, which is a harder challenge with the SDLC method.

Adding Security to SDLC

Security is a crucial element of the SSDLC. Developers implement security-related practices into the SDLC right from the start and through to the end. To deliver a stronger, more secure product, developers conduct security testing right from the earliest phases of the SDLC. This gives them the time to implement fixes, so the flaw doesn't travel down the SDLC and end up in the final product. Thus, the SSDLC focuses on security that's baked in from the beginning rather than tacked on at the end. That's why the SSDLC is vital for modern software development.

The SSDLC effectively incorporates security into every stage of the development process, an idea that's known as **shifting left**. Early and frequent testing enables teams to find security risks in the code and detect security vulnerabilities in their components. Since security is no longer an afterthought but, rather, a deliberate choice, it minimizes the number of security flaws that go into production and ensures that the final product is more secure.

Another benefit of the SSDLC is that it unites the development and security teams, so they have a shared goal to deliver a high-quality and secure product to end users as soon as possible. They can collaborate better to shorten the development timeline and ensure that the application and its code are always protected.

Finally, when security is baked into the application from the beginning, vulnerabilities become clearer faster. Teams can then act quickly to remediate gaps, thus avoiding the costs and delays of rework. Moreover, they can leverage many automated tools and frameworks such as the NIST **Secure Software Development Framework (SSDF)** (`https://csrc.nist.gov/projects/ssdf`) to reduce the need for the manual effort of security testing and coding.

> **Note**
>
> You can read the NIST SSDF reference document at `https://csrc.nist.gov/projects/ssdf`.

By now, you are aware that organizations today adopt cloud-native tools, microservice architectures, and open source components to their benefit, but these technologies can also introduce new vulnerabilities and risks to the software. The traditional SDLC has a serious drawback for not focusing on security but on completing development and bringing a software-based product to market. On the other hand, the SSDLC implements security-related practices and conducts security testing from the earliest phases to find and remediate vulnerabilities early and deliver a more secure product to the end user.

The next section will explain how security gets embedded into every phase of the SDLC to evolve into the SSDLC.

Utilizing the SSDLC in Cloud Projects

It is crucial for the SSDLC to incorporate security measures at the outset of the development process rather than as an afterthought. This proactive approach ensures that security is a fundamental consideration throughout all phases of software creation. This section will delineate the specific steps necessary to integrate security effectively from the initial stages. These steps include defining security requirements along with functional requirements, conducting threat modeling to assess potential vulnerabilities early on, and embedding regular security reviews and testing into each phase of the SSDLC. By following these guidelines, organizations can significantly reduce vulnerabilities and enhance the overall security posture of their software products. This section will focus on the steps required to do it successfully.

SSDLC Activities and Phases

In addition to security testing, the SSDLC also includes activities related to the following:

- Secure coding

- Writing security and functional requirements simultaneously

- Performing architecture risk analyses

- Leveraging vulnerability-scanning tools

- Implementing automated workflows to fix vulnerabilities quickly and early

All these activities happen at various phases of the SDLC and contribute to the application's security. Now, let's learn how security gets embedded into every phase of the SDLC to create the SSDLC.

Phase 1: Planning

During planning in the traditional SDLC, development teams and their stakeholders plan key aspects of the project such as resource and capacity allocation, timeline and schedules, quality goalposts, and budget. In the SSDLC, they also brainstorm and incorporate security considerations right from this first phase.

Thus, while creating regular deliverables such as cost estimates, task schedules, and project plans that provide clarity on the project, teams add the security aspects that could influence the project to these deliverables.

Phase 2: Requirements and Analysis

During requirements gathering in the SDLC, the team will identify the security considerations for functional requirements, create a plan to incorporate security into the final design, and decide which frameworks, languages, and technologies to use during development.

For example, if one functional requirement is that a user should be able to verify their credit card information in the application, a security consideration could be that they should only be able to view and verify their own information. This requirement for security mapping is a crucial part of the SSDLC and can determine the ultimate security posture of the application.

In this phase, teams will also try to determine whether any particular vulnerabilities may crop up during development or whether certain coding practices may introduce insecurities into the application. They must consider such aspects and their implications to ensure that they don't inadvertently adopt insecure practices or tools.

Phase 3: Design

The design phase is where in-scope requirements are translated into a plan of action. The plan should be detailed enough to show what the requirements will look like in the final application and how the development team will get to that point.

Teams also create design documents, such as a list of the components and patterns for the project, and a functional design that describes what should happen in the application. In some projects, they may create a rapid prototype that will be used as a starting point during development.

In the SSDLC, the design and prototyping phase always begins with security in mind. Thus, the team would also describe the security concerns and requirements applicable to the design.

Phase 4: Development

Security considerations are particularly vital during the development phase of the SSDLC. So, instead of simply generating code, developers follow secure coding guidelines to create vulnerability-free code. They will also confirm that open source or third-party components are secure before incorporating them into the application.

During the development phase, the testing team will create coding guidelines to secure the code. They regularly review the code as it is being developed. Code reviews help ensure that the application matches the planned features and functionalities. In addition, they will conduct security testing to identify code weaknesses and open source vulnerabilities. Identified issues then go back to developers so they can address them quickly and correctly before the application reaches production.

To accomplish these objectives, teams often use automated **Static Application Security Testing** (SAST) tools. SAST tools analyze coding-in-progress applications from the **inside out** so that developers don't have to manually analyze their source code to find vulnerabilities. The tool can run throughout the SSDLC to provide an ongoing way to identify and fix vulnerabilities proactively and continually.

Applications that incorporate open source components can significantly benefit from the use of **Software Composition Analysis** (SCA) tools, which are specifically designed for rigorous testing environments. These tools play a critical role in ensuring the security and integrity of the application by scanning, tracking, and analyzing open source components throughout the development process.

SCA tools enable developers to thoroughly inspect the application for any dependencies and potential vulnerabilities associated with these components. By providing a comprehensive overview of the open source libraries used, SCA tools empower developers to proactively secure their code base. This proactive approach not only helps in enhancing the security of the application but also contributes to improving the overall health and reliability of open source components integrated within the software. Testing happens during the verification and maintenance phases to ensure that no flaws fall through the cracks.

Phase 5: Verification and Deployment

Some organizations deploy the application as soon as it is ready. However, this approach should only be adopted if the appropriate security tools and practices are in place to accommodate speedy deployment without sacrificing security.

In the SSDLC, the application is only deployed after it is thoroughly tested to confirm that it matches the original design and security requirements. A pre-deployment round of automated security testing helps to catch and remediate any remaining vulnerabilities.

In DevOps and other agile methodologies, automated tools such as **Continuous Integration (CI)/Continuous Deployment (CD)** pipelines are used to control app verification and release, while automated deployment tools help to dynamically swap application secrets for the production environment.

Phase 6: Maintenance

The main goal of the SSDLC and its **shift-left** approach is to find and address as many vulnerabilities as possible before application release. But despite the enhanced focus on security and security testing, some vulnerabilities may still creep through into the live app. The hypothetical application in development might also behave differently in the production environment. For these reasons, security testing should continue on the software and its production environment even after deployment.

In the maintenance phase of the SSDLC, establishing a direct feedback loop between the security and development teams is indispensable. This seamless communication channel ensures that issue management and remediation processes are executed swiftly and efficiently. Such high-velocity responses are paramount to maintaining the security integrity of the application, as they significantly reduce the window of opportunity for adversaries to exploit any security gaps. Rapid remediation is essential not only for protecting the application itself but also for safeguarding the broader organization from potential cyber threats that could exploit these vulnerabilities. During this phase, DevOps teams may also do the following:

- Plan to accommodate production issues in future releases.

- Verify the security configuration of cloud environments.

- Check the resources affecting application functionality, such as container engines and orchestration tools.

DevOps

DevOps is a set of practices and cultural philosophies that aims to unify software **Development (Dev)** and software **Operation (Ops)**. The primary goal of DevOps is to shorten the system's development life cycle while delivering features, fixes, and updates frequently in close alignment with business objectives. It emphasizes people and culture and seeks to improve collaboration between development and operations teams to enable faster, more efficient production cycles along with ensuring the quality and security of the software.

DevOps can be considered both a development methodology and an organizational culture:

- As a development methodology:

  - **CI/CD**: DevOps introduces practices such as CI/CD that automate the integration of code changes from multiple contributors and ensure that the application is systematically tested and deployed.

  - **Automation**: DevOps heavily relies on automation to speed up routine operational and development tasks. This includes everything from testing software to setting up and tearing down environments.

  - **Monitoring and performance**: DevOps encourages proactive monitoring of the performance of applications and infrastructure to ensure they are running efficiently and to quickly rectify issues as they arise.

- As an organizational culture:

  - **Collaboration and communication**: DevOps breaks down the traditional silos between development, operations, and other teams within an organization. It fosters a culture of transparency and sharing where all members are aligned toward the common goal of delivering quality software rapidly and reliably.

  - **Shared responsibility**: DevOps promotes the idea that everyone is responsible for the stability and performance of the software, not just the operations team. This shared responsibility helps to achieve higher quality and more secure deployments.

  - **Continuous learning and adaptation**: DevOps culture emphasizes continuous improvement and encourages experimentation and learning from failures. It supports an adaptive planning process, flexible to changes in user demands or market conditions.

In essence, DevOps transcends being merely a methodology focused on technical processes and tools; it encapsulates a broader cultural shift that integrates and aligns the goals of development and operations teams. It is about building a cooperative and adaptive environment where building, testing, and releasing software can happen rapidly, frequently, and more reliably.

DevOps is built upon a set of guiding principles that shape its practices and methodologies, aiming to improve collaboration between development and operations teams and enhance the entire software delivery process. Here are the seven core principles of DevOps:

- **CI**: Developers regularly merge their code changes into a central repository, after which automated builds and tests are run. CI aims to improve software quality and reduce the time taken to validate and release new software updates.

- **CD**: This extends CI by ensuring that software can be released to production at any time. It involves automating the entire software release process, from build to deployment, ensuring that the software is always in a deployable state throughout its life cycle.

- **Microservices architecture**: This principle involves structuring applications as a collection of loosely coupled services, which improves modularity and makes the application easier to understand, develop, and test, and become more resilient to architecture erosion. It allows individual service scalability and faster development cycles.

- **Infrastructure as Code (IaC)**: This practice involves managing and provisioning computing infrastructure through code, rather than through manual processes. It enables consistent environments to be deployed more rapidly and at scale by leveraging modern automation tools that apply the same configurations to every environment.

- **Monitoring and logging**: Continuous monitoring of the application and its underlying infrastructure is crucial to immediately react to issues or failures and to optimize performance. Logging and monitoring systems help in understanding how application changes or updates impact users, providing insights into the root causes of problems or unexpected changes.

- **Communication and collaboration**: Enhanced communication and collaboration are often considered the most critical aspects of the DevOps philosophy. This principle encourages the breaking down of silos among teams, promoting a culture where everyone works toward a common goal and shares responsibilities for the success of the software.

- **Continuous feedback and improvement**: DevOps relies on continuous feedback loops to improve products and processes. By continuously measuring and receiving feedback on performance, teams can understand their effectiveness, uncover areas for improvement, and innovate more rapidly. This iterative process of learning and adaptation is key to DevOps success.

Why the SSDLC Is Vital for Modern Software Development

The SSDLC is vital for modern software development primarily because it integrates security directly into the development process, ensuring that security considerations are embedded from the beginning rather than being an afterthought. This approach helps in identifying and mitigating security vulnerabilities early, significantly reducing the risk and impact of security breaches. Early integration of security not only enhances the overall security of software products but also aligns with regulatory compliance requirements, protecting the organization from potential legal and financial penalties.

Additionally, the SSDLC promotes a security-aware culture within development teams, making security a fundamental part of the development ethos and decision-making process. Ultimately, incorporating the SSDLC leads to more secure, robust, and reliable software, which is essential in today's digital landscape where cyber threats are increasingly sophisticated and pervasive.

In this section, you evaluated the reasons why the SSDLC is vital for modern software development and how security gets embedded into every phase of the SDLC to create the SSDLC. You also identified how DevOps improves the pace and efficiency of software delivery while maintaining high-quality, high-security, and high-dependability standards.

In the next section, you will learn about a process that is used in software development and security not only to recognize potential security risks but also to develop strategies for eliminating those risks.

Threat Modeling

Threat modeling is a process used in software development and security to identify potential security risks and vulnerabilities in a system, as well as build strategies for minimizing or removing those risks. It involves evaluating the system architecture and identifying potential attack vectors, determining the probability and effect of these threats, and implementing mitigation techniques to counteract them.

Diverse threat modeling approaches exist, including the Microsoft Threat Modeling Framework (https://www.microsoft.com/en-us/securityengineering/sdl/threatmodeling), the **Spoofing, Tampering, Repudiation, Information disclosure, Denial of service, and Elevation of privilege (STRIDE)** method, and the **Process of Attack Simulation and Threat Analysis (PASTA)** method. Typically, these approaches involve the following steps:

1. Determine the assets and data that must be safeguarded within the system.
2. Identify the potential threats to these assets and data, such as internal and external adversaries as well as their potential attack tactics.
3. Assess the probability and impact of each hazard, then rank them according to their risk level.
4. Develop measures to minimize or limit the risk posed by each danger, such as adopting security controls or developing incident response plans.
5. Evaluate and verify the effectiveness of the mitigation techniques and make any necessary modifications.

Modeling threats should be performed continuously throughout the SDLC, from design to deployment through maintenance. It ensures that security is included in the system from the beginning, as opposed to being an afterthought. It also provides a method for communicating security threats to stakeholders in a language they can comprehend, as well as a shared knowledge of the risks the system is exposed to and the countermeasures implemented.

From the exam perspective, you need to first learn about each of the threat modeling approaches and determine the differences between them. So, first, let's look at each of the components of the STRIDE model (with a few examples).

STRIDE

The STRIDE model is a mechanism for identifying and classifying potential software security concerns. It is an acronym representing the six sorts of threats: spoofing, tampering, repudiation, information disclosure, **Denial of Service (DoS)**, and elevation of privilege. Since STRIDE is one of the more well-known threat models, you need to look at each threat and review examples of the threat in a scenario. Looking at examples will help solidify the concepts and help you remember them for the exam:

- **Spoofing**: An attacker impersonates a genuine user or system in order to gain unauthorized access to sensitive data or resources. Examples of spoofing attacks are phishing, IP spoofing, ARP spoofing, DNS spoofing, and man in the middle:

  - **Phishing attack example**: An attacker sends an email or text message that looks to be from a reputable source, such as a bank or online merchant, in an attempt to deceive the user into divulging their login credentials, personal information, or financial information. Here, the attacker is impersonating a legitimate organization to obtain access to sensitive data.

  - **IP spoofing example**: An attacker modifies the source IP address of a packet or connection so that it appears to have originated from a different network or device. In other words, the attacker impersonates a different device or network to get access to sensitive data or do other harmful actions.

  - **DNS spoofing example**: An attacker updates the **DNS** records to route users to a malicious website that looks like a legal site, thus impersonating a legitimate website to deceive users into divulging personal information or login credentials.

  - **Man-in-the-middle attack example**: An adversary modifies and intercepts the communication between two parties. In this way, the attacker impersonates a genuine party to obtain access to sensitive data or do other destructive acts.

- **Tampering**: This form of threat involves the modification or corruption of data in transit or at rest by an adversary. Examples of tampering attacks are the rewrite SQL injection, file tampering, tampering with software updates, and man in the middle:

  - **Rewrite SQL injection attack example**: An attacker adds malicious code into a SQL query to change or damage the database data. This is an example of tampering, as the attacker is modifying data without permission.

  - **File tampering example**: An attacker edits a file on the target system, such as a configuration file, in order to alter the settings of a service running on the system. Here, the attacker is manipulating data in an unauthorized way.

- **Tampering with software updates example**: Before the software update package is installed on the target system, the attacker alters the package on a public repository and gets access to the target system or installs malicious software on it.

- **Man-in-the-middle attack example**: An adversary modifies and intercepts the communication between two parties, thus manipulating data in an unauthorized way while in transit.

- **Repudiation**: This sort of threat comprises an attacker denying having undertaken a certain action or denying the existence of particular information. Examples of repudiation attacks are digital signature spoofing, session hijacking, IP spoofing, log tampering, and email spoofing:

  - **Digital signature spoofing example**: An attacker modifies a document's digital signature to make it appear as if a different person signed it, thus denying the signature's legitimacy.

  - **Session hijacking example**: An adversary takes control of a user's session and acts on their behalf. The attacker essentially denies responsibility for the activities.

  - **IP spoofing example**: An attacker modifies the source IP address of a packet or connection so that it appears to have originated from a different network or device. This way, the attacker disputes the IP address's legitimacy.

  - **Log tampering example**: An attacker modifies a system's logs to make it appear as if certain events never occurred. Here, the attacker denies the existence of specific occurrences.

  - **Email spoofing example**: An attacker modifies the email headers to make it look as if the message was sent by a different sender, thus denying the email's legitimacy.

- **Information disclosure**: This sort of threat involves unauthorized access to or capture of sensitive information. Examples of such attacks include eavesdropping, social engineering, vulnerability scanning, **Cross-Site Scripting** (**XSS**), and malware:

  - **Eavesdropping example**: An attacker intercepts and monitors network traffic to have unauthorized access to sensitive data, such as login passwords or personal information.

  - **Social engineering example**: An attacker manipulates a user into divulging sensitive information, such as login credentials or personal data. This way, the attacker is getting information with authorization.

  - **Vulnerability scanning example**: An attacker searches for flaws in a system and sensitive files or data on a server using automated tools to gain access to unauthorized information.

  - **XSS example**: An attacker injects malicious code into a web page to steal sensitive information, such as login credentials or personal data.

  - **Malware example**: An attacker installs malware on a system to steal sensitive data, such as login passwords and personal information.

- **DoS**: This threat involves an adversary rendering a system or service inaccessible to authorized users, thus preventing authorized users from accessing the system or service. Examples of such attacks include **Distributed Denial of Service (DDoS)**, resource depletion, amplification, **Synchronize (SYN)** flood, and application layer attacks:

  - **DDoS attack example**: An attacker overwhelms or floods a server or network with traffic from multiple sources in order to overload or block access.

  - **Resource depletion attack example**: An attacker exhausts all available system resources, such as CPU and memory, rendering the machine inoperable.

  - **Amplification attack example**: Utilizing a botnet or amplification techniques such as NTP amplification, an attacker increases the traffic to a targeted site or network.

  - **SYN flood attack example**: In this attack, an attacker sends a large number of SYN packets with a spoofed source IP address to a targeted server, causing the server to exhaust its resources while attempting to complete the handshake procedure.

  - **Application layer attack example**: An attacker sends a high number of requests to a server or service, leading it to become overloaded and unable to respond to genuine requests.

- **Elevation of privilege**: This type of threat is characterized by an adversary getting unauthorized privileges or access. Here, the attacker is getting more privileges or access beyond what is intended. Examples of such attacks include privilege escalation assaults, pass the hash, pass the ticket, malware, and misconfigured permissions:

  - **Privilege escalation example**: An attacker gains elevated access to a system by exploiting a vulnerability. This is an instance of elevation of privilege since the attacker is getting more privileges or access beyond what is intended.

  - **Pass-the-hash attack example**: An attacker gets a user's password hash and uses it to impersonate that user and gain access to sensitive data or services. This is an instance of elevation of privilege since the attacker is getting more privileges or access beyond what is intended.

  - **Pass-the-ticket attack example**: An attacker steals a user's Kerberos ticket in order to impersonate that user and obtain access to sensitive data or resources. This is an instance of elevation of privilege since the attacker is getting more privileges or access beyond what is intended.

  - **Malware example**: An attacker installs malware on a machine that grants them elevated privileges. This is an instance of elevation of privilege since the attacker is getting more privileges or access beyond what is intended.

- **Misconfigured permissions example**: An attacker identifies and exploits incorrectly configured permissions on a system or its resources in order to gain access to higher levels of access. This is an instance of elevation of privilege since the attacker is getting more privileges or access beyond what is intended.

The STRIDE model offers a systematic approach to detecting and classifying security threats, which can aid in the development of effective mitigation techniques and the identification of locations where additional security controls may be required. It ensures that all potential dangers are examined and provides a means of communicating risks to stakeholders in a manner they can appreciate.

PASTA

The PASTA methodology is a detailed, seven-step process designed to enhance software security by aligning it with business objectives. Here's a concise overview of each step:

- **Define business objectives**: Identify and outline the key business goals that the software supports, establishing a foundational understanding crucial for effective threat modeling.

- **Define technical scope**: Determine the technical elements crucial to the application, including software, hardware, and networks, to provide a clear scope for the threat model.

- **Application decomposition**: Break down the application into its fundamental components and workflows to understand the architecture, data flows, and potential security weaknesses.

- **Threat analysis**: Identify potential threats using structured frameworks such as STRIDE, focusing on vulnerabilities specific to the application's architecture and workflows.

- **Vulnerability analysis**: Pinpoint specific vulnerabilities within the application that could be exploited, using tools such as static and dynamic analysis, manual reviews, and penetration testing.

- **Attack modeling**: Simulate potential attacks based on the identified threats and vulnerabilities, outlining how an attacker could exploit the system, the tools they might use, and the potential impact.

- **Risk and impact analysis**: Evaluate the risks identified during the attack modeling phase in terms of their potential impact on the business and likelihood of occurrence, allowing for prioritization and targeted mitigation efforts.

This structured approach not only helps in identifying and mitigating risks effectively but also ensures that security efforts are prioritized according to their potential impact on the organization's strategic goals.

By simulating various attack scenarios and examining how the system would respond to them, PASTA enables teams to gain a more comprehensive understanding of prospective attack scenarios. This can aid in identifying potential system vulnerabilities that may not be apparent using conventional threat modeling techniques, and in developing more effective mitigation solutions.

In addition, PASTA provides a method for communicating security concerns to stakeholders in a language they can comprehend, as well as a shared knowledge of the risks to which the system is exposed and the mitigation actions that have been implemented.

PASTA is not a standalone approach; when combined with STRIDE and DREAD (you will cover DREAD next), it can be utilized throughout the SDLC. The next section will explain the DREAD threat modeling methodology.

DREAD

Damage, Reproducibility, Exploitability, Affected users, and Discoverability (DREAD) is a threat modeling methodology that assesses the risk of prospective security risks. The five elements of DREAD are as follows:

- Damage is the severity of the threat's potential impact on the system or its users.
- Reproducibility is the ease with which an attacker can duplicate an assault.
- Exploitability is the ease with which an attacker may execute an attack.
- Affected users, as the name implies, relates to the number of users or systems that would be affected by the assault.
- Discoverability is the ease with which an attacker can identify the exploited vulnerability.

The DREAD approach comprises the following steps:

1. Determine the assets and data that must be safeguarded within the system.
2. Determine the potential threats to the assets and data in question.
3. Assess the risk associated with each danger by assigning a score between 1 and 10 for each of the five DREAD elements.
4. Rank the discovered threats according to their total DREAD score.
5. Develop and implement methods to remove or lessen the risk posed by each hazard.
6. Evaluate and verify the effectiveness of the mitigation techniques and make any necessary modifications.

The DREAD technique provides a means of prioritizing risks by evaluating the entire impact and likelihood of an attack. In addition, it provides a straightforward method for communicating risks to stakeholders and making data-driven decisions regarding where to concentrate resources to manage risks. As with other approaches, DREAD should be included in the broader threat modeling process and utilized throughout the SDLC.

Using STRIDE in conjunction with DREAD provides a comprehensive approach to threat modeling:

- **Identification with STRIDE**: First, use STRIDE to identify and categorize potential threats to the system. This involves examining the software architecture and identifying where and how these types of attacks could occur.

- **Prioritization with DREAD**: Once threats are identified, use DREAD to assess and prioritize them based on how critical they are to address. This helps in allocating resources effectively to where they are most needed.

For instance, if an analysis with STRIDE reveals a potential for an **elevation of privilege** attack on a system component, DREAD can be used to assess the potential damage, how easy the attack would be to reproduce and exploit, how many users would be affected, and how easily the vulnerability could be discovered. Based on the scores, if the threat is rated highly across most or all DREAD categories, it should be prioritized highly in the mitigation plans.

Integrating these models helps organizations effectively manage their security by ensuring that the most significant risks are identified and addressed promptly, thereby enhancing the overall security posture of their software systems.

In this section, you learned about the diverse threat modeling approaches—namely, the STRIDE method, the PASTA method, and the DREAD method. While the STRIDE model is a mechanism for identifying and classifying potential software security concerns, the PASTA model focuses on finding and assessing potential attack scenarios. DREAD, on the other hand, assesses the risk of prospective security risks. The next section will list ways to practically safeguard code from being vulnerable to attacks.

Avoiding Vulnerable Code

Now that you have discovered how frameworks and models can be used to avoid vulnerable code, this section will reinforce some practical steps that can be undertaken to accomplish the same. The practical steps are as follows:

1. **Integrate security concerns into each phase of the SDLC**: Starting with the requirements collecting phase, security issues should be included in every phase of the SDLC. This guarantees that security threats are discovered and handled throughout the development process.

2. **Use threat modeling with the STRIDE technique to detect potential security threats**: Using the STRIDE methodology, the development team may determine the potential security dangers that the software could face such as spoofing, tampering, repudiation, information leakage, DoS, and elevation of privilege.

3. **Apply mitigation techniques**: Once the possible security threats have been identified, the development team can develop and apply mitigation techniques to mitigate their risk. Implementing security controls such as access controls, encryption, and input validation may be part of this process.

4. **Utilize automated testing tools to detect and mitigate security flaws**: Automated testing techniques can assist in identifying security issues in code. This can include static analysis tools that check the code for vulnerabilities such as SQL injection and XSS. These tools can also assist in identifying flaws in third-party libraries and frameworks used by the software.

5. **Regular security reviews and penetration testing**: This can help uncover any remaining security flaws in the code. This includes discovering potential misconfigurations or vulnerabilities in the software or its supporting infrastructure.

By adhering to these methodologies, development teams can utilize frameworks such as the SDLC and threat modeling with STRIDE to avoid developing insecure code and produce secure software. In addition, it is essential to stay abreast of the most recent security best practices and recommendations and to provide frequent security training for the development team.

In the *DevOps* section, you learned how DevOps integrates operation considerations into the software development process. In the same way, DevSecOps integrates security considerations into the process. It strives to ensure that security is an intrinsic component of the software development process from start to finish.

DevSecOps improves the security of the software development process in the following ways:

- **Automating security testing**: The development teams may quickly uncover and address code security problems by using automated techniques. While tools such as static analysis and dynamic analysis scan the code for vulnerabilities, automated penetration testing tools find weaknesses in the deployed program.

- **Integrating security into the pipeline for CI/CD**: By integrating security into the CI/CD pipeline, development teams can ensure that all code modifications undergo security testing. This helps to detect security flaws early in the development process when they are less expensive and simpler to remedy.

- **Creating a security culture**: DevSecOps encourages a security culture inside the development team. This involves providing the development team with regular security training, integrating security considerations into the team's everyday work, and involving security experts in the software development process.

- **Communication and collaboration**: DevSecOps improves communication and collaboration between development, security, and operations teams. This ensures that security issues are detected and addressed early in the software development process and that security considerations are incorporated into the software development process.

By implementing DevSecOps principles, development teams may ensure that security is incorporated throughout the whole software development process—from development to deployment—making it more successful at finding and mitigating software vulnerabilities.

Since this section is dedicated to writing less vulnerable code, it is critical to grasp the importance of securing the CI/CD pipeline.

The CI/CD Pipeline Concept

The CI/CD pipeline is a set of automated processes that enables software development teams to integrate code changes and create, test, and deploy software applications continuously and efficiently. The CI/CD pipeline is intended to offer rapid and consistent feedback on code changes, eliminate errors and manual intervention, and enhance the quality of software applications overall.

The usual CI/CD pipeline includes the following stages:

1. **Push code**: Developers push code modifications to a centralized source code repository, such as Git.

2. **Continuous integration**: Changes to the code base are automatically built, tested, and integrated.

3. **Continuous testing**: Automated tests are executed to verify the quality and functionality of code modifications.

4. **Continuous deployment**: Code modifications are deployed automatically to the production environment or other pre-production environments, such as staging.

5. **Continuous monitoring**: Deployed software is continuously monitored to detect and address issues and performance difficulties.

The CI/CD pipeline enables software development teams to produce software applications more quickly, with more quality and dependability, and at a lower risk of errors and security vulnerabilities.

Securing a CI/CD pipeline is a crucial component of software development that necessitates a comprehensive strategy for identifying and mitigating potential security issues. The following measures are taken to secure a CI/CD pipeline:

- Implement access control, and restrict pipeline and resource access to just authorized users. This involves implementing **Role-Based Access Control** (**RBAC**) to guarantee that users have the bare minimum of privileges required to complete their tasks.

- Use encryption to safeguard sensitive pipeline data, such as access keys and passwords with encryption. This involves employing secure protocols, such as HTTPS and **Secure Shell** (**SSH**), to encrypt data in transit and using encryption at rest to safeguard data kept on disk.

- Adopt a secrets management solution to store and manage secrets, as well as regularly rotate secrets, such as access keys and passwords, in a secure manner.

- Monitor and audit the pipeline to discover and respond to possible security incidents. This involves logging and monitoring tools to record and analyze system and application logs, as well as to detect and report suspicious activities.

- Utilize automated security testing to assist in identifying and resolving code security problems. This uses automated techniques designed to rigorously evaluate the security of software. These include static analysis tools, which scrutinize source code without executing it to uncover potential vulnerabilities, and dynamic analysis tools, which analyze running programs to detect runtime vulnerabilities.

- Continuously patch and update the pipeline to maintain the pipeline and its components with the most recent security updates and fixes. This involves the use of automatic patch management technologies to guarantee that the pipeline and its components are functioning properly.

Summary

In this chapter, you grasped the methodologies that enable writing secure code, which is essential to how modern cloud-native applications are written. You came to know the phases of the SDLC and elements of the STRIDE threat model from an exam's perspective. You also learned that to design software code that is largely free of security flaws, it is essential to adopt a comprehensive strategy that includes the following:

- Adopting the SSDLC and threat modeling techniques, such as PASTA or STRIDE, to identify and mitigate potential security issues throughout the development process.

- Using the DevSecOps approach to ensure that security is an inherent component of the software development process—from development to deployment—by integrating security considerations into the CI/CD pipeline.

- Implementing access control and encryption to safeguard critical pipeline data, such as access keys and passwords.

- Using secrets management to store, maintain, and rotate pipeline secrets in a secure manner.

- Monitoring and auditing the pipeline in order to discover and address possible security incidents.

- Using automated security testing to find and address security flaws in the code and maintain the pipeline and its components with the most recent security patches and upgrades.

By adhering to these principles, development teams ensure that security is integrated across the whole software development process, making it more efficient and effective in finding and addressing software vulnerabilities.

In the next chapter, you will evaluate the importance of assurance, validation, and verification. You will look at functional and non-functional testing, security testing concepts, third-party testing concepts, and how to secure APIs.

Exam Readiness Drill – Chapter Review Questions

Apart from a solid understanding of key concepts, being able to think quickly under time pressure is a skill that will help you ace your certification exam. That is why working on these skills early on in your learning journey is key.

Chapter review questions are designed to improve your test-taking skills progressively with each chapter you learn and review your understanding of key concepts in the chapter at the same time. You'll find these at the end of each chapter.

> **How to Access These Materials**
>
> To learn how to access these resources, head over to the chapter titled *Chapter 25, Accessing the Online Resources.*

To open the Chapter Review Questions for this chapter, perform the following steps:

1. Click the link – `https://packt.link/CCSPE1_CH13`.

    Alternatively, you can scan the following **QR code** (*Figure 13.1*):

Figure 13.1 – QR code that opens Chapter Review Questions for logged-in users

2. Once you log in, you'll see a page similar to the one shown in *Figure 13.2*:

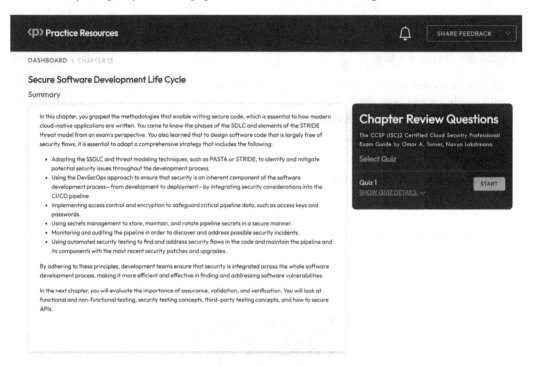

Figure 13.2 – Chapter Review Questions for Chapter 13

3. Once ready, start the following practice drills, re-attempting the quiz multiple times.

Exam Readiness Drill

For the first three attempts, don't worry about the time limit.

ATTEMPT 1

The first time, aim for at least **40%**. Look at the answers you got wrong and read the relevant sections in the chapter again to fix your learning gaps.

ATTEMPT 2

The second time, aim for at least **60%**. Look at the answers you got wrong and read the relevant sections in the chapter again to fix any remaining learning gaps.

ATTEMPT 3

The third time, aim for at least **75%**. Once you score 75% or more, you start working on your timing.

> Tip
>
> You may take more than **three** attempts to reach 75%. That's okay. Just review the relevant sections in the chapter till you get there.

Working On Timing

Target: Your aim is to keep the score the same while trying to answer these questions as quickly as possible. Here's an example of how your next attempts should look like:

Attempt	Score	Time Taken
Attempt 5	77%	21 mins 30 seconds
Attempt 6	78%	18 mins 34 seconds
Attempt 7	76%	14 mins 44 seconds

Table 13.1 – Sample timing practice drills on the online platform

> Note
>
> The time limits shown in the above table are just examples. Set your own time limits with each attempt based on the time limit of the quiz on the website.

With each new attempt, your score should stay above **75%** while your "time taken" to complete should "decrease". Repeat as many attempts as you want till you feel confident dealing with the time pressure.

14

Assurance, Validation, and Verification in Security

This chapter delves into the significance of assurance, validation, and verification within the context of cloud security, focusing on their application in cloud-native systems. Assurance in cloud security provides stakeholders with confidence that security measures are appropriately designed and implemented to protect cloud-based resources. Validation involves testing these security measures to ensure they function as expected and effectively mitigate identified risks. Verification goes a step further by continuously monitoring and ensuring that all components of the cloud-native application adhere to specified security standards and protocols throughout their life cycle. Together, these three forms of security are fundamental in building and maintaining robust defenses against evolving threats, thereby protecting sensitive data and ensuring the reliability and performance of cloud-native applications. This is crucial not only for maintaining operational integrity but also for ensuring compliance with stringent regulatory requirements, as well as preserving the trust of customers and partners.

In this chapter, you will start with some basic definitions and then take a deep dive into the following topics:

- The three primary forms of security – assurance, validation, and verification
- Functional and non-functional testing
- Security testing and quality assurance
- Software verification
- Third-party review
- Securing APIs

First, you will learn about the three primary forms of security.

The Importance of Assurance, Validation, and Verification

As you are aware, cybersecurity is a massive concern to both businesses and individual users in the digital era. Millions of lives have been affected and billions of dollars have been lost because of cyber breaches. Internal controls to prevent fraud and limit cyberattacks are the best way to ensure digital security. This requires ensuring that any software used in an environment where data or proprietary information is used or stored is safe to be deployed, and also while it interacts with other pieces of technology that it performs in the right way. As a CCSP-certified individual, your knowledge in this area will be critical for conversations with stakeholders as well as for being successful in the exam.

While the terms **assurance**, **validation**, and **verification** are often used interchangeably, they each have distinct meanings that are crucial to understand for their effective application in a cloud-native environment. It is essential to grasp these differences to ensure that each process is correctly implemented, thereby guaranteeing the security and integrity of cloud-native systems. Clear understanding and proper execution of assurance, validation, and verification are fundamental to maintaining robust security measures, achieving compliance with regulatory standards, and building trust with users and stakeholders. This differentiation not only enhances the security posture but also optimizes operational performance within the cloud:

- **Assurance**: This is a measure of confidence in the cloud environment or the application itself that measures how well it can keep itself, its users, and the systems it connects to and that are dependent on it safe from any threats it might encounter over the course of its use. It covers everything from the architecture, procedures, and practices to the security features of an application.

- **Validation**: This is the process of confirming that an application meets both the expectations and the needs of its users. Its purpose is to test the reliability, functionality, and usability of the product, and it does this by executing the application's code. Its functions include but are not limited to usability testing, functionality testing, security testing, system testing, and performance testing.

- **Verification**: This is the process of making sure an application is developed as per the specifications it was designed for. It tests the code, design, and architecture of the application but does not need to execute code to do so. This process commonly includes code verification, design verification, and requirement verification. During code verification, specific parameters such as coding standard compliance, code structure, security vulnerabilities, and the integration of different code modules are scrutinized. This thorough inspection helps identify discrepancies early in the development process, ensuring the application's architecture and design are correctly implemented before moving forward.

You have now grasped the definitions of assurance, validation, and verification. In the next section, you will learn about the two types of testing – functional and non-functional.

What Are Functional and Non-Functional Testing?

Functional and non-functional testing occur at different times and have different expectations. Functional testing typically occurs after an application has been deployed, with the release management team putting the application through thorough testing in order to identify and correct small errors. The application's development team is in charge of performing functional testing after each new section of code is completed or a section of code is taken out and upgraded or altered.

In the development life cycle, functional testing normally occupies the longest stretch of time. It is a much smarter idea to perform functional testing in piecemeal mode, where each new section of code undergoes testing before its integration with the next section begins. Some organizations put all the code together and then perform functional testing on the entire application to see what is discovered. This can very easily lead to a lot more of a time sink as well as delays, as errors in one section of code might cause false positive errors in other sections.

The purpose of the essential functional test is to analyze a specific function or a part of an application. This is in contrast to full testing, where programmers check out the entirety of the application to see how it functions in every instance.

Functional testing enables programmers to evaluate whether the outputs produced by an application align with the expected results. This approach offers a holistic view of the application's performance, going beyond isolated module testing. By observing how different modules interact within the application, programmers can more effectively identify and address errors. This integrated testing environment often makes discrepancies more apparent, facilitating quicker and more efficient debugging and validation of the software's functionality.

In software development, a module refers to a self-contained unit of code that performs a specific function within a larger system. Each module is designed to execute a part of the software's overall functionality independently while maintaining the ability to interact with other modules via well-defined interfaces. This design allows for encapsulation, where modules conceal their internal data and expose only necessary elements to other parts of the application. Modules are reusable across different parts of an application or in different projects, enhancing development efficiency. They also facilitate easier and more effective testing because they can be tested individually before integration. Overall, the modular approach simplifies the complexity of developing, maintaining, and testing software systems.

Functional testing can be broken down into several main types, one of which is also a form of non-functional testing (also known as performance testing, which you will get on to later). The four types of functional testing are as follows:

- **Unit testing**: Unit testing involves executing individual components of a software system in a controlled testing environment to validate their behavior against predefined specifications outlined in the original documentation. These tests are designed to be clear-cut and straightforward, assessing whether the system's responses match expected outcomes. Unit tests are particularly crucial because they help ensure that each part of an application functions correctly on its own. However, their effectiveness can diminish if the application's functionality changes unexpectedly or if it regresses from its intended design, making it essential to keep unit tests updated alongside the software's development.

- **Integration testing**: Integration tests check out how well different software modules sync together. When app developers write code, the code's components rely on specific conditions to determine how they interact with each other. These tests ensure that each piece of the app correctly performs its task or gives a warning if it causes a regression.

- **User acceptance testing**: At this stage of functional testing, developers use test subjects or real end users to perform real-world tests on how well an application functions. It is viewed in some circles as taking too much time, costing too much, and being unreliable, but in most applications, it continues to have its uses.

- **Closed-box testing**: Closed-box tests look only at the outputs of an application without considering how it functions internally. Only the outermost layers of the code are tested, and the output is evaluated.

Non-functional testing evaluates aspects of an application that, while not critical to the core functionality required for operation in a cloud environment, are significant to the user experience. This form of testing focuses on the system's attributes such as performance, usability, reliability, and security. Although identifying an issue through non-functional testing does not necessarily imply an immediate impact on the end user's interaction with the application, it can still highlight potential problems that may affect the application's performance or user satisfaction at a later stage in its life cycle. Such insights are crucial for ensuring that the application not only functions correctly but also delivers a quality experience to users consistently. The four main types of non-functional tests include the following:

- **Performance testing**: Performance testing is fundamentally concerned with assessing the speed and efficiency of an application's system to ensure that it operates effectively under expected workloads. This type of testing can be likened to checking a car's engine and other critical components to confirm they are functioning optimally before driving on the road. If an application responds slowly or performs poorly, it can significantly hinder the user experience, making swift and reliable functionality crucial for a positive interaction with the software.

- **Load testing**: The purpose of load testing is to find out how much traffic an application can handle in the cloud environment. Every application works amazingly when handling only one request per second. But how about when the usage is increased to 10,000 requests per second? Load testing lets you see what your app's peak load and the breaking point are on heavy-duty workloads.

- **Usability testing**: Performing a non-functional usability test is not something you can do repeatedly for every user, as it's too time-consuming and focused to be scalable. However, when you create your app for the first time, usability is a good way to make sure that the interface is intuitive and makes sense to a user before pushing out more code.

- **Security testing**: Testing for security purposes is essential in the cloud environment, and in a non-functional capacity it includes everything from automatic scanning to periodic penetration tests that see how well the app responds to different levels of exposure to threats.

Now, you'll see the difference between functional and non-functional testing.

Comparing Functional and Non-Functional Testing

Non-functional testing is committed to seeing whether your code does things the correct way, while functional testing checks whether the code performs the right functions. Both forms of testing are concerned with backend and frontend elements equally, and there is also some overlap between them.

Functional tests can be deemed the more necessary of the two, since they deal with the actual use and output of the application. Non-functional is also necessary, but most of the issues are secondary. Functional testing in the cloud environment has even more weight, since there are typically more regulatory and legal ramifications to tackle when moving an app to the cloud.

Assurance and validation are, thus, big parts of functional testing, while verification – that the app is built correctly – falls under the gaze of non-functional testing. Validation of the entire build package and the steps to deploy the application must be gained before it is put into the cloud environment.

In this section, you learned that while non-functional testing is committed to seeing whether your code does things the correct way, functional testing checks whether the code performs the right functions. The next section will take a closer look at testing with a security-centric approach.

Security Testing and Quality Assurance

Security testing in cloud-native apps is a massive concern that must constantly pass assurance and validation checks in order to keep users secure and happy. Security testing often overlaps with the sort of testing done in a data center. Both environments promote themselves as utterly secure for third-party computing, so ensuring that "nothing gets in, nothing gets out" is imperative. The following details the different types of security testing along with the appropriate circumstances where they can be used:

- **Penetration testing**: Penetration testing is a form of black-box testing, meaning an application's functionality is tested without any knowledge of the structure of its internal code, paths, or implementation details. It is focused on how the input matches the output. A penetration test emulates how a cybercriminal or hacker would attempt to get around the security of the app, using the same tools that the attacker would use. The purpose is to see how vulnerable the app is in the cloud environment while running it live in a typical scenario. This sort of testing allows the development team to find any potential weak points or vulnerabilities in the application and correct them before deployment. This is a test that will continue to be run once the application goes live as new threats and new tools and protocols become available to the bad actors of the world. This falls directly into the assurance form of security.

- **Dynamic application security testing (DAST)**: Another form of black-box testing called DAST uses a live system. The programmers running it do not have any knowledge of the system that they will be testing. The testing must use its own resources to test the paths and interfaces that it is designed to test. DAST is often used in conjunction with **Static Application Security Testing (SAST)** so that developers can test both the inside and outside of an app, providing the most comprehensive report possible and ensuring future success.

- **Runtime application self-protection (RASP)**: This is a good test for systems that can change their own security settings, based on the kind of environment they are in and the sort of attack they measure as being used against them. RASP is an excellent way to evaluate the flexibility of an application against various types of threats and check whether one or more threats might pose a larger-than-acceptable danger, based on its assurance measure.

- **Static application security testing (SAST)**: Touched upon briefly in DAST, SAST is a white-box form of testing that takes a closer look at the code and various components of each application. It is the opposite of DAST because SAST has full knowledge of what's going on inside the application, and it is an effective way of measuring verification by finding errors made by programs or specific vulnerabilities such as SQL injection.

- **Vulnerability scanning**: This sort of testing goes along with the white-hat hacker theory of an organization going up against its own systems, using known attacks favored by actual hackers and cybercriminals. Doing so will produce a rating of risk or vulnerability that coincides with the level of assurance that an organization has, related to how effective its application is at defending itself from the ulterior motives of outside forces. The scanning is not a one-time occurrence but, rather, done repeatedly to produce a rating for risk for the entire app.

Quality assurance (QA) can be tough to negotiate in the world of dynamic cloud-native applications. The challenge of any cloud environment is how to adapt to its native architecture. Using QA tools and processes can help meet these challenges head-on and ensure that you are maintaining the big three primary forms of security – that is, assurance, validation, and verification – every time. Services inside a cloud-native app are simple, but when you start mixing services together, the requirements for infrastructure and interactions between the said services get complicated quickly. If you add distributed hosting environments, there are more variables to be tested. QA teams employ parallel testing and automated testing to tackle how complex some cloud-native apps can be.

Having explored how QA tackles the complexities of cloud-native applications and ensures robust security through assurance, validation, and verification, let's now turn our focus to the specific strategies and methodologies employed in software verification to further enhance the reliability and performance of these applications.

Two Approaches to Software Verification

Having applications that are verified is important in the cloud environment, as it ensures that the code, architecture, and design are all done correctly. If you've ever heard the phrase that something is only as strong as its weakest link, this also very much applies to cloud-native software. Any loose ends or faulty bits of code can lead to easy access for bad actors or a simple failure to perform as intended, which is disastrous for businesses and their customers. A weakness in one component can be disastrous for an entire application. If there is no verification process, applications can particularly be at risk of attacks where a user can force their way into the administrator role, gain access to control the application and the underlying infrastructure, and expose data as well. These are the sort of data breaches that wind up making headline news when companies aren't aware anything is wrong, until someone dumps all of their user data on the internet.

There are two main ways that software can be verified – dynamic and static verification. Read on to learn more about them.

Static verification, much like SAST in the *Security Testing and Quality Assurance* section, checks out the software to ensure that it meets the requirements by inspecting the actual code before the software is deployed. Techniques of static verification include the following:

- **Coding conventions verification**: Coding conventions are the guidelines for the particular programming language recommended for an application. All parts of that language and how it is used are checked in this stage.

- **Anti-pattern detection**: Anti-pattern detection looks for any part of the application structure that appears to be effective but usually has more bad results than good ones. If left in the existing code, it could work against the overall goals of the code.

- **Software metrics calculation**: Software metrics refer to the degree to which an app contains a certain property – in this case, how much functionality or consistency it contains.

- **Formal verification**: A formal verification takes on the algorithms contained in an app using a formal proof of the mathematical model. The purpose is to either prove or disprove that the algorithms successfully function as intended.

- **Verification by analysis**: Some or all of these can be incorporated into the analysis verification method, which conducts an entire investigation using logical evaluation, standard textbook calculations, and an investigation methodology to measure and observe the expected results of the application.

Dynamic verification is similar to software testing in that it is done in conjunction with running an application and seeing how it works inside a live environment. The behavior of the application is important, and experimentation with different inputs to see what sort of outputs are generated is commonplace. To comprehensively assess the application's performance, three distinct methods are commonly utilized, each representing a specific type of test:

- **A test in the small**: This is a test that examines a single function of an application or a single class of code. It is also known as a unit test because it focuses on a single unit of source code.

- **A test in the large**: This test takes on a group of classes at once. A module test will take on one module, while an integration test tackles more than one module. In integration testing, modules are put together as a group and measured for compliance or to ensure they have the correct functionality. The biggest format, known as the system test, quite logically looks over the entire system as one unit.

- **An acceptance test**: An acceptance test formally checks the acceptance criteria of software, namely its stress test, which pushes it beyond its normal parameters to see how it handles things such as error handling, availability, and robustness under a heavier-than-normal load, along with a functional test that ensures that the interface works correctly across the board.

Dynamic verification is designed to find any errors that happen as the result of an activity taken by the application or a repetitive action.

Having discussed how dynamic verification involves real-time testing to analyze an application's behavior in a live environment, let's now explore how the third-party review process can complement these efforts, by providing an unbiased assessment of the application's security and functionality from an external perspective.

Third-Party Review Processes

The COVID-19 pandemic woke up a lot of people and raised huge concerns about cybersecurity in cloud environments. With people working from home and leaving their traditional on-premises jobs in great numbers at the same time, the demand for cloud technology spiked lots of lapses in security that led to major issues. Different software publishers have different ideas about what security should look like, how much testing needs to be done, and what standards they should be beholden to. That will not work when you are trying to have everyone work together in one cloud environment that a company uses for all of its business.

Therefore, the future of the cloud must be a trusted third-party review entity to look over all public APIs, as well as any third-party code that is going to be deployed into the cloud environment. This will convince software publishers that their work might be subject to an audit if they do not comply with what is expected of them. This will quell validity and verification concerns for users who fear that their data will suffer leaks, as it moves from one application to another in the same cloud environment. All developers need to be held to the same minimum standard of security and data protection.

The cloud has numerous points of entry that can double as places to attack for those with malicious intent. Different industries are in the process of compiling industry-specific compliance requirements. Companies are also aware that their brand reputations can be on the line if they do business in a cloud environment that doesn't offer robust security practices.

Reviewing software programs by a third party is essential to ensure their security and quality. This procedure entails evaluating the security posture of an application produced by a third-party vendor to ensure that it fulfills the organization's security requirements and standards. The processes to be undertaken while doing a third-party assessment of software applications are as follows:

- **Establish the review's scope**: The first step is to determine the review's scope. This includes identifying the application that requires assessment, the business processes it supports, and the sort of data it processes. In addition, it is crucial to identify the systems that interact with the application.

- **Identify the review team**: A review team comprised of professionals from various disciplines, such as software development, security, infrastructure, and project management, must be assembled. The team should have the requisite abilities and experience to conduct a thorough examination of the application.

- **Create a review framework**: The scope of the review should guide the establishment of the review framework. It should contain the review's objectives, techniques, and instruments.

- **Gather information**: The review team should collect application-related information, such as the application architecture, data flow diagrams, user access restrictions, data security measures, and user roles and permissions.

- **Examine the source code**: The review team should examine the application's source code to find any security flaws or coding problems. They should utilize code review tools that can detect vulnerabilities such as SQL injection, **Cross-Site Scripting (XSS)**, and buffer overflow.

- **Perform a vulnerability assessment**: A vulnerability assessment should be conducted in order to uncover any security flaws in the application. The review team can uncover vulnerabilities using either automated vulnerability assessment technologies or human testing procedures.

- **Perform penetration testing**: Penetration testing should be conducted to evaluate the efficacy of the application's security mechanisms. This entails replicating a real-world attack against the program in order to find any vulnerabilities in its defenses.

- **Evaluate security controls**: The review team should review the application's security controls, such as access controls, data encryption, and data backup and recovery. They should ensure that these controls comply with the security rules and requirements of the firm.

The team conducting the review should communicate its results to the organization. The report should provide a summary of the review procedure, its results, and its suggestions. This report should be used to identify areas in need of improvement and to implement the necessary changes.

In order to conduct a third-party evaluation of cloud-native software applications, you can use a range of technologies to uncover vulnerabilities, evaluate security controls, and verify compliance with industry standards. The following are some of the most important technologies to conduct third-party reviews of cloud-native software applications:

- **Cloud security posture management**: The **Cloud Security Posture Management (CSPM)** tools can help you evaluate the security posture of cloud-native applications by discovering misconfigured and underused resources, as well as other security issues. These solutions can also help you ensure compliance with regulatory standards such as the **Health Insurance Portability and Accountability Act (HIPAA)**, the **Payment Card Industry Security Standards Council (PCI DSS)**, and the **General Data Protection Regulation (GDPR)**.

- **Container security tools**: These tools can assist in evaluating the security posture of cloud-native application containers, identifying vulnerabilities in container image files, scanning for malware, and monitoring the behavior of running containers.

- **Cloud infrastructure security tools**: These tools help you evaluate the security of the cloud infrastructure that supports cloud-native apps. They also assist you in identifying cloud infrastructure security concerns, such as unauthorized access, misconfigured access controls, and unsecured APIs.

- **Web applications**: These tools, built with cloud-native technologies, can help you identify web application vulnerabilities and uncover typical vulnerabilities in online applications, including SQL injection, XSS, and broken authentication and session management.

- **Configuration auditing and compliance tools**: These tools can assist in ensuring that cloud-native apps are correctly set up and adhere to industry standards. They also assist you in identifying configuration errors, such as incorrectly configured security groups, non-compliant access rules, and non-compliant encryption techniques.

- These tools can assist you in identifying typical security flaws, such as buffer overflows, SQL injection, and XSS.

- Building on the insights gained from third-party review processes, which provide an external evaluation of application security, you will now shift our focus to the specific measures necessary to secure APIs, a critical component that interfaces directly with other applications and services.

Measures to Secure APIs

In the term **Application Programming Interface** (**API**), "application" refers to any software that has a specific function. The interface is the point of contact and the exchange between two different applications, with the point of contact being how the two pieces of software interact with each other, based on the information requested and the response to those requests. API documentation refers to the information that developers use to structure and maintain those requests and responses. APIs work much like a client and a server. The application that makes the request is the client and the application that gives the response is the server.

The biggest reason for stringent API security is that companies and individuals use APIs every minute of the day to transfer data and connect different services. If an API is vulnerable to data loss, it can have disastrous consequences. API breaches are not something that only happens to small businesses with lax security. In the past decade, Yahoo, LinkedIn, Facebook, and the United States Post Office have all been victimized by cybercriminals exploiting vulnerable APIs; Yahoo's was among the most negligent, taking three years to be revealed. In 2016, Yahoo admitted that more than 3 billion user accounts had been compromised, with a lot of personal information exposed.

Having an insecure API is a huge threat to cloud environments because cloud services are excessively dependent on APIs for their operation and functionality. It is fair to imagine APIs as the backbone of the cloud environment as well as its deployment. Many functions in the cloud would not work correctly without APIs, including provisioning, autoscaling, authentication, authorization, and the operations of the cloud application itself.

The danger increases when you consider that many cloud applications that employ APIs are within reach of public use or are depended on by other applications. In order to reduce the risk, there needs to be a commitment from the application owner as well as the cloud provider to make sure that tight, dedicated security controls are in place. This includes utilizing powerful encryption and ensuring strict access to authorization to use APIs.

Emerging technologies are something to keep an eye on when it comes to protecting APIs. Newer APIs have more features and robust capabilities for management, along with new ways to perform rapid provisioning and autoscaling, along with news offerings such as virtual image security.

There are two types of APIs usually used in cloud systems. The first is known as **Representational State Transfer (REST)** and the second is **Simple Object Access Protocol (SOAP)**:

- REST is a type of software architecture that describes how components, connectors, and data points are used for various web applications. They use the HTTP protocol and support numerous data formats, including XML and JSON.

- SOAP is both a protocol and a standard for exchanging data between two or more web services. The most common is HTTP, but **File Transfer Protocol (FTP)** is also popular. SOAP only allows data that is XML-formatted, and there is no data caching involved. It typically has worse performance and not as much scalability as REST.

If unsecured APIs are introduced to an environment, particularly if those APIs are integrated outside of an application's security protocol, then all data used by that application risks being either exposed, stolen, corrupted, or lost. When external APIs are involved, any cloud security worker must make sure that the external applications receive the same total evaluation and testing as the applications that exist inside the cloud environment. That means that all three steps of assurance, validation, and verification must be taken and completed with satisfactory results. Depending on whether the API uses REST or SOAP, everything has to be tested and validated to meet the security requirements of both an organization and the industry.

A lot of cloud-based systems and web applications are routinely composed of lots of different types of software leveraging API calls, external data sources, different components from different manufacturers, and so on. Integrating external API calls and other web services allows apps to access large numbers of data sources from outside the cloud environment, while not having to worry about storing or taking care of data in local portals. However, as stated previously, that flexibility comes with added concerns, most notably that the API relies on code that has not been checked out by the organization running the cloud environment and integrates data that originates outside of its parameters.

That means that cloud service professionals don't know or perhaps even cannot know how these sources test and validate their code, what the data sources actually are, what their security models look like, and what their design architecture appears to be. This is nothing new; open source software has been around for decades with the idea that the entire community using them does so with good intentions, and alterations are constantly made for the improvement of everyone.

Lots of developers who operate in cloud environments rely heavily on open source applications, frameworks, and tools to make their jobs easier and give them more access to what they want to build. The element of mutual trust is built into open source software at an intrinsic level, as most are used in lots of different industries with a lot of specific testing and analysis by the organizations employing them. With rare exceptions, this means open source software not only is very secure but also, when a vulnerability or a bug is recognized, the community will embrace the spirit of collaborative coding to find the errors, get them patched or replaced, and re-upload the software so that it can be confirmed as successful for everyone involved.

As a cloud security professional, you will have to adjust to the wants and needs of the organization you work for when it comes to open source applications. Those operating on a fixed budget, such as nonprofits and academic institutes, typically see open source software as a boon to be harnessed time and again. Conversely, government agencies and large corporations are more likely to see open source as a problem waiting to happen, given that they are constantly preaching to their customers about their security, professionalism, and threat-level analysis. Revealing that one or more of the items inside a cloud environment are also being used by companies around the world and are frequently being shared may not be acceptable to most stockholders and boards of directors.

APIs expose an application's functionality to external parties, making API security essential for application security. A vulnerable API can result in data breaches, illegal access, and other security problems. The following are the measures an expert in application security would take to secure APIs:

- **Authentication**: This is the process of confirming the identity of an API user. By requiring legitimate credentials to access the API, authentication can prevent unwanted access. Several technologies, including API keys, OAuth 2.0, and OpenID Connect, can be used to enable API authentication.

- **Authorization**: This is the process of allowing access to particular resources or actions depending on a user's role and permissions. Authorization can ensure that users only have access to the resources they require and prevent them from gaining unauthorized access to other resources. Authorization can be done with either **Role-Based Access Control** (**RBAC**) or **Attribute-Based Access Control** (**ABAC**).

- **Encryption**: Encrypted sensitive data transferred between the client and the API encryption safeguards data against illegal access, interception, and modification. **Transport layer security** (**TLS**) can be used to implement API encryption.

- **Input validation**: Implementing input validation is essential to prevent malicious data from exploiting API vulnerabilities. By enforcing input validation, you ensure that the API only accepts appropriate and expected inputs, thereby mitigating risks associated with attacks such as SQL injection, XSS, and buffer overflow. Utilizing frameworks and libraries that offer built-in input validation features can simplify the process of securing APIs against such threats. This approach helps maintain the integrity and security of the API by filtering out potentially harmful data before it can cause any damage.

- **Rate restriction**: This can prevent malicious actors from overwhelming an API with requests. Rate restriction can restrict the amount of API calls made by a user or IP address to prevent the API from becoming overloaded. By using API management platforms or bespoke code, rate limitations can be established.

- **Logging and monitoring**: This can provide visibility into API usage and detect questionable activities. API queries and responses can be logged, while monitoring can notify security personnel of potential security vulnerabilities. Using logging frameworks, monitoring tools, and **Security Information and Event Management (SIEM)** systems, logging and monitoring can be deployed.

- **Routine API security audits**: These audits can detect API vulnerabilities and weaknesses. Using vulnerability scanners, penetration testing, and code reviews, API security evaluations can be carried out.

APIs must be secured to ensure application security. Incorporating authentication, authorization, encryption, input validation, rate restriction, logging, and monitoring, as well as conducting routine API security assessments, can protect APIs from unwanted activities and assure their security.

API security is a very comprehensive and important topic, and you have covered a lot over the last few pages. Let's summarize and provide a list of how to approach API security in your organization. Securing APIs is crucial for maintaining the integrity and security of cloud services. A cloud security architect can follow these steps to ensure robust API security:

1. Define security requirements:

    - Start by understanding and defining the specific security needs related to the APIs. This includes identifying the data sensitivity handled by the APIs and the compliance requirements that need to be met.

2. Implement strong authentication and authorization:

    - Use strong authentication mechanisms to verify the identity of users and systems interacting with APIs. OAuth, OpenID Connect, and **JSON Web Tokens (JWT)** are common methods for securing API access.

    - Implement robust authorization practices to ensure that authenticated users have appropriate permissions. This often involves RBAC or ABAC.

3. Secure data:

    - Encrypt sensitive data transmitted to and from APIs using TLS to prevent interception during transmission

    - Ensure that any sensitive data stored by the API is encrypted at rest to protect it from unauthorized access

4. Use API gateways:

    - Deploy API gateways to manage API traffic, which can provide an additional layer of security such as rate limiting, IP whitelisting, and attack detection

    - API gateways can also handle authentication and authorization, offloading these tasks from individual services

5. Implement input validation:

    - Validate all input received through APIs to prevent common vulnerabilities such as SQL injection, XSS, and command injection. Ensure that inputs are checked against a strict type, length, format, and range.

6. Audit and logging:

    - Maintain comprehensive logs of all API interactions, including access logs and transaction logs. This can help to identify and respond to security incidents effectively.

    - Ensure that logs are stored securely and are only accessible by authorized personnel.

7. Regular security testing:

    - Conduct regular security assessments and audits of the APIs, including penetration testing and vulnerability scanning

    - Use automated tools to continuously test and monitor for new vulnerabilities

8. API version management and deprecation strategy:

    - Regularly update and version APIs to incorporate security patches and improvements. Clearly communicate API deprecation policies and timelines to consumers.

    - Ensure backward compatibility or provide migration paths to newer API versions to avoid exposing old, less secure APIs.

9. Rate limiting and throttling:

    - Implement rate limiting and throttling to protect APIs from being overwhelmed by too many requests, which can lead to **denial of service (DoS)** attacks

10. Educate and train developers:

    - Educate developers about the best practices in API security. Encourage secure coding practices and awareness of the latest security threats and mitigation techniques.

By following these steps, a cloud security practitioner can significantly enhance the security of APIs, protecting both the data they handle and the services they interact with from potential threats and breaches.

Summary

Regardless of the software, component, or interface being brought into your organization's cloud environment, it is imperative that the three primary forms of security – assurance, validation, and verification – need to be checked and completed every time. While different organizations have different opinions on how much security is necessary for various components that link together to make a company operate efficiently, it is imperative that any cloud security professional employed by an organization running its own cloud-native environment always be on the lookout for warning signs of improper access or unchecked code coming into the said environment. The smallest change can result in easy avenues for bad actors to gain entry to the cloud environment, wreaking havoc on your organization.

The CCSP certification covers essential API security and software validation and assurance topics. To pass the CCSP exam, a candidate must be familiar with the various types of APIs, API security controls, and API security testing methodologies, as well as the **Software Development Life Cycle** (**SDLC**) fundamentals (covered in *Chapter 13, Secure Software Development Life Cycle*, software testing methodologies, software security testing methodologies, software security best practices, and software security compliance requirements.

The next chapter will cover the important specifics of traditional cloud application architecture, with a focus on essential security components such as WAF, DAM, and API gateways, as well as cryptography, sandboxing, and securing virtualized applications.

The next chapter will cover the important specifics of traditional cloud application architecture, with a focus on essential security components such as **Web Application Firewall** (**WAF**), **Data Access Management** (**DAM**), and **Application Programming Interface** (**API**) gateways, as well as cryptography, sandboxing, and securing virtualized applications.

Exam Readiness Drill – Chapter Review Questions

Apart from a solid understanding of key concepts, being able to think quickly under time pressure is a skill that will help you ace your certification exam. That is why working on these skills early on in your learning journey is key.

Chapter review questions are designed to improve your test-taking skills progressively with each chapter you learn and review your understanding of key concepts in the chapter at the same time. You'll find these at the end of each chapter.

> **How to Access These Materials**
>
> To learn how to access these resources, head over to the chapter titled *Chapter 25, Accessing the Online Resources*.

To open the Chapter Review Questions for this chapter, perform the following steps:

1. Click the link – `https://packt.link/CCSPE1_CH14`.

    Alternatively, you can scan the following **QR code** (*Figure 14.1*):

Figure 14.1 – QR code that opens Chapter Review Questions for logged-in users

2. Once you log in, you'll see a page similar to the one shown in *Figure 14.2*:

Figure 14.2 – Chapter Review Questions for Chapter 14

3. Once ready, start the following practice drills, re-attempting the quiz multiple times.

Exam Readiness Drill

For the first three attempts, don't worry about the time limit.

ATTEMPT 1

The first time, aim for at least **40%**. Look at the answers you got wrong and read the relevant sections in the chapter again to fix your learning gaps.

ATTEMPT 2

The second time, aim for at least **60%**. Look at the answers you got wrong and read the relevant sections in the chapter again to fix any remaining learning gaps.

ATTEMPT 3

The third time, aim for at least **75%**. Once you score 75% or more, you start working on your timing.

> **Tip**
> You may take more than **three** attempts to reach 75%. That's okay. Just review the relevant sections in the chapter till you get there.

Working On Timing

Target: Your aim is to keep the score the same while trying to answer these questions as quickly as possible. Here's an example of how your next attempts should look like:

Attempt	Score	Time Taken
Attempt 5	77%	21 mins 30 seconds
Attempt 6	78%	18 mins 34 seconds
Attempt 7	76%	14 mins 44 seconds

Table 14.1 – Sample timing practice drills on the online platform

> **Note**
> The time limits shown in the above table are just examples. Set your own time limits with each attempt based on the time limit of the quiz on the website.

With each new attempt, your score should stay above **75%** while your "time taken" to complete should "decrease". Repeat as many attempts as you want till you feel confident dealing with the time pressure.

15

Application-Centric Cloud Architecture

This chapter highlights the various ways to keep cloud applications safe and the tools needed to do so. The security for cloud apps should work the same regardless of where they're hosted, including different platforms, deployment models, and service models. The diverse nature of applications introduces complexities and uncertainties regarding their behavior and potential vulnerabilities. This highlights the critical importance of sandboxing and application virtualization for safety in cloud application environments. Additionally, the chapter discusses additional safety tools such as **Web Application Firewalls (WAFs)**, **Database Activity Monitoring (DAM)**, XML firewalls, and **Application Programming Interface (API)** gateways. These tools are important for ensuring that cloud apps are protected from various cyber threats. Furthermore, it delves into how cryptography can keep data safe, both at rest and in transit.

By the end of this chapter, you will be able to confidently answer questions on the following:

- Supplemental security components such as WAF, DAM, XML firewalls, and API gateways
- Cryptography
- Sandboxing
- Application virtualization and orchestration

Supplemental Security Components

When developing applications, relying solely on the application itself for security is insufficient, and you often need extra (or supplemental) security components. Supplemental security components, such as WAF, DAM, and XML firewalls, are necessary to address specific vulnerabilities or threats that may not be adequately covered by primary security measures alone. They add redundancy to the organization's defense mechanisms, enhancing its resilience against evolving cybersecurity challenges. You will now study in detail some of the prominent supplemental security components with respect to the CCSP exam.

WAFs

A WAF has similarities to a traditional network firewall, but they differ in scope and purpose. A network firewall protects the broader network by filtering traffic based on the source IP address, destination IP address, port number, and protocol type. Conversely, a WAF, an OSI layer-7 firewall, is dedicated to safeguarding a web application from specific application-layer threats.

A WAF is typically deployed at the perimeter of a network as a proxy between the web server of a web application and the client (users or devices accessing the application). Most major cloud service providers offer WAF services as part of their cloud security offerings. Cloud-based WAF services offer advantages such as scalability, ease of deployment, and integration with other cloud services. You can configure rules, monitor traffic, and mitigate threats to ensure the security of your web applications.

Operating at the application layer of the network stack, a WAF specifically focuses on HTTP and HTTPS traffic, analyzing the content of web requests and responses. It can effectively detect and mitigate attacks such as SQL injection, **Cross-Site Scripting (XSS)**, **Cross-Site Request Forgery (CSRF),** directory traversal, buffer overflows, cookie-based attacks, and **Distributed Denial-Of-Service (DDOS)** attacks.

DAM

DAM focuses on monitoring and analyzing activities that occur within a database system. The primary purpose of DAM is to provide real-time visibility into database transactions, detect suspicious or unauthorized activities, and help organizations ensure the security and integrity of their sensitive data. DAM may extend its monitoring capabilities to network connections, inspecting and analyzing the traffic between clients and database servers.

DAM solutions log and audit database activities, providing a detailed record of user logins, queries, data modifications, and more. This comprehensive monitoring allows organizations to swiftly identify and respond to unauthorized or suspicious activities through real-time alerts and notifications, making it an effective detective control measure. DAM systems play a role in ensuring that only authorized users with appropriate roles and privileges access the database, safeguarding data integrity, security, confidentiality, and compliance with industry regulations such as HIPAA, GDPR, and SOX.

DAM tools often include the ability to analyze SQL queries and detect anomalies or patterns indicating potential security risks or malicious intent. This helps in identifying potential SQL injection attacks or other malicious activities. While SQL databases are commonly associated with DAM, these tools are adaptable to various database types, including NoSQL, big data platforms, cloud-based databases, and file-based databases. DAM's goal is to offer comprehensive monitoring and security for sensitive data across diverse database technologies.

Advanced DAM solutions employ behavior analysis to establish baseline activity patterns, enabling the detection of unusual deviations that may signal potential security threats. In the event of security incidents, DAM logs are invaluable for forensic analysis, helping organizations understand the nature and cause of breaches and ultimately strengthening their data security measures.

XML Firewalls

XML is used for structured data representation and interchange between systems. It is used in web services, configuration files, document formats, and various domains where organized and standardized data exchange is essential. XML firewalls play a role in ensuring the security, integrity, and reliability of XML-based communication.

XML firewalls carry out several functions, starting with **content filtering**. They scrutinize XML data payloads within messages, requests, and responses, allowing for the filtering, validation, and transformation of XML content to ensure compliance with security policies and standards. These firewalls are also equipped to detect and prevent common security threats associated with XML, such as XML injection attacks, schema validation issues, and malicious payloads.

Access control is another security feature of XML firewalls. They enforce access control policies to ensure that only authorized users or systems are permitted to access or interact with specific XML services or APIs. This includes mechanisms for user authentication and authorization to ensure that users have the necessary permissions to access and modify XML data.

One key function of XML firewalls is **threat protection**. By validating and sanitizing incoming and outgoing XML data, they safeguard against XML-specific vulnerabilities and attacks, such as XML injection, ensuring the integrity of the data exchanged.

These firewalls also play a pivotal role in **transformation**. They ensure that XML messages adhere to specified schemas and standards, enabling them to modify or transform content according to security policies or to facilitate interoperability between systems with different data formats. Encryption and decryption capabilities are integrated into XML firewalls to secure the transmission of sensitive data within XML messages, safeguarding confidentiality and preventing unauthorized interception.

Monitoring and logging functionalities provided by XML firewalls are essential for compliance, troubleshooting, and security incident detection. By tracking and auditing XML traffic, organizations can gain insights into potential security threats and maintain a record of activities for regulatory compliance purposes. Additionally, XML firewalls contribute to mitigating DDoS attacks by intelligently filtering and prioritizing incoming XML requests, ensuring the continued availability of web services.

Throttling and rate limiting mechanisms implemented by XML firewalls control the rate of incoming requests, preventing abuse and optimizing resource allocation. They also enforce security policies, ensuring that web services comply with industry standards, access controls, and data validation requirements. In the context of modern API-based architectures, XML firewalls play a crucial role in securing APIs that utilize XML as the data format. They protect against API-specific threats, ensure data integrity, and enforce security policies, contributing to the overall security posture of web services relying on XML-based communication.

XML firewalls also employ predefined threat signatures or patterns to identify, and block known XML-based attacks and vulnerabilities, strengthening the overall security posture of XML-based communication. They can offer load balancing and routing capabilities, distributing incoming XML traffic across multiple servers or service instances, thereby improving performance and system reliability.

API Gateways

An API gateway serves as an intermediary between client applications and a collection of microservices, web services, or APIs. It acts as a central point of control, management, and security for all API interactions. Its primary role is to streamline and optimize API interactions. One of its core functions is API routing, where it directs incoming requests to the appropriate microservices or backend services based on specific criteria such as URL or path, simplifying client access to various services and enhancing the overall efficiency of the system.

API gateways also play a pivotal role in load balancing, ensuring that incoming requests are evenly distributed across multiple instances of a microservice, thereby improving performance and enhancing system reliability. API gateways also implement caching mechanisms, storing frequently requested data to reduce the load on backend services and enhance response times. Scalability is a core feature, ensuring that they can seamlessly manage increased API traffic as a system grows.

Security is a paramount concern, and API gateways address this very well:

- **Authentication and authorization**: API gateways verify the identity of clients through methods such as API keys or tokens and ensure that authenticated clients possess the necessary permissions to access specific resources. This helps prevent unauthorized access and ensures a controlled and secure interaction with the API.

- **Rate limiting and throttling**: These mechanisms control the rate of incoming requests, preventing abuse and ensuring fair resource usage. By setting limits on the number of requests a client can make within a specified timeframe, API gateways mitigate the risk of **denial-of-service (DoS)** attacks and protect against resource exhaustion.

- API gateways contribute to **data security** through data validation and transformation. They validate incoming data to ensure it conforms to the expected format, guarding against common security vulnerabilities such as injection attacks. Additionally, gateways perform data transformations as needed, facilitating compatibility with backend services while addressing security concerns.

- **Encryption**: secures the transmission of data between clients and servers. When a client sends an HTTPS request to the API gateway, the data is encrypted using TLS.

- **TLS termination**: This can be set up at the API gateway, allowing it to decrypt incoming HTTPS requests. Acting as a TLS termination point, the gateway decrypts the encrypted data received from the client, allowing it to inspect the content of the request. After inspection, the gateway may re-encrypt the data before forwarding it to backend services. TLS termination enables the gateway to perform functions such as content inspection, modification, and routing based on the decrypted data while maintaining secure communication between clients and the gateway.

- **Logging and monitoring**: These are integral to API security, and API gateways generate logs capturing details of API requests and responses. These logs are instrumental in monitoring and auditing API activity, helping to detect and investigate security incidents. Real-time insights into traffic patterns and potential threats contribute to a proactive security posture. API gateways also address **Cross-Origin Resource Sharing (CORS)** concerns by enforcing policies that control which domains are permitted to make requests to the API. This prevents unauthorized cross-origin requests, fortifying the security of web applications that consume the API. Integration with WAFs further strengthens API security by providing protection against common web application attacks. This includes safeguards against SQL injection, XSS, and other vulnerabilities identified in the OWASP Top Ten.

Additionally, API gateways play a role in enforcing compliance with security standards and industry regulations. This includes adherence to authentication protocols, data encryption standards, and privacy regulations, contributing to a comprehensive security framework for APIs. In summary, API gateways function as central security perimeters for APIs, implementing a suite of protective measures to enhance the overall security and reliability of API interactions.

Cryptography

Cryptography involves both the practice and study of safeguarding communication and information through the transformation of data into an unreadable format, accessible solely to authorized individuals possessing the requisite decryption keys. Utilizing mathematical algorithms and principles, cryptography converts **plaintext** (readable data) into **ciphertext** (unreadable data) and vice versa. Its core objectives include ensuring the confidentiality, integrity, and authenticity of data, making it the backbone for securing sensitive information and communication channels against unauthorized access:

- **Encryption**: Converts plaintext into ciphertext.
- **Decryption**: Converts ciphertext back into plaintext.

As cloud services often involve multiple tenants sharing the same infrastructure, cryptography helps in ensuring the security and isolation of data between different tenants, preventing unauthorized access.

Integrating cryptography across all phases of the software development life cycle is a fundamental practice to establish comprehensive security. Cryptography should not be viewed as a singular, isolated step; rather, it should seamlessly integrate into the entire development process. It protects information, both when it's at rest (data-at-rest encryption) residing on storage devices and in databases, and when it's in motion (data-in-motion encryption), traversing networks and communication channels.

Data-at-Rest Encryption

Data security is a paramount concern in cloud computing, and encrypting data at rest is a fundamental practice to ensure the confidentiality and integrity of sensitive information stored on cloud platforms. **Cloud service providers (CSPs)** play a pivotal role in offering native encryption features that safeguard data stored in the cloud environment. These features often include **server-side encryption (SSE)**, where the CSP takes responsibility for encrypting the data at rest, utilizing encryption keys managed by the provider.

Major cloud providers also provide the option of customer-managed keys, where encryption keys are controlled and managed by the cloud customers, rather than by the cloud service provider. It provides the customers with better control over their data security and access.

In addition, organizations may opt for client-side encryption, where users encrypt their data locally before uploading it to the cloud. This approach grants users greater control over the encryption keys and adds an extra layer of security. Client-side encryption is particularly beneficial when handling highly sensitive information.

Data-in-Motion Encryption

Encrypting data in motion, or during transit, is a fundamental practice to fortify the confidentiality and integrity of information as it traverses various components within cloud environments and extends beyond the cloud's boundaries. A cornerstone of this approach is the implementation of well-established protocols such as **Transport Layer Security (TLS)** or **Secure Sockets Layer (SSL)**. These protocols establish encrypted channels, shielding data from eavesdropping and unauthorized tampering during communication between clients and cloud services.

Furthermore, the adoption of end-to-end encryption is pivotal, ensuring that data remains protected throughout its entire journey, from its source to its ultimate destination. This approach guarantees that only the intended recipient possesses the necessary decryption keys to access the encrypted data. **Virtual Private Networks (VPNs)** add an extra layer of security by creating secure communication channels, particularly valuable when data traverses public networks.

The selection of robust encryption algorithms such as **Rivest-Shamir-Adleman (RSA)** and **Elliptic Curve Cryptography (ECC)** is instrumental in enhancing the overall security of data in motion. Cloud service providers offer built-in encryption services tailored for data in motion, facilitating secure communication between various components of the cloud infrastructure, both within and outside the cloud. Monitoring and logging mechanisms play a role, enabling organizations to track and analyze network traffic, promptly identifying potential security incidents or anomalies in data transmission.

Key Management

The prevention of unauthorized access is a direct outcome of robust key management practices. If cryptographic keys were to fall into the wrong hands, the security of the encrypted data would be compromised. Therefore, secure **key management** is a frontline defense against data breaches and unauthorized disclosures.

Managing encryption keys involves the secure generation, storage, distribution, rotation, and disposal of cryptographic keys used in various processes such as data encryption, authentication, and digital signatures. CSPs often provide key management services, and some allow users to bring their own encryption keys (BYOK), offering an added layer of control.

While many CSPs offer the convenience of managing keys, situations demanding heightened security necessitate user-generated and controlled keys. Recognizing the diverse security needs of their clients, CSPs provide tools and infrastructure for managing encryption keys, empowering cloud customers to maintain a higher degree of control over the encryption process.

The process begins with the secure generation of cryptographic keys, employing robust algorithms to create keys in an unpredictable manner that resists potential attacks. Equally critical is the secure storage of these keys, often facilitated by cloud providers through dedicated key management services and **hardware security modules (HSMs)**. However, HSMs are not commonly used by cloud customers due to their high cost and challenges with implementation and management. Their usage is primarily driven by regulatory requirements.

Efficient key distribution mechanisms must be established to provide authorized users and systems with secure access to cryptographic keys while avoiding insecure practices. Regular key rotation is essential, mitigating risks associated with compromised keys, and cloud environments often offer automated tools for seamless rotation processes. The ability to promptly revoke and replace compromised or obsolete keys is crucial in maintaining the security of encrypted data.

Role-Based Access Control (RBAC) principles play a pivotal role in controlling access to cryptographic keys, ensuring that permissions are granted based on the principle of least privilege. Robust monitoring and auditing practices help track key usage, detect anomalies, and ensure compliance, providing insights into potential security incidents.

Lastly, secure procedures for the disposal of keys that are no longer needed must be in place. This includes practices for securely deleting or archiving keys to prevent potential misuse. In essence, effective key management demands secure practices at every stage of the key lifecycle, including defining clear policies for key practices, including generation, usage, rotation, and retirement.

Separation of Duties

The separation of duties between the service handling encrypted data and the service managing encryption keys is a fundamental aspect of enhancing security. It ensures that no single entity holds all the keys to the data kingdom. Even if a security breach occurs, the compromise of encryption keys is mitigated by this separation, enhancing the overall security posture.

Sandboxing

Sandboxing, a security mechanism, is designed to confine and execute untrusted or potentially malicious code within a controlled and isolated environment, commonly referred to as a "sandbox." A sandbox functions as a secure and restricted area for running programs, applications, or processes, ensuring that they cannot adversely affect the broader system.

The primary characteristic of sandboxing lies in its ability to isolate code execution to a confined environment. This strict containment ensures that the code cannot access or modify resources beyond the predefined boundaries of the sandbox.

You are probably already aware of the various types of sandboxing, but it's worth reviewing them as each is tailored to specific use cases and security requirements:

- **Hardware-based sandboxing** involves physical isolation using dedicated hardware, suitable for high-security environments where complete physical separation is crucial.

- **Software-based sandboxing** achieves isolation through software mechanisms, enabling testing in controlled environments without requiring dedicated hardware.

- **Network sandboxing** isolates network traffic to prevent unauthorized access, restricting communication within a network and safeguarding data.

- **Operating system-level sandboxing** isolates processes and applications at the OS level, enhancing security by confining applications within designated spaces.

- **Application sandboxing** confines the actions of specific applications, preventing untrusted apps from impacting the system at large.

- **Web browser sandboxing** isolates browser processes, preventing web-based threats from compromising the entire system during online browsing.

Sandboxing finds widespread use in security testing, particularly for applications, software, or code. You can leverage sandboxes to analyze and observe the behavior of potentially harmful code without exposing the actual system to any associated risks.

In secure SDLC, developers benefit from sandboxes mainly in the development and testing phases. It provides a controlled environment for testing and debugging code, enabling experimentation with new features or changes without impacting the production environment.

Sandboxing serves as a critical tool for the analysis and understanding of malware. You can use sandboxes to execute malicious code in an isolated environment, allowing the study of its behavior without jeopardizing the overall system.

Email security solutions often integrate sandboxes to open and analyze email attachments within a controlled environment. This proactive approach aids in identifying and containing potential threats before they can harm the recipient's system.

Sandboxing becomes especially crucial in cloud computing due to multitenancy. Sandboxing is essential for isolating workloads, ensuring that the activities of one customer do not impact the performance or security of another. Sandboxing provides a safe learning environment for users to experiment with cloud services without impacting the operational stability of the shared infrastructure.

Sandboxing in the cloud can be achieved through a variety of technologies, including virtualization, containers, microservices and **function-as-a-service (FaaS)**. Virtualization creates isolated instances known as **Virtual Machines (VMs)** within a single physical server. Each VM operates independently, running its own operating system and applications, allowing multiple sandbox environments to coexist on the same hardware without interference. Network isolation is achieved by assigning each VM its own virtual network, enhancing the security of sandboxed activities. Containers provide lightweight and portable environments to isolate applications and services, whereas microservices architectures decompose applications into smaller, autonomous components. Additionally, FaaS platforms such as AWS Lambda and Azure Functions facilitate serverless code execution within isolated environments.

Application Virtualization and Orchestration

Application virtualization and orchestration, demonstrated by solutions such as Kubernetes, Docker Swarm, and Amazon ECS, are reshaping the landscape of cloud computing. Application virtualization is a technique that enables applications to run in isolated environments, abstracted from the underlying operating system and hardware. This abstraction allows applications to be portable and independent of the host system, offering several benefits in cloud computing. Virtualization can run applications on different operating systems or versions, providing compatibility and flexibility for cloud customers.

Virtual containers provide a controlled execution environment, enabling strict management of permissions, access controls, and interactions with the host system. Applications operate within predefined boundaries, limiting interference with other applications and access to sensitive resources for enhanced security.

Application virtualization isolates applications, preventing conflicts and dependencies, reducing the attack surface, and minimizing the impact of security breaches within one virtual container on other applications or the host system. Virtual containers serve as secure sandboxes for security testing, allowing isolation of applications. Application virtualization streamlines patching and updates by allowing them to be applied within the virtual container. This reduces the risk of introducing vulnerabilities or disruptions to the host system, maintaining a more secure software environment with the latest security patches.

In the development phase, application virtualization ensures consistency in environments. Developers work with standardized virtual containers, minimizing security issues due to differences between development and production environments.

In **Mobile Device Management (MDM)**, application virtualization allows for the creation of isolated environments, separating corporate and personal data on mobile devices. This ensures that business applications and data are kept secure and separate from the user's personal applications, reducing the risk of data leaks and unauthorized access.

Application orchestration involves the automated management, deployment, and scaling of applications in a cloud environment. It ensures that applications are provisioned, configured, and scaled based on defined policies and requirements. Orchestration tools can dynamically scale applications in response to changing workloads. This elasticity is essential for maintaining performance and cost-efficiency. Orchestration streamlines the deployment and management of complex applications, reducing manual intervention and potential errors. It enforces consistency in application deployments, configurations, and updates, enhancing reliability and reducing the risk of misconfigurations.

In cloud computing, application virtualization and orchestration work hand in hand to enhance the deployment and management of applications. Here are some practical applications:

- **Containers**: Utilizing application virtualization, technologies such as Docker create portable, lightweight containers capable of running anywhere. These containers encapsulate applications and dependencies, providing a consistent and portable runtime environment. Operating in isolated user spaces, containers efficiently share the host system's operating system kernel, optimizing resource utilization and enhancing portability. For streamlined deployment and management of containerized applications, orchestration platforms such as Kubernetes automate tasks such as scaling. Kubernetes, a container orchestration tool, plays a crucial role in coordinating and managing containerized applications, ensuring efficient operation. Containers, integral to DevOps practices and CI/CD pipelines, contribute to modern software development by offering efficiency, security, and consistency in deployment.

- **Microservices**: Application virtualization forms the groundwork for creating microservices, which are self-sufficient components of an application operating independently and with loose coupling. Orchestration tools play a pivotal role in managing the deployment and scaling of these microservices, fostering agility and rapid development. In the microservices architecture, applications are broken down into small, independently deployable services that communicate through well-defined interfaces. This approach enhances flexibility, scalability, and ease of maintenance, empowering teams to develop, deploy, and scale services independently.

In conclusion, application virtualization and orchestration are fundamental to the modern cloud computing landscape. They empower cloud users with flexible, efficient, and secure application management, enabling organizations to harness the full potential of cloud resources and deliver reliable, scalable applications.

Summary

In this chapter, you learned about the critical aspects of securing cloud application architecture within the cloud computing's dynamic landscape. You explored multifaceted dimensions, including confidentiality, integrity, and availability. The chapter emphasized the importance of a defense-in-depth approach.

The chapter discussed supplemental security components as layers fortifying cloud applications. It detailed the functions of WAF, DAM, XML firewalls, and API gateways. These components worked synergistically to enhance system resilience and protect against various threats, from application-layer attacks to unauthorized database activities.

The chapter highlighted cryptography's role in maintaining data security at rest and in transit. It emphasized the integration of encryption across all phases of the software development life cycle, encompassing encryption of data at rest and data in motion. Key management was underscored as a frontline defense against data breaches, focusing on secure key generation, distribution, rotation, and disposal.

The chapter provided insights into sandboxing, a security mechanism confining and executing untrusted code within controlled environments. It elucidated sandboxing's applications in security testing, development, malware analysis, and email security. The chapter emphasized virtualization's role in simplifying and enhancing sandboxing in cloud environments, including hardware-based and software-based approaches.

The chapter explored application virtualization and orchestration as integral concepts reshaping the cloud computing landscape. It explained how application virtualization enabled the creation of isolated environments for portable and independent application execution, enhancing security through controlled permissions and access. The chapter clarified that application orchestration automated the deployment and scaling of applications, ensuring efficiency, reliability, and adherence to policies.

The chapter discussed practical applications of application virtualization and orchestration, such as containers and microservices. It highlighted container technologies such as Docker for their portability, consistency, and efficient resource utilization. The chapter portrayed microservices architecture, enabled by application virtualization, as a paradigm fostering agility, scalability, and independent service deployment.

The next chapter will discuss how **Identity and Access Management (IAM)** solutions play a pivotal role in securing organizations. You will cover key areas such as identity providers, federated identities, secret management, and other crucial IAM solutions.

Exam Readiness Drill – Chapter Review Questions

Apart from a solid understanding of key concepts, being able to think quickly under time pressure is a skill that will help you ace your certification exam. That is why working on these skills early on in your learning journey is key.

Chapter review questions are designed to improve your test-taking skills progressively with each chapter you learn and review your understanding of key concepts in the chapter at the same time. You'll find these at the end of each chapter.

> **How to Access These Materials**
>
> To learn how to access these resources, head over to the chapter titled *Chapter 25, Accessing the Online Resources*.

To open the Chapter Review Questions for this chapter, perform the following steps:

1. Click the link – https://packt.link/CCSPE1_CH15.

   Alternatively, you can scan the following **QR code** (*Figure 15.1*):

Figure 15.1 – QR code that opens Chapter Review Questions for logged-in users

2. Once you log in, you'll see a page similar to the one shown in *Figure 15.2*:

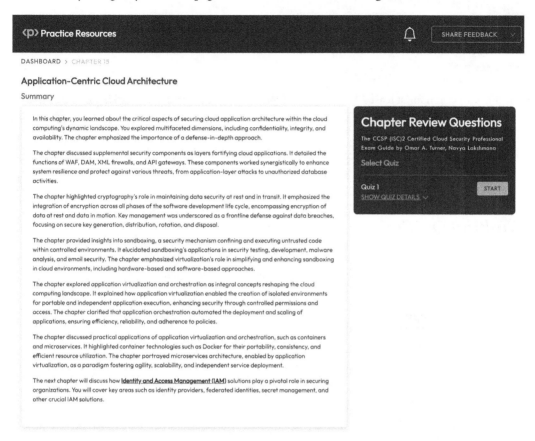

Figure 15.2 – Chapter Review Questions for Chapter 15

3. Once ready, start the following practice drills, re-attempting the quiz multiple times.

Exam Readiness Drill

For the first three attempts, don't worry about the time limit.

ATTEMPT 1

The first time, aim for at least **40%**. Look at the answers you got wrong and read the relevant sections in the chapter again to fix your learning gaps.

ATTEMPT 2

The second time, aim for at least **60%**. Look at the answers you got wrong and read the relevant sections in the chapter again to fix any remaining learning gaps.

ATTEMPT 3

The third time, aim for at least **75%**. Once you score 75% or more, you start working on your timing.

> **Tip**
>
> You may take more than **three** attempts to reach 75%. That's okay. Just review the relevant sections in the chapter till you get there.

Working On Timing

Target: Your aim is to keep the score the same while trying to answer these questions as quickly as possible. Here's an example of how your next attempts should look like:

Attempt	Score	Time Taken
Attempt 5	77%	21 mins 30 seconds
Attempt 6	78%	18 mins 34 seconds
Attempt 7	76%	14 mins 44 seconds

Table 15.1 – Sample timing practice drills on the online platform

> **Note**
>
> The time limits shown in the above table are just examples. Set your own time limits with each attempt based on the time limit of the quiz on the website.

With each new attempt, your score should stay above **75%** while your "time taken" to complete should "decrease". Repeat as many attempts as you want till you feel confident dealing with the time pressure.

16

IAM Design

This chapter on **Identity and Access Management** (IAM) will unravel the intricacies of processes and technologies designed to ensure secure access to organizational resources. You will first explore the components of **Identity Management** (IDM) and **Access Management** (AM), emphasizing user provisioning, role definition, and password management. The chapter then moves on to **privileged user management**, shedding light on the oversight of elevated access privileges, with a focus on **Multi-Factor Authentication** (MFA) and audit considerations. You will thoroughly examine centralized directory services and their role as repositories for efficient user IDM. You will finaly have a comprehensive overview of **Federated Identity**, **Single Sign-On** (SSO), and the crucial security measures provided by MFA.

By the end of this chapter, you will be able to confidently answer questions on the following topics:

- IAM
- Federated Identity
- **Identity Providers (IdPs)**
- **SSO**
- **MFA**
- **Cloud Access Security Brokers (CASBs)**
- Secrets management

IAM

IAM is similar to a security guard for digital resources, deciding who gets access to what resources in a secure and organized way. *Figure 16.1* shows the various objectives of IAM. The primary objectives of IAM are to securely manage and govern user identities, authenticate individuals, and control their access to systems, applications, and data. IAM comprises two interconnected components: IDM and AM. IDM focuses on the creation, maintenance, and lifecycle management of digital identities within an organization. Key aspects of IDM include user provisioning and deprovisioning, defining roles and permissions, password management, and managing the entire identity lifecycle. IDM ensures that individuals, devices, or systems are uniquely identified, authenticated, and granted appropriate access based on their roles and attributes. AM, operating in conjunction with IDM, is concerned with controlling and regulating the access rights of authenticated users or systems. It involves defining access control policies, determining resource permissions, and implementing SSO for streamlined user access. AM ensures that access is aligned with the individual's or system's identity, providing the right level of permissions and preventing unauthorized access.

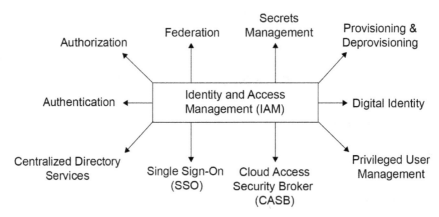

Figure 16.1 – Objectives of IAM

Digital Identity

Digital identity is the representation of an individual, organization, or device within the digital realm. In the online landscape, it encompasses a unique set of attributes and credentials that distinguish one entity from another. One fundamental aspect of digital identity involves identifiers, which are distinctive labels or codes assigned to entities in the virtual space. Examples of identifiers include usernames, email addresses, and account numbers. These identifiers form the basis for recognizing and distinguishing digital entities in the vast online landscape.

Authentication

Authentication is the verification process of confirming the identity of an individual, system, or entity seeking access to a specific resource, system, or service, ensuring that it is genuinely who or what it claims to be. For instance, when you want to log in to your online banking, you put in your username (digital identity) and a password (authentication factor). Some systems may also do extra checks (MFA), such as sending a one-time code to your phone. Your digital identity is essentially your account, and the authentication process makes sure that only the right person, with the correct username and password, can see your important banking details.

Authentication factors and MFA will be discussed later in this chapter.

Authorization

Authorization operates hand in hand with authentication. Once an entity's identity has been successfully authenticated, authorization comes into play to determine the specific access rights and permissions that the authenticated entity should be granted. This pivotal step ensures that individuals, systems, or entities gain access only to the resources or functionalities for which they are authorized, aligning with the principle of least privilege.

At its core, authorization involves the definition and enforcement of access control policies. These policies serve as rules and regulations dictating the actions or operations permitted for each authenticated identity within a system. Permission levels play a crucial role in the authorization process by assigning specific rights and roles to authenticated entities. For instance, administrators may have elevated privileges compared to regular users, allowing them to perform administrative tasks.

An example of authentication and authorization at a banking website is shown in *Figure 16.2*.

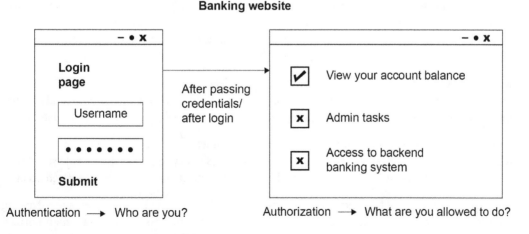

Figure 16.2 – Authentication and authorization at a banking website

Provisioning and Deprovisioning

Provisioning and deprovisioning are fundamental processes in AM, involving the granting and revoking of access rights for users. Provisioning refers to the process of setting up and providing users with the necessary resources, permissions, and credentials to access specific systems or services within an organization's infrastructure. Provisioning streamlines user onboarding, enhances security by enforcing standardized authentication and least privilege principles, maintains audit trails for accountability, and ensures consistency through automated workflows.

Deprovisioning, on the other hand, is the process of revoking or disabling access rights when a user no longer requires them. This could be due to various reasons, such as an employee leaving the organization or a change in their role. Deprovisioning is a crucial security measure aimed at mitigating risks. It ensures that individuals only have access to the resources and data necessary for their current job responsibilities. This helps prevent unauthorized access and reduces the chances of security breaches.

> **Note**
>
> **Authorization creep** refers to the gradual accumulation of excessive access rights or privileges beyond what is necessary for an individual's role or responsibilities. This phenomenon poses security risks as it may result in individuals having more permissions than required, potentially leading to unauthorized access, data breaches, or misuse of resources. Preventing authorization creep involves regular reviews, proper provisioning and deprovisioning processes, and adherence to the principle of least privilege to ensure that access rights align with job responsibilities.

Privileged User Management

Privileged user management oversees and regulates users with elevated access privileges, such as administrators and system engineers, to ensure responsible and secure utilization of these rights. It aims to mitigate security risks, prevent unauthorized access, and establish robust protocols for authentication, monitoring, and periodic reviews. Enforcing MFA adds an extra layer of security, enhancing protection against unauthorized access.

Key audit items for privileged users include elevated privilege usage, adherence to the principle of least privilege, login times, abnormal behavior monitoring, and access log reviews for potential security incidents. Regular audits strengthen security and facilitate early detection of unauthorized activities.

Password management for privileged users involves implementing strong password policies, regular rotations, using MFA, and regular audits of password usage and policy adherence. Privileged users should employ unique and complex passwords stored securely to mitigate the risk of unauthorized access.

Segregation of Duties (SoD) in privileged users is vital for preventing conflicts of interest and enhancing security. It divides tasks among different individuals, reducing the risk of fraud and unauthorized activities. SoD in privileged users may involve separating roles related to system administration, data management, and security controls.

Centralized Directory Services

Centralized directory services play a critical role in IAM by serving as a consolidated repository for storing and managing user identities and associated attributes within an organization. This directory service, commonly based on protocols such as **Lightweight Directory Access Protocol** (**LDAP**) or Active Directory, acts as a centralized source of truth for user information, enabling efficient authentication, authorization, and access control.

In IAM, a centralized directory service streamlines user provisioning and deprovisioning processes, ensuring consistent granting or revoking of access rights across different systems and applications. It facilitates the enforcement of access-related policies and rules, allowing organizations to implement security measures effectively. This approach simplifies the administration of user accounts, group memberships, and access permissions, reducing the complexity associated with managing identities in a distributed environment.

LDAP, crafted with a focus on the X.500 standard, stands as a widely used protocol for accessing and managing directory information services. Tailored for querying and modifying hierarchical directory services, LDAP organizes data in a tree-like structure known as the **Directory Information Tree** (**DIT**). Each entry, representing an object such as a user, is identified by a **Distinguished Name** (**DN**), providing a unique identifier in the LDAP directory. For example, the DN for a user might be something such as `uid=jdoe,ou=users,dc=example,dc=com`.

Entries in an LDAP directory possess attributes storing information about the represented object, such as usernames, email addresses, or group memberships. LDAP supports various operations, such as search, add, modify, delete, and bind. It can operate over secure channels using protocols such as **LDAP Secure** (**LDAPS**) or StartTLS, ensuring encryption for data in transit. Authentication mechanisms, including simple password-based authentication or more secure options such as **Simple Authentication and Security Layer** (**SASL**), can be employed.

Notable LDAP directories include Microsoft's Active Directory, the open source implementation OpenLDAP, and various directory services offered by cloud platforms.

Federated Identity

Federated Identity involves linking and coordinating a user's digital identity and attributes across multiple IDM systems or organizations. It establishes a framework where a user's identity is recognized and trusted across various domains, allowing for the efficient and secure exchange of identity information. This interconnected identity model enables users to access resources and services without the need for separate credentials for each domain. Users manage a single set of credentials, simplifying the login process and encouraging strong password practices. Organizations can collaborate seamlessly, allowing users to access resources beyond organizational boundaries with their trusted identities.

Here are some of the key components of federation:

- **IdPs**: Organizations maintain their IdPs responsible for authenticating users and asserting their identity.

- **Service Providers (SPs) or relying parties**: These are entities or applications that rely on the federated identity system to grant users access based on authenticated identities.

- **Attributes and claims**: Federated Identity involves the exchange of user attributes and claims between IdPs and SPs, enabling informed access decisions.

Federation Standards

These protocols and specifications define the framework for secure and interoperable authentication and authorization across different domains or organizations. These standards facilitate the establishment of trust relationships between IdPs and SPs in a federated environment. Some key Federated Identity standards are the following:

Security Assertion Markup Language (SAML)

SAML is a standard using XML. It's like a digital language that lets different websites and services exchange and verify login details, especially between an IdP and an SP. It enables SSO and supports the secure transfer of assertions about a user's identity.

The SAML workflow involves an authentication request initiated by the SP, leading to user authentication at the IdP. Upon successful authentication, the IdP generates a SAML assertion containing relevant user information, such as identity and attributes. This assertion is then sent back to the SP, allowing the user seamless access to the requested service.

SAML assertions are digitally signed by the IdP to ensure their integrity and authenticity. Additionally, encryption can be applied to protect sensitive information within assertions, mitigating the risk of tampering and eavesdropping during data transmission.

The latest version of SAML, SAML 2.0, represents a significant evolution, introducing enhancements and features that make it widely adopted for secure and interoperable SSO implementations.

OpenID Connect

OpenID Connect serves as an authentication protocol, allowing clients to request and obtain information about the end user's identity. It builds on OAuth 2.0's authorization capabilities, extending them to include identity information.

The primary innovation introduced by OpenID Connect is the ID token. After successful authentication, the IdP issues an ID token containing claims about the end user, such as their unique identifier, name, and email address. This token is digitally signed by the IdP for integrity and authenticity. ID tokens are typically formatted as **JSON Web Tokens (JWTs)**, providing a compact and self-contained way to convey information between parties. JWTs are digitally signed and, optionally, encrypted to ensure the security of the identity information.

OAuth 2.0 (Open Authorization)

OAuth 2.0, an open standard for authorization, facilitates secure and delegated access to user data by third-party applications without the need to expose sensitive credentials. The protocol defines distinct roles, including the resource owner (typically the end user), the client (requesting access), the authorization server (responsible for authentication and token issuance), and the resource server (hosting protected resources).

Access tokens, at the core of OAuth 2.0, represent the authorization granted to the client and are crucial for accessing protected resources. Bearer tokens are used for authentication in OAuth 2.0, where possession of the token is sufficient for access. However, implementing proper security measures, such as token validation, is imperative to prevent misuse.

OAuth 2.0 introduces security measures to safeguard the authorization process. This includes using HTTPS for secure communication, protecting client secrets, and validating tokens to prevent unauthorized access. Refresh tokens are employed to obtain new access tokens without requiring the resource owner's reauthorization, contributing to a smoother user experience.

OAuth 2.0 finds widespread adoption across industries for various purposes, ranging from social media logins to mobile app integrations and **Application Programming Interface (API)** access control.

Web Services Federation (WS-Federation)

WS-Federation stands as a specification integral to enabling secure and interoperable identity federation and SSO within web services architectures. At its core, WS-Federation operates on a token-based authentication model, where security tokens are exchanged to convey information about a user's identity, attributes, and authentication status. Central to the WS-Federation architecture is the **Security Token Service (STS)**, responsible for issuing and validating these security tokens across participating domains.

Federation Metadata, another key component, takes the form of XML-based information describing the capabilities and configurations of IdPs, SPs, and STS within the federation. This Metadata plays a crucial role in establishing trust relationships and facilitating secure communication.

IdPs

In a federation flow, IdPs play a pivotal role in facilitating secure and streamlined authentication and authorization across multiple services or domains. Google, Microsoft Azure AD, Okta, OneLogin, Ping Identity, and Salesforce Identity are all notable examples of IdPs.

The user initiates the authentication process by accessing a service in one domain. The service, acting as an SP, recognizes the need for authentication and redirects the user to the designated IdP. The IdP authenticates the user, often by validating their credentials through a username and password, MFA, or other secure means. Once authenticated, the IdP generates a security token or assertion that vouches for the user's identity. The token contains information about the user (claims) and the authentication event. Common token formats include **Security Assertion Markup Language** (**SAML**) assertions or JWTs.

The user is redirected back to the original service (SP) with the issued token. The SP can then validate the token by communicating with the IdP or using shared keys or certificates. The SP authorizes access to its resources based on the validated token. The user gains access to the requested services without needing to log in separately, promoting an SSO experience.

In some federated systems, the IdP may include additional attributes about the user in the token. The SP can use these attributes to personalize the user's experience or make access decisions based on specific user attributes. Federation flows often include mechanisms for **Single Logout** (**SLO**), enabling users to log out from all federated services in a single action. The IdP plays a key role in coordinating and initiating the logout process.

Throughout this flow, the IdP acts as a trusted entity responsible for authenticating users, issuing secure tokens, and conveying the user's identity to various services within the federation. This centralized authentication mechanism streamlines user access, enhances security, and simplifies IDM across multiple domains or services in a federated environment. Now, you will see how IdPs function in some of the primary federation models:

- **Web of Trust (WoT) federation**: In a WoT federation, each IdP within the federation not only provides identity information about its users but also relies on identity information shared by other members. Every participant acts as both an IdP, asserting the identities of its users, and a relying party, depending on the assertions made by other entities. Users or IdPs trust each other based on their own assessments of reliability and trustworthiness. This model is often decentralized and relies on the reputation of the entities involved. Each participant is responsible for managing its trust relationships, and trust is transitive, meaning if A trusts B and B trusts C, then A can also trust C. WoT is commonly associated with peer-to-peer networks and decentralized systems.

- **Trusted Third Party (TTP)**: In a TTP federation model, trust is mediated through a central authority or trusted intermediary. This third party, often referred to as an IdP or authentication service, is responsible for verifying and validating the identities of users. Other SPs in the federation trust this central authority to authenticate users and provide necessary identity attributes. The TTP model simplifies trust management as entities only need to establish trust with the central authority rather than with each other.

SSO

SSO is a user authentication process that streamlines access to multiple applications or services by allowing users to log in only once, providing them with subsequent access without the need to re-enter their credentials. In an SSO system, a user's initial authentication is validated by an IdP, and this authenticated session is then shared across various applications or services within the same security domain. This eliminates the need for users to remember multiple sets of usernames and passwords for different systems, improving user convenience and reducing the risk associated with password fatigue.

IT administrators can manage user authentication from a centralized location, simplifying user provisioning and deprovisioning and enforcing security policies consistently across multiple applications. Organizations can realize cost savings associated with reduced helpdesk support for password-related issues, as users are less likely to forget or need assistance with their credentials.

SSO enhances both security and user experience within a specific organization or domain, as users can seamlessly move between various applications without repetitive logins, contributing to increased efficiency and a more user-friendly authentication process.

SSO introduces a single point of access, and if the credentials for this SSO are compromised, attackers could gain access to multiple services, making the protection of initial authentication crucial to preventing unauthorized access. Additionally, the security of the session token or cookie used in SSO is paramount, as mishandling could render it vulnerable to session hijacking or replay attacks. Trust in the IdP is fundamental in SSO systems, and if compromised, it could lead to unauthorized access across multiple services. SSO poses a potential single point of failure in a system, as reliance on a central authentication mechanism means that any disruption or compromise could lead to widespread denial of access. To address this vulnerability, organizations implement redundancy measures and robust security practices to ensure system resilience and uninterrupted service availability.

> **Note**
>
> SSO simplifies user access within a specific organization by allowing users to log in once and seamlessly access various services without repeated authentication. It is designed for a single security domain, enhancing user convenience and reducing the need for multiple logins within the organization. On the other hand, federation extends the concept of SSO beyond organizational boundaries. It involves collaboration between multiple IdPs and SPs across different security domains, enabling users to access resources in a federated environment. Federation is particularly beneficial when users need to interact with services or resources distributed across various organizations, fostering seamless authentication and access in a collaborative and multi-domain setting.

MFA

MFA uses various authentication factors to confirm a user's identity. These factors typically fall into three categories:

- **Knowledge factors**: This involves information only the user should know, such as a password, PIN, or answers to security questions.

- **Possession factors**: This involves something the user possesses, adding a physical element to the authentication process. Common examples include receiving a one-time code on a mobile device or using a physical security token.

- **Inherence factors**: Biometric data, such as fingerprints or facial recognition, is an example of something inherent to the user. Biometrics add a unique and personal dimension to the authentication process.

The strength of MFA lies in its ability to create a layered defense, where compromising one factor does not grant unauthorized access. Even if a password is stolen, an additional verification step is required, providing an additional barrier against unauthorized entry. This is particularly important as cyber threats continue to evolve, and traditional authentication methods alone may not be sufficient to protect sensitive information and systems.

In addition to the three main categories of factors (knowledge, possession, and inherence), MFA can incorporate various additional factors for enhanced security. These factors contribute to a more comprehensive and diverse approach to user verification. Here are some examples:

- **Location factors**: Geographical information or the user's specific location can be considered. If a login attempt occurs from an unusual or unexpected location, it may trigger additional verification steps.

- **Time factors**: Authentication may consider the time of day or the regular patterns of user activity. Unusual login times or deviations from established usage patterns can trigger additional scrutiny.

- **Behavioral factors**: User behavior analytics can be employed to assess the typical patterns of user interaction with systems. Deviations from established behavior may prompt additional authentication steps.

- **Device factors**: Information about the device used for authentication, such as device fingerprints or security features, can be considered. This helps ensure that the user is accessing the system from a recognized and/or secure device.

- **Risk-based factors**: Dynamic risk assessment involves continuously evaluating the risk associated with a particular login attempt. Factors such as the user's history, the device used, and the context of the request contribute to risk analysis.

CASB

A CASB is a security solution that serves as an intermediary between cloud customers and cloud SPs to ensure the secure and compliant use of cloud applications and services, with the flexibility of deployment as either software or hardware-based solutions. CASBs address the challenges related to the adoption of cloud computing, providing visibility, control, and security for data and applications in the cloud.

Their primary function is to guarantee that the utilization of cloud services aligns with the intentions and policies set forth by the organization. CASBs play a pivotal role in enforcing security measures, monitoring activities, and implementing controls to ensure that cloud usage complies with organizational standards, maintaining a secure and policy-consistent cloud environment. CASBs offer visibility into the cloud applications and services being used within an organization, even those adopted without IT approval. This helps in discovering potential shadow IT and assessing the usage of unauthorized or unmanaged cloud applications.

CASBs implement **Data Loss Prevention (DLP)** policies to monitor and control the transfer of sensitive data, preventing unauthorized access or accidental exposure. They can encrypt data, apply access controls, and monitor for policy violations. CASBs can inspect and analyze encrypted traffic to ensure that sensitive data is not being transferred maliciously or against organizational policies. This includes the ability to decrypt, inspect, and re-encrypt traffic without compromising security.

CASBs enforce access controls and authentication policies to ensure that only authorized users can access specific cloud resources. This includes enforcing MFA and SSO for cloud applications. CASBs protect against various cloud-specific threats, including malware, data breaches, and insider risks. They leverage threat intelligence and advanced security measures to detect and mitigate risks.

CASBs help organizations maintain compliance with industry regulations and internal policies. They provide tools for auditing and reporting on user activities, ensuring that cloud usage aligns with established governance standards. CASBs assist in configuring cloud services securely, aligning with best practices and security standards. This includes enforcing encryption, securing APIs, and implementing secure configurations for cloud applications.

CASBs continuously monitor user activities and data transactions in real time. They generate alerts and notifications for potential security incidents, allowing organizations to respond promptly to threats or policy violations. CASBs often integrate with cloud SPs' APIs to extend security controls directly into cloud applications. This enables more granular security policies and visibility into cloud-native environments.

CASBs can be classified into three main types based on their deployment methods:

- **API-based CASBs**: API-based CASBs integrate directly with the APIs of cloud SPs, enabling real-time communication. This method offers seamless visibility into user activities and data configurations, allowing for effective policy enforcement and anomaly detection. While non-intrusive, the efficacy of API-based CASBs relies on the features and granularity of APIs provided by cloud platforms.

- **Agent-based CASBs**: Agent-based CASBs require software agents installed on end user devices, facilitating direct connections to the CASB platform. This approach allows for granular control over data flow, enabling real-time monitoring and security policy enforcement. Despite offering detailed control, the need for software installations on each device can be challenging, especially in environments with diverse or numerous endpoints.

- **Agentless CASBs**: Agentless CASBs operate without installing software on end user devices, using alternative methods such as proxy-based architectures. This quicker deployment is suitable for scenarios where agent installations are impractical. Agentless CASBs still provide essential security functions, including policy enforcement and threat detection. However, their effectiveness depends on intercepting and managing traffic without dedicated software agents on endpoints.

CASBs can be classified into two types based on their approach to securing cloud environments:

- **Inline CASBs**: Inline CASBs function in real time within the data flow between users and cloud services, acting as an intermediary for immediate inspection and policy application. While offering instant security controls, they may introduce latency, potentially impacting data transaction performance. Organizations typically deploy inline CASBs for critical data protection needs requiring real-time monitoring and enforcement.

- **Out-of-band CASBs**: Out-of-band CASBs operate non-disruptively, away from the live data flow between users and cloud services. Instead of intercepting real-time transactions, they analyze logs or copies of data post-transaction. This minimizes impact on data transaction performance but may lead to delays in detecting and responding to security incidents. Out-of-band CASBs are favored when organizations prioritize minimal interference with data flows and can tolerate delayed response times for monitoring and enforcing security policies.

Secrets Management

Secrets management involves the secure creation, storage, retrieval, and handling of sensitive information, often referred to as "secrets," within an organization's computing environment. These secrets can include credentials, encryption keys, API keys, passwords, and other confidential information.

The lifecycle of secrets and effective secrets management involves several critical stages to ensure the security, confidentiality, and proper handling of sensitive information. The process begins with the creation of secrets, emphasizing the use of secure methods to generate strong and unique values while avoiding easily guessable or default options. Secure storage practices are then implemented, including encryption at rest, the use of access controls, and the consideration of hardware security modules for enhanced protection. Access control is crucial, and organizations should enforce strict policies based on the principle of least privilege, utilizing IAM tools for effective rule enforcement.

Regularly rotating secrets is a fundamental practice to reduce the risk of compromise, and automation is recommended to ensure consistency and timeliness in the rotation process. Monitoring and auditing play a crucial role in secrets management, involving continuous monitoring of access, detailed logging of activities, and alerts for suspicious behavior. In the event of a compromise or unused secrets, a well-defined revocation process is essential, and automation should be leveraged to minimize response time. Secure deletion procedures must be in place for retired or obsolete secrets, ensuring that they are irrecoverable and removed from configurations and repositories.

Encryption techniques, such as strong encryption algorithms and end-to-end encryption for secrets in transit, contribute to a robust secrets management strategy. User education and awareness programs are essential components, fostering a culture of security and accountability through regular training and encouraging the reporting of any suspected breaches. Finally, a well-defined incident response plan is crucial for addressing secrets exposure promptly, with procedures for containment, investigation, and remediation.

The following are some of the pointers to keep in mind while managing secrets with respect to the CCSP exam:

- Instead of hardcoding secrets in code, employ dynamic configuration management tools or environment variables to securely handle sensitive information. This ensures that secrets are not exposed in the source code and can be easily updated without modifying the code base.

- Configure logging mechanisms to exclude sensitive information, ensuring that secrets are not stored in logs.

- Enforce the use of expiration dates for secrets to mitigate the risk of prolonged exposure in the event of a security breach.

- Prioritize the implementation of high availability for the secrets management system. Ensure continuous access to secrets, especially in dynamic operational environments, to maintain the application's functionality and security.

Summary

In this chapter, you learned about the intricate processes, policies, and technologies that have been designed to ensure the appropriate access of resources to individuals or systems within an organization.

You delved into IDM, concentrating on the creation, maintenance, and lifecycle management of digital identities. The chapter discussed key components of IDM, including user provisioning and password management, to underscore the importance of a well-organized and controlled identity ecosystem. It also emphasized the critical role of IDM in uniquely identifying and authenticating individuals, devices, or systems, thus establishing a secure foundation for AM.

Collaborating seamlessly with IDM, AM has emerged as the guardian of authenticated access rights. Through the definition of access control policies and resource permissions and the implementation of SSO, AM ensures that the principle of least privilege is upheld. The discussion on provisioning and deprovisioning highlighted the foundational processes that dictate the granting and revoking of access rights, emphasizing the need to mitigate authorization creep—a phenomenon where individuals accumulate unnecessary access privileges over time.

The chapter also delved into privileged user management, focusing on the oversight of users with elevated access. It stressed the importance of MFA, audit items, and SoD to fortify security measures. The significance of centralized directory services in streamlining user provisioning and deprovisioning processes was underscored, emphasizing their role as a consolidated repository for managing user identities.

The chapter further discussed Federated Identity, illuminating the interconnectedness of user identities across multiple systems or organizations. You also explored SSO and learned its efficiency in streamlining user access within a specific security domain, acknowledging its vulnerabilities and the paramount importance of trust in the IdP. The chapter provided a holistic perspective on MFA and IAM, emphasizing proactive digital IAM to enhance overall organizational security.

In the next chapter, you will learn about crucial configuration needs for both physical and logical infrastructure in cloud environments, along with common controls for operational and maintenance tasks.

Exam Readiness Drill – Chapter Review Questions

Apart from a solid understanding of key concepts, being able to think quickly under time pressure is a skill that will help you ace your certification exam. That is why working on these skills early on in your learning journey is key.

Chapter review questions are designed to improve your test-taking skills progressively with each chapter you learn and review your understanding of key concepts in the chapter at the same time. You'll find these at the end of each chapter.

> **How to Access These Materials**
>
> To learn how to access these resources, head over to the chapter titled *Chapter 25, Accessing the Online Resources.*

To open the Chapter Review Questions for this chapter, perform the following steps:

1. Click the link – `https://packt.link/CCSPE1_CH16`.

    Alternatively, you can scan the following **QR code** (*Figure 16.3*):

Figure 16.3 – QR code that opens Chapter Review Questions for logged-in users

2. Once you log in, you'll see a page similar to the one shown in *Figure 16.4*:

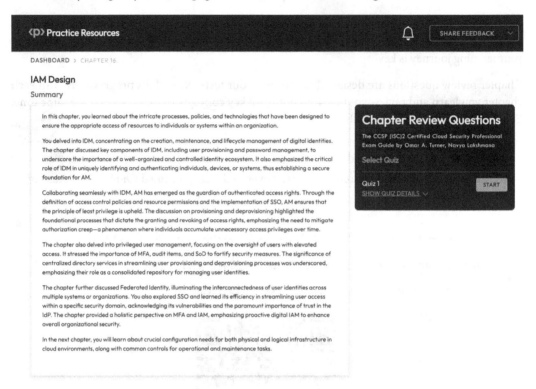

Figure 16.4 – Chapter Review Questions for Chapter 16

3. Once ready, start the following practice drills, re-attempting the quiz multiple times.

Exam Readiness Drill

For the first three attempts, don't worry about the time limit.

ATTEMPT 1

The first time, aim for at least **40%**. Look at the answers you got wrong and read the relevant sections in the chapter again to fix your learning gaps.

ATTEMPT 2

The second time, aim for at least **60%**. Look at the answers you got wrong and read the relevant sections in the chapter again to fix any remaining learning gaps.

ATTEMPT 3

The third time, aim for at least **75%**. Once you score 75% or more, you start working on your timing.

> **Tip**
>
> You may take more than **three** attempts to reach 75%. That's okay. Just review the relevant sections in the chapter till you get there.

Working On Timing

Target: Your aim is to keep the score the same while trying to answer these questions as quickly as possible. Here's an example of how your next attempts should look like:

Attempt	Score	Time Taken
Attempt 5	77%	21 mins 30 seconds
Attempt 6	78%	18 mins 34 seconds
Attempt 7	76%	14 mins 44 seconds

Table 16.1 – Sample timing practice drills on the online platform

> **Note**
>
> The time limits shown in the above table are just examples. Set your own time limits with each attempt based on the time limit of the quiz on the website.

With each new attempt, your score should stay above **75%** while your "time taken" to complete should "decrease". Repeat as many attempts as you want till you feel confident dealing with the time pressure.

17
Cloud Physical and Logical Infrastructure (Operationalization and Maintenance)

In this chapter, you will delve into crucial skills for constructing and implementing the physical and logical infrastructure of a cloud environment. You will learn about hardware-specific security measures such as **Hardware Security Modules (HSMs)** and **Trusted Platform Modules (TPMs)**. The chapter will also cover operational and maintenance aspects by exploring access controls, secure network configurations, network security controls, and the hardening of **Operating Systems (OSs)**. Additional topics will include patch management, the **Infrastructure as Code (IaC)** strategy, the availability of clustered hosts and guest OSs, as well as performance and capacity monitoring. The chapter will conclude with insights into hardware monitoring and the configuration of backup and restore functions for both host and guest OSs. Lastly, it will explore the intricacies of the management plane.

By the end of this chapter, you will be able to confidently answer questions on the following:

- Hardware-specific security configuration requirements
- Virtual hardware-specific security configuration requirements
- Installation of guest OS virtualization toolsets
- Access controls for local and remote access
- Secure network configuration and network security controls
- OS baselining
- Patch management

- IaC strategy

- Availability of clustered hosts guest OS

- Performance and capacity monitoring and hardware monitoring

- Configuration of host and guest OS backup and restore functions

- Management plane

Hardware-Specific Security Configuration Requirements

Establishing a secure cloud environment commences with a focus on security throughout the planning and implementation of both physical and logical infrastructures. This involves careful consideration of components such as servers, networking, and storage devices. Despite unique cloud-specific requirements, the principles for securely configuring systems remain similar to those for on-premises systems, although the techniques may differ. In the shared responsibility model of the cloud, the responsibility for physical security often falls on the cloud provider.

HSMs

An HSM is a dedicated hardware device or appliance that provides secure key management, cryptographic operations, and protection of sensitive data. HSMs are designed to offer a high level of security and tamper resistance, making them suitable for safeguarding cryptographic keys and performing cryptographic functions in a secure environment. The key management capabilities of HSMs encompass various aspects, including key generation, secure distribution, rotation, and destruction, ensuring the confidentiality and integrity of cryptographic keys.

HSMs perform cryptographic operations within a secure and isolated environment, shielding sensitive processes from potential attacks. They are equipped with a high-quality random number generator, essential for generating secure and unpredictable random numbers crucial to cryptographic applications. The secure hardware design of HSMs incorporates features such as tamper-evident seals and secure key storage, aiming to resist physical tampering and unauthorized access. In addition to cryptographic operations, HSMs offer mechanisms for securely backing up and recovering cryptographic keys, ensuring business continuity in case of hardware failure.

Moreover, HSMs provide well-defined APIs that allow applications and systems to interact with cryptographic functions securely. Certification processes, such as FIPS 140-2 and Common Criteria, validate many HSMs for compliance with industry standards and regulations, attesting to their security and reliability.

Cloud Service Providers (**CSPs**) often offer **virtual Hardware Security Modules** (**vHSMs**) as part of their cloud services to address the cryptographic and key management needs of cloud-based applications. vHSMs replicate the functionalities of traditional physical HSMs but operate in a virtualized environment within the cloud.

TPMs

A TPM is a dedicated hardware component integrated into the motherboard of computing devices to enhance the overall security of the platform. Functioning independently of the **Central Processing Unit (CPU)**, the TPM plays a pivotal role in establishing a hardware-based root of trust. This root of trust is fundamental for ensuring the integrity and security of the computing environment.

One of the primary functions of the TPM is to provide secure key storage. It acts as a vault for cryptographic keys. In addition to key management, the TPM is integral to the Secure Boot process. It ensures that only authorized and digitally signed firmware and OS components are loaded during the boot sequence. This protects against the execution of unauthorized or malicious code, reinforcing the platform's security from the early stages of operation.

The TPM measures the integrity of the platform by creating a unique hash value, known as a **Platform Configuration Register (PCR)**. This value reflects the state of the system and can be used to verify the platform's integrity at different points in time. The attestation capability of the TPM allows the system to prove its integrity and configuration to external entities, contributing to establishing trust in a distributed environment. While TPMs are commonly found in consumer devices such as laptops and desktops, they are a crucial component in creating a trusted computing environment.

Just as there are vHSMs, there are also **virtual Trusted Platform Modules (vTPMs)** to bolster the security of virtualized environments, particularly within hypervisors. These vTPMs serve a similar function to physical TPMs but are implemented in software and operate within the virtualized infrastructure rather than on dedicated hardware. vTPMs play a crucial role in extending the security benefits of TPM technology to **Virtual Machines (VMs)** and hypervisor-based environments.

Storage Controllers

Storage controllers, which are integral components in computing systems, manage and regulate data exchange between a computer and its storage devices. The hardware or software components are pivotal for the seamless operation of storage systems, overseeing functions such as data transfer, read and write operations, error handling, caching, encryption support, access control, and communication with diverse storage devices. Storage controllers encompass a range of types, fulfilling distinct roles in storage management. These include **disk controllers**, overseeing data flow for both **Hard Disk Drives (HDDs)** and **Solid-State Drives (SSDs)**; **RAID controllers**, implementing RAID configurations for enhanced data protection and performance; **Storage Area Network (SAN) controllers**, managing communication between servers and networked storage devices; and **Network-Attached Storage (NAS) controllers**, governing data access and storage functions within NAS systems.

Beyond traditional storage controllers, here are some of the specialized protocols and technologies that have emerged:

- **iSCSI controllers** enable the utilization of the iSCSI protocol to transmit SCSI commands over **Internet Protocol** (**IP**) networks, facilitating SAN connections through existing Ethernet infrastructure.

- **Fibre Channel/Fibre Channel over Ethernet (FCoE) controllers** are designed for high-performance, low-latency environments. Fibre Channel serves as a dedicated storage networking protocol, and FCoE encapsulates Fibre Channel frames within Ethernet, enabling the convergence of storage and regular network traffic.

Network Controllers

Network controllers, also known as **Network Interface Controllers** (**NICs**) or network adapters, are essential components within computing systems, acting as the intermediary for communication between the computer and the network. Serving as a gateway for data transmission, these controllers enable devices to connect and communicate seamlessly over networks. There are distinct types of network controllers including Ethernet controllers, wireless network adapters, and onboard controllers.

Operating based on standardized communication protocols such as Ethernet and Wi-Fi, network controllers manage data transmission using protocols such as **Transmission Control Protocol** (**TCP**) and IP. Network controllers play a crucial role in modern networking, supporting advanced features such as **Quality of Service** (**QoS**), VLAN tagging, and energy-efficient Ethernet. They may also include support for security protocols in wireless networks, ensuring the integrity and confidentiality of transmitted data.

Network controllers play a pivotal role in the paradigm of **Software-Defined Networking** (**SDN**), reshaping traditional network architectures by centralizing control and emphasizing programmability. In traditional networks, the control plane functions were distributed across individual devices, such as routers and switches. SDN shifts this control to a centralized entity—the network controller. In an SDN environment, the control plane is abstracted from the data plane, allowing network controllers to manage and orchestrate communication across the network.

Installation and Configuration of Management Tools

Management tools enable heightened operational efficiency, optimized resource usage, streamlined collaboration, fortified security measures, and improved cost control in cloud computing.

Scheduling within the management plane involves the coordination and allocation of resources to execute tasks efficiently. This includes assigning computing resources, managing workloads, and optimizing the utilization of available infrastructure based on predefined schedules and priorities.

Orchestration involves the coordination and automation of various tasks and processes to achieve specific business objectives. In the management plane, orchestration streamlines the deployment and configuration of complex applications, ensuring seamless integration and interaction between different components.

Maintenance operations within the management plane focus on ensuring the health, security, and performance of the cloud infrastructure. This includes applying software updates, patches, and security fixes, as well as performing routine tasks such as system checks and diagnostics.

The management plane facilitates **dynamic resource scaling** to adapt to changing workloads. This includes auto-scaling mechanisms that automatically adjust the number of resources based on demand, optimizing resource usage and providing a responsive and cost-effective environment.

Synchronization of tasks is crucial for managing parallel operations and maintaining consistency within the cloud environment. The management plane orchestrates tasks to avoid conflicts, optimize resource usage, and ensure the smooth execution of operations.

Policies governing resource allocation, security, and compliance are enforced through the management plane. This includes configuring and maintaining access controls, implementing security measures, and ensuring adherence to organizational policies across the cloud infrastructure.

The management plane offers **user interfaces and APIs** that enable administrators to interact with and control the cloud environment. Through graphical interfaces or programmable APIs, users can manage, monitor, and configure resources within the cloud infrastructure.

Securing management tools within an organization's IT infrastructure is paramount to ensuring the overall integrity, confidentiality, and availability of critical functions. Management tools often handle sensitive information and, as the command center for essential operations, these tools require robust security measures to prevent unauthorized access and control, including access credentials and configuration data, necessitating strong authentication mechanisms and access controls to safeguard against unauthorized access. Additionally, securing these tools is vital to prevent tampering with configurations that could lead to vulnerabilities or unauthorized changes, compromising system stability and security. Mitigating insider threats, whether intentional or unintentional, is another critical aspect of securing management tools through continuous monitoring and access controls.

You will now see some of the best security practices to harden management tools:

- Implement **strong authentication mechanisms**, including **Multi-Factor Authentication (MFA)**, to ensure that only authorized personnel can access management tools. Employ **granular access controls** to restrict privileges based on roles and responsibilities.

- **Encrypt** data in transit and at rest to protect sensitive information handled by management tools. Use secure communication protocols, such as HTTPS, and encrypt stored data to prevent unauthorized access.

- Keep management tools and underlying infrastructure **up to date with the latest security patches and updates**.

- Enable robust **audit logging** to track user activities within management tools. Implement real-time monitoring to detect anomalous behavior and potential security incidents.

- Follow the **principle of least privilege** when configuring management tools. Disable unnecessary features and services to reduce the attack surface. Regularly review and update configurations based on security best practices.

- **Isolate management tools** from other parts of the network through network segmentation. This helps contain potential security breaches and prevents lateral movement within the network.

- Develop and regularly update an **incident response plan** specific to the management tools.

- Conduct regular **security audits and assessments** of management tools to identify vulnerabilities and ensure compliance with security standards.

- Provide comprehensive **training for employees** who have access to management tools. Educate them on security best practices, the importance of strong passwords, and the potential risks associated with their roles.

- Implement regular backups of critical configurations and data handled by management tools. Develop a robust **disaster recovery plan** to ensure business continuity in the event of a security incident or system failure.

- Implement **high-availability architectures** to ensure uninterrupted access to management tools. This involves redundant systems, load balancing, and failover mechanisms to maintain operations even in the face of hardware or software failures.

Virtual Hardware-Specific Security Configuration Requirements

Securing virtual hardware in the cloud presents unique challenges primarily because of the shared nature of underlying physical hardware and the dynamic, multi-tenant environment. In cloud computing, multiple VMs often share the same physical hardware resources, such as CPU, memory, and storage. This shared infrastructure introduces a challenge in implementing traditional air-gap designs, where physical separation ensures security, as VMs from different users coexist on the same servers. The concept of "noisy neighbors" refers to situations where the performance of one VM on shared hardware impacts the performance of neighboring VMs. This dynamic resource sharing can lead to security challenges, especially if one VM poses a security risk to others.

In cloud environments, the effective provisioning of virtual hardware plays a pivotal role in ensuring optimal performance and resource utilization. Striking the right balance between allocating sufficient storage, CPU, and memory resources for all VMs is crucial for meeting the demands of diverse workloads. Overprovisioning, where resources are assigned in excess of actual requirements, poses concerns related to cost inefficiency and underutilization. On the other hand, underprovisioning, or allocating insufficient resources, can lead to performance bottlenecks and diminished user experience. Achieving an optimal resource allocation strategy requires careful consideration of application requirements, workload patterns, and adherence to vendor recommendations and best practices. The availability of resources is a key focal point, and organizations must proactively monitor usage patterns, making adjustments as needed to maintain a robust and available infrastructure.

IaC is a key methodology in cloud computing that enables the automated and consistent deployment of infrastructure resources. IaC involves describing and managing infrastructure components, including VMs, networks, and storage, using code or configuration files rather than manual processes.

Hypervisor

The hypervisor, responsible for managing multiple VMs on a host, becomes a critical security point. Vulnerabilities in the hypervisor could potentially lead to security breaches across multiple VMs. Ensuring the security of the hypervisor is crucial in a shared environment and it is the responsibility of the cloud provider in the cloud.

Regularly applying security patches and updates ensures that known vulnerabilities are addressed promptly, reducing the risk of exploitation. Enabling Secure Boot features is another crucial step, ensuring that only signed and trusted components are loaded during the hypervisor's boot process. This proactive measure helps prevent the execution of malicious code during startup.

By following hypervisor-hardening guidelines provided by the vendor, including the deactivation of unnecessary ports, services, and protocols, organizations can strengthen their security posture. Implementing strong access controls, such as **Role-Based Access Control** (**RBAC**), restricts access to the hypervisor management interface, ensuring that users have only the minimum necessary privileges.

Segregating the management network for the hypervisor from other networks adds an extra layer of security, preventing unauthorized access. Securing the management interfaces with strong authentication mechanisms, such as MFA, and using encrypted connections for remote management interfaces further safeguards against potential threats such as eavesdropping. Robust monitoring and logging, combined with regular reviews of logs and prompt incident response, contribute to a proactive security approach. Conducting regular security audits and assessments collectively establishes a comprehensive security framework for the hypervisor environment. Ensuring strong isolation between VMs, setting resource allocation limits, and employing network segmentation within the hypervisor contribute to maintaining a secure virtualized environment.

Network

In cloud computing, both physical and virtual networking play crucial roles. Physical networking in the cloud involves the infrastructure that enables data transfer between servers, storage, and other hardware components. Cloud providers maintain physical data centers with a complex network infrastructure, including routers, switches, and physical cabling. Physical networking, while managed by the cloud provider, forms the foundational layer upon which virtual networking is built.

Virtual networking in the cloud leverages software-defined concepts to create isolated and customizable network environments. The **Virtual Private Cloud** (**VPC**) in AWS and GCP and VNet in Azure are prime examples, offering users the ability to provision logically isolated sections of the cloud where they can deploy resources. It provides an isolated space where users can launch resources, such as VMs, and define their own network configurations. VPCs and VNets enable logical isolation and segmentation within a cloud environment, allowing users to create multiple VPCs or VNets for different projects or departments. Users can define and manage IP address ranges for their VPCs, facilitating seamless integration with on-premises networks if needed. VPCs and VNets support custom routing tables, allowing users to define how traffic is routed within the virtual network. VPCs and VNets offer various connectivity options, such as **Virtual Private Network** (**VPN**) connections and private connectivity, enabling secure communication between on-premises data centers and cloud resources. This virtualized approach allows for greater flexibility, scalability, and customization compared to traditional physical networking.

Security groups in AWS and Azure and firewall rules in GCP are a form of virtual firewall that controls inbound and outbound traffic for VMs within a VPC. These groups act as a virtualization of traditional firewalls, allowing users to define rules that permit or deny traffic based on protocols, ports, and IP addresses. Security groups enhance the security posture of virtualized environments by restricting unauthorized access and facilitating a least-privileged access model.

A security group is a fundamental element of a VPC. It acts as a virtual firewall for resources within a VPC, controlling both inbound and outbound traffic based on specified rules. The security group is a critical component that contributes to the overall security and isolation within the VPC environment.

In a VPC, users can create and configure security groups to define access rules for their VMs and other resources. Each security group operates at the instance level, and users can associate multiple instances with a specific security group. The rules within a security group dictate what type of traffic is allowed or denied, based on factors such as protocols, ports, and source/destination IP addresses. By associating resources with specific security groups, users can enforce a fine-grained security posture, contributing to the principle of least privilege.

> **Note**
>
> The configuration of inbound and outbound rules depends on whether the service is stateful or stateless. In stateful services, the firewall automatically allows return traffic for outbound connections initiated from within the network. Therefore, you typically only need to define rules for inbound traffic. On the other hand, stateless services require explicit rules for both inbound and outbound traffic because the firewall does not automatically allow return traffic. So, you need to define rules for both directions to ensure proper communication.

Installation of Guest OS Virtualization Toolsets

Guest OS virtualization toolsets are software packages that optimize the performance, integration, and functionality of guest OSs within VMs, providing features such as guest OS-specific packages (e.g., Windows and Linux), device drivers, display enhancements, security enhancements, performance monitoring, filesystem sharing, and time synchronization. In IaaS, the responsibility for managing guest OSs typically lies with the customer, whereas in SaaS, it's usually the responsibility of the provider. In PaaS, it can vary depending on the specific service, but it often lies with the provider.

Before installing guest OS virtualization toolsets, ensure compatibility with the specific hypervisor in use. Confirm that the guest OS is supported by the virtualization toolset. Guest additions are typically available for a range of OSs, including various versions of Windows, Linux distributions, and others.

Download toolsets only from official and trusted sources, such as the hypervisor vendor's website. When transferring or distributing installation packages, use secure and encrypted channels. Keep the virtualization toolset up to date by checking for and applying any updates released by the hypervisor vendor.

Prioritize the security configuration of the guest OS. Apply security best practices such as regular patching, disabling unnecessary services, configuring firewalls, and employing strong authentication mechanisms. A secure guest OS foundation enhances the overall security posture.

Access Controls for Local and Remote Access

Access controls in the cloud are crucial measures regulating local and remote access, forming a multi-layered security approach to prevent unauthorized entry, safeguard against potential threats, and ensure secure interactions with cloud resources. These controls span various protocols and mechanisms to ensure a robust security posture.

Remote Desktop Protocol (**RDP**), a proprietary Microsoft protocol, stands as a widely utilized means for remote access to VMs and desktops. RDP incorporates robust security features, including encryption, authentication, and **Network-Level Authentication** (**NLA**). To bolster RDP security, administrators can configure Group Policy settings, applying restrictions on user access, session timeouts, and other security-related configurations aligned with organizational policies. Firewall considerations play a vital role, as configuring firewalls to permit RDP traffic exclusively from trusted IP addresses reduces the attack surface. Implementing **Two-Factor Authentication** (**2FA**) provides an additional layer of security, requiring users to provide a second form of identification beyond passwords. Regular updates and patching are essential elements of RDP security, addressing vulnerabilities and ensuring resilience against emerging threats. While native to Windows, RDP is available for other OSs, including Linux and Unix, extending its utility across diverse environments.

Secure terminal access involves using protocols such as **Secure Shell** (**SSH**) to establish secure command-line connections to remote servers or VMs in the cloud. SSH is a cryptographic network protocol designed for secure communication over potentially insecure networks. Serving as a secure alternative to unencrypted protocols such as Telnet, SSH ensures encrypted connections for remote access, command-line execution, and secure file transfers between clients and servers. At its core, SSH employs robust encryption algorithms to maintain the confidentiality of data during transmission, thwarting unauthorized interception and eavesdropping. To fortify SSH security, it offers various authentication mechanisms, including password-based, public-key, and certificate-based authentication. Public-key authentication, a key component, enhances security by eliminating the need for passwords, thereby reducing the risk of credential-based attacks. 2FA can be implemented to add an extra layer of security to SSH access. Regular software updates are crucial to addressing vulnerabilities and introducing new security features.

> **Note**
> It's advisable to avoid using RDP or SSH for remote access to VMs in the cloud. Instead, consider alternative solutions such as private connectivity (such as AWS Direct Connect), VPN tunnels, or dedicated options provided by CSPs such as AWS Session Manager or Azure Bastion. These alternatives offer enhanced security and better align with modern cloud security practices.

CSPs deliver **console-based access** as a pivotal element in their service offerings, providing customers with a versatile interface to manage and oversee their cloud resources. It encompasses both a **Graphical User Interface** (**GUI**) for intuitive point-and-click resource management and a **Command-Line Interface** (**CLI**) for advanced users and IaC practitioners, providing versatility. Console-based access facilitates comprehensive resource management, allowing users to create, modify, monitor, and delete a spectrum of cloud resources, including VMs, storage, databases, and networking components. Security is paramount, with features such as MFA, RBAC, and audit trails embedded within the console interface to ensure secure and authorized access. Real-time monitoring tools and analytics dashboards enhance visibility into resource performance and usage metrics, enabling proactive management and optimization.

Jumpboxes, also known as bastion hosts or jump servers, play a crucial role as secure intermediary systems within a network, providing an additional layer of security by serving as a single point of entry. These dedicated servers act as gateways, controlling and managing access to other machines, and require robust authentication measures such as MFA to establish connections. Access controls for jumpboxes prevent unauthorized users from directly accessing sensitive systems, enforcing a controlled and monitored access paradigm. Organizations deploy jumpboxes strategically to enhance security, ensuring that users must authenticate through the jumpbox before reaching critical resources. This intermediary approach not only helps monitor and control remote access but also contributes to a defense-in-depth strategy.

Virtual client solutions are platforms or applications facilitating secure and flexible remote access to virtualized desktop environments from diverse locations, empowering users to log in and access their virtual desktops securely. Access controls, including strong authentication measures and authorization protocols, are employed to verify the identity of users and limit access to authorized individuals. Users can access their virtual desktops without compromising sensitive information, as the actual processing occurs on servers rather than individual devices.

Keyboard, Video, Mouse (KVM) is a hardware device or technology that allows users, primarily CSPs, to control multiple computers or servers using a single set of peripherals. The KVM switch enables users to switch between different computers without the need for additional keyboards, monitors, and mice. KVM switches are commonly used in data centers, server rooms, and other environments where multiple computers need to be accessed and controlled by a single user. Stringent controls are implemented to limit interactions solely to keyboards and mice, while other USB and I/O functions are deliberately restricted. Firmware in secure KVMs is securely locked, reducing the risk of unauthorized modifications. The pushbuttons on secure KVMs are designed with air-gapped technology, adding an extra layer of security by physically isolating the control mechanism from electronic components. Components within the secure KVM are tamper-proof, meaning that any attempt to physically manipulate or compromise the device will be detectable. Secure KVMs isolate channels for keyboard, video, and mouse signals, preventing potential crosstalk or interference between connected systems and maintaining data integrity.

Secure Network Configuration

A well-designed, configured, and secured network is instrumental in fostering resilience, scalability, and the overall robustness of cloud-based infrastructures.

Virtual Local Area Networks (VLANs)

VLANs are a network segmentation technique that allows the partitioning of a physical network into multiple logical networks. By grouping devices logically rather than physically, VLANs enhance network efficiency, security, and management. Devices within the same VLAN communicate seamlessly, regardless of their physical location, while traffic between VLANs requires routing. VLANs are widely used in enterprises to isolate broadcast domains, improve network performance, and enforce security policies by creating distinct broadcast domains within a single physical network infrastructure. Key uses include network segmentation for enhanced performance, isolation of broadcast domains for reduced network congestion, and improved security by logically segregating different types of traffic.

Virtual Extensible LAN (VXLAN) extends the concept of VLANs to cloud environments, providing network segmentation and isolation functionalities at scale. Similar to VLANs, VXLAN allows for the logical partitioning of network resources into multiple virtual networks. However, VXLAN uses encapsulation techniques to encapsulate Layer 2 Ethernet frames within Layer 3 IP packets, enabling the creation of overlay networks that can span across physical network boundaries and data centers. This overlay network architecture enables seamless movement of workloads between different physical hosts and data centers, without being constrained by the underlying physical network topology. In cloud environments such as AWS, Azure, and GCP, VXLAN-like functionality is often implemented using technologies such as VPC in AWS and GCP and VNet in Azure.

Transport Layer Security (TLS)

TLS, the successor of **Secure Sockets Layer (SSL)**, is a cryptographic protocol designed to secure communication over a computer network. It operates at the transport layer, ensuring the privacy and integrity of data exchanged between applications, such as web browsers and servers, or within other network services.

The TLS protocol consists of two primary sub-protocols:

- **Record Protocol**: The Record Protocol is responsible for fragmenting, compressing (optional), encrypting, and authenticating the data to be transmitted. It ensures the confidentiality and integrity of the information exchanged between the client and server.

- **Handshake Protocol**: The TLS Handshake Protocol is crucial for establishing a secure connection between the client and server. It involves the following key steps:

    I.   The client initiates the handshake by sending a message indicating supported cryptographic algorithms and other parameters.

    II.   The server responds by selecting appropriate parameters from the client's list and providing its own certificate.

    III. Next is key exchange. The client and server agree on a premaster secret, from which they derive the session keys for encrypting and decrypting data.

IV.   The server presents its digital certificate to prove its identity, and the client may also provide one if requested.

V.   Both parties confirm the successful completion of the handshake, and the encrypted session begins.

Dynamic Host Configuration Protocol (DHCP)

DHCP is a network protocol used to automatically assign and manage IP addresses and configuration information to devices within a network. Key features of DHCP include the following:

- DHCP eliminates the manual configuration of IP addresses by automatically assigning unique addresses to devices within the network. This dynamic allocation promotes efficient resource utilization.

- In addition to IP addresses, DHCP can provide devices with essential configuration parameters such as subnet mask, default gateway, and **Domain Name System** (**DNS**) server addresses. This ensures that devices have the necessary information for seamless network communication.

- DHCP leases IP addresses to devices for a specific duration known as the lease period. The lease can be renewed or released, allowing for efficient IP address reuse and management.

- DHCP centralizes the administration of IP address allocation, making it easier for network administrators to manage and control IP address assignments within the network.

- By automating the IP address assignment process, DHCP reduces the likelihood of configuration errors that can arise from manual entry, promoting network stability and accuracy.

DNS and Domain Name System Security Extensions (DNSSEC)

DNS serves as the backbone of the internet, translating human-readable domain names into IP addresses. DNSSEC adds a layer of security to the current DNS structure. When a client initiates a DNS query (e.g., resolving a domain name to an IP address), it sends a DNS request to a DNS resolver or directly to a DNS server. The DNS resolver or server processes the request, either responding with the IP address associated with the queried domain or forwarding the request to other DNS servers if necessary.

DNS uses a hierarchical structure, with authoritative DNS servers responsible for specific domains. The resolver navigates this hierarchy to obtain the required information.

Responses from DNS servers include the requested data and are sent back to the client. DNS messages contain various types of **Resource Records** (**RRs**) providing information such as IP addresses, name servers, and more.

DNS primarily utilizes UDP on port 53 for quick and lightweight query-response transactions, while employing TCP on port 53 for larger data transfers and scenarios where responses may exceed the size accommodated by a single UDP packet, such as in zone transfers and handling large DNS responses.

The inherent vulnerabilities in traditional DNS create opportunities for malicious activities. To address these concerns, DNSSEC was introduced. Here are some of the key components of DNSSEC:

- **Digital signatures**: DNSSEC employs digital signatures to sign DNS RRs, offering a mechanism for verifying the authenticity and integrity of DNS data.

- **Public Key Infrastructure (PKI)**: DNSSEC relies on PKI principles, utilizing public keys for signature verification. The **DNSKEY** RR contains the public key used in the authentication process.

- **Chain of trust**: DNSSEC establishes a chain of trust, linking the authenticity of one domain's DNS data to the cryptographic signatures of its parent domain.

You also must be aware of the major DNS attacks that DNSSEC prevents:

- **Cache poisoning**: In cache poisoning attacks, attackers manipulate DNS caches to redirect users to malicious websites by injecting false information into the cache.

- **DNS spoofing**: DNS spoofing involves providing false DNS responses to redirect users to malicious sites, exploiting the trust placed in the DNS system.

- **Man in the Middle (MitM)**: In a DNS-based MitM attack, an adversary intercepts and alters DNS requests and responses, enabling them to eavesdrop on or manipulate communications.

VPN

A VPN is a technology that establishes a secure and encrypted connection over the internet, creating a private network for users to access remotely. The primary purpose of a VPN is to ensure the confidentiality and privacy of data transmission, allowing users to securely connect to a private network from various locations. This technology is vital for remote access, enabling users to interact with corporate resources as if they were physically present within the network.

In a **split-tunnel VPN** configuration, the network traffic is divided between the VPN tunnel and a direct connection to the internet. Only specific data destined for the corporate network traverses the VPN, optimizing bandwidth usage and reducing latency for internet-bound traffic. This configuration is advantageous for organizations seeking a balance between security and efficient bandwidth utilization. Users can simultaneously access local resources and the internet securely through the split-tunnel VPN.

Contrastingly, a **full-tunnel VPN** directs all network traffic, including both corporate and internet-bound, through the VPN tunnel to the corporate network. This ensures that all data is encrypted and passes through the secure VPN connection, providing a higher level of security. Full-tunnel VPNs are often chosen for scenarios where centralized control and monitoring of all user traffic are crucial. This configuration facilitates the consistent application of security policies and offers enhanced visibility into user activities.

Remote-access VPNs, also known as client-side VPNs or point-to-site VPNs, cater to individual users, allowing them to securely connect to a private network from remote locations. This form of VPN is particularly relevant to employees working from home or traveling, providing a secure channel to access corporate resources. Remote access VPNs ensure that users can maintain productivity and connectivity while preserving the confidentiality and integrity of data transmitted over the internet.

In contrast, **site-to-site VPNs** establish secure connections between different physical locations of a network, connecting entire networks together. This configuration is commonly utilized for interconnecting branch offices or geographically dispersed locations. Site-to-site VPNs facilitate seamless communication between multiple sites, allowing for the secure exchange of data and resources across the connected networks.

Network Security Controls

Network security controls are critical measures implemented to safeguard a computer network infrastructure from unauthorized access, cyber threats, and data breaches.

Firewalls

Firewalls are fundamental network security devices designed to monitor, filter, and control incoming and outgoing network traffic based on predetermined security rules. They act as a barrier between a trusted internal network and untrusted external networks, establishing a secure perimeter. You must know the different types of firewalls as listed below:

- **Hardware firewalls**: These are physical devices positioned at the network perimeter to filter traffic between the internal network and the internet.

- **Software (host-based) firewalls**: These are software applications or programs installed on individual devices to monitor and control network traffic at the device level.

- **Stateless firewalls**: These network traffic filters operate based solely on predefined rules and criteria, such as source and destination IP addresses, ports, and protocols. They lack awareness of the state or context of the active connections; each packet is evaluated in isolation.

- **Stateful inspection firewalls**: These maintain awareness of the state of active connections, making decisions based on the context of the traffic. They track the state of connections, allowing for a more sophisticated analysis of packet content and context.

- **Proxy firewalls**: These act as intermediaries between internal and external systems, forwarding requests and responses. They can modify and filter traffic for enhanced security. They mask the identity of internal devices, inspect and filter application-layer data, and provide an additional layer of privacy.

- **Application firewalls**: These focus on the application layer of the OSI model, allowing or blocking traffic based on the specific application or service. They provide granular control over applications, enabling security policies based on the characteristics of each application.

- **Next-Generation Firewalls (NGFWs)**: These integrate traditional firewall features with advanced capabilities such as intrusion prevention, application awareness, and user identity tracking. They offer a holistic approach to security by combining traditional firewall functionalities with advanced threat detection and mitigation features. They are layer 7 firewalls operating at the application layer of the OSI model, allowing them to analyze traffic at a deeper level. This enables them to detect attacks and control traffic based on the actual protocol being used, even if they use non-standard network ports for communication.

Intrusion Detection System (IDS) and Intrusion Prevention System (IPS)

An IDS monitors and analyzes network or system activities for signs of unauthorized access, malicious activities, or security policy violations. An IDS operates by employing various detection methods, including signature-based detection (the system compares observed patterns against known signatures of known threats), anomaly-based detection (establishes a baseline of normal behavior and flags any deviations that may indicate an intrusion), and heuristic-based detection (identifies novel threats based on behavioral deviations). An IDS can be deployed as either a **Network-Based IDS (NIDS)**, monitoring network traffic, or a **Host-Based IDS (HIDS)**, focusing on individual devices or hosts by analyzing log files and system activities.

Building upon the capabilities of an IDS, an IPS takes a more proactive approach by not only detecting but also actively preventing or blocking malicious activities within a network or system. An IPS also can be deployed as a **Network-Based IPS (NIPS)**, operating at the network level to actively block malicious traffic, or a **Host-Based IPS (HIPS)**, installed on individual devices to prevent malicious activities at the host level.

While IDSs and IPSs play pivotal roles in network security, they are not without their challenges. Both systems may generate false positives, triggering alerts or blocking legitimate traffic. Additionally, the continuous monitoring and active prevention features of IDSs and IPSs can consume system resources, requiring careful consideration during implementation. Despite these challenges, the combined strengths of IDSs and IPSs contribute to a robust defense strategy against a dynamic and ever-evolving threat landscape.

Honeypots

A honeypot is a security mechanism designed to deceive and lure attackers into a controlled environment, allowing security professionals to study their tactics, techniques, and motives. Honeypots are intentionally made vulnerable to attract malicious actors who eventually provide insights into the methods they use and their potential targets. Honeypots provide valuable threat intelligence by capturing the tactics and tools used by attackers. They can detect attacks in their early stages before they reach critical systems. By diverting attackers to a controlled environment, honeypots deceive them, reducing the risk to actual production systems.

A **honeynet** is a network of interconnected honeypots and other security devices designed to work together. It provides a more comprehensive view of an attacker's activities across multiple systems.

However, honeypots have their own challenges that you must be aware of. Here are the major obstacles faced while implementing a honeypot project:

- Honeypots can be resource-intensive to deploy and maintain. This may include significant resource investments.

- The management of honeypot-generated alerts may divert attention from genuine security events, impacting incident response efficiency.

- In some cases, honeypots may inadvertently disclose details such as IP addresses, hostnames, or other attributes that could be used by attackers to gather intelligence about the organization's infrastructure.

- Deploying honeypots without proper authorization or failing to adhere to legal boundaries can lead to legal consequences. The concept of entrapment, where an organization lures attackers into illegal activities, may raise ethical concerns and legal challenges.

Despite these challenges, strategically deploying and managing honeypots can greatly contribute to threat intelligence, early detection, security research, and understanding the tactics of adversaries.

OS Baselining

A **baseline** refers to a predefined set of configurations, settings, or standards that represent the desired and secure state of a system, application, or environment. It serves as a reference point or starting point for ensuring consistency, security, and compliance across multiple instances or components within an organization's IT infrastructure. The purpose of a baseline is to establish a known and secure configuration that aligns with industry best practices, regulatory requirements, and organizational security policies. Baselines are commonly used for OSs, software applications, network devices, and other IT components. Key characteristics of a baseline include the following:

- Baselines promote standardization by defining a common set of configurations that should be applied consistently to all relevant systems.

- Security baselines are designed to meet security standards and mitigate known vulnerabilities, reducing the risk of security incidents.

- They ensure that all instances or components within an IT environment share similar configurations, minimizing variations that could introduce security risks.

- They provide a basis for auditing and assessing compliance with security policies, regulatory requirements, and industry standards.

- They simplify the management and maintenance of IT systems by offering a unified and secure starting point for deployments.

Organizations often use frameworks or guidelines, such as those provided by the **Center for Internet Security (CIS)** or the **Defense Information Systems Agency (DISA)**, to create and maintain security baselines tailored to their specific needs. However, there are some best practices for OS baselining that are valid across different requirements. These are listed below:

- Definition of comprehensive security policies and guidelines governing user access, authentication, authorization, and data protection.

- Configuration of user account settings, including removal of default accounts, implementation of robust password policies, account lockout thresholds, and permissions.

- Settings related to network interfaces, firewall rules, and communication protocols to ensure secure and efficient network operations. Elimination of unnecessary open ports.

- Procedures for applying OS patches and updates to address known vulnerabilities, ensuring the removal of potential security risks.

- Configuration of logging and auditing mechanisms to capture security-related events for monitoring, analysis, and compliance purposes.

- Installation and configuration of antivirus software and endpoint protection measures to detect and mitigate malware threats.

- Settings to control and manage external devices, peripherals, and removable media, removing unnecessary services and libraries to prevent unauthorized access and data breaches.

- Definition of rules for allowing or restricting the execution of specific applications, removing unnecessary or potentially risky apps to enhance application security.

- Implementation of encryption measures for data at rest and data in transit.

- Establishment of backup and recovery procedures to ensure data integrity and facilitate system restoration in case of failures or incidents.

- Thorough documentation of the baseline configurations, including versioning and change history, to support auditing and accountability.

- Implementation of monitoring tools and processes to detect deviations from the baseline and respond to potential security incidents.

- Integration with configuration management tools to automate the deployment, maintenance, and tracking of baseline configurations across various systems.

- Integrate automated vulnerability scanning tools into the baseline for regular assessments and proactive identification of security risks.

- Consider adopting an immutable infrastructure approach where VMs are treated as disposable and are replaced with new instances rather than being modified. This reduces the risk of unauthorized changes.

- Integrate **policy-as-code** solutions such as AWS **Service Control Policies** (**SCPs**), Azure policies, or GCP organization policies to establish guardrails, such as prohibiting resource deployments with public IPs, enforcing encryption at rest, and ensuring auditing and key rotation.

Any proposed changes or deviations from the established baseline should undergo a meticulous review and approval process within the change and configuration management framework. This comprehensive approach ensures that modifications align with established security policies and standards. Changes approved in this process should be promptly reflected in the **Configuration Management Database** (**CMDB**), contributing to accurate documentation of the current state of configurations and supporting informed decision-making.

It is imperative to document any approved exceptions or modifications to the baseline configuration. This documentation should include a clear rationale for each exception, an assessment of associated risks, and the implementation of mitigating controls. Regular reviews of baseline configurations are essential to assess their effectiveness and relevance in the evolving security landscape. If a consistent need for exceptions or changes arises during these reviews, organizations should consider adjusting the baseline to accommodate emerging security requirements.

Patch Management

Patches are software updates designed to address specific issues, enhance functionality, or, most critically, fix vulnerabilities in software, OSs, or applications. These updates are released by software vendors to improve security, performance, and overall reliability.

Patch management is the systematic process of planning, implementing, monitoring, and reviewing the deployment of patches across an organization's IT infrastructure. The primary goal is to keep software and systems up to date, secure, and resilient by addressing known vulnerabilities and ensuring that the latest improvements are applied.

Now, you will go through the key aspects of the patch management process:

- Regularly scan and assess systems to **identify vulnerabilities** that require patching.

- **Evaluate the criticality** of each patch based on the severity of the vulnerability and the importance of the affected systems.

- **Test patches** in a controlled environment to ensure they do not cause conflicts or disruptions.

- Obtain the **necessary approvals** for deploying patches, considering factors such as potential impact on operations.

- **Deploy** approved patches to systems according to a schedule that minimizes impact on users and operations.

- **Monitor** the deployment process to ensure that patches are successfully applied and systems remain operational.

- **Verify** that the applied patches have effectively addressed the identified vulnerabilities.

- Maintain **documentation** of applied patches, including version details and deployment dates.

- **Communicate** with users to inform them about upcoming patches, their importance, and any potential impact on system availability.

- Generate **reports** on the status of patch deployments, system compliance, and any issues encountered.

Automated and Manual Patching

Automated patching involves using software tools to automatically identify, download, and deploy patches to systems. It streamlines the patch management process, reduces human error, and ensures consistency. It is well suited for large-scale deployments and environments with numerous endpoints.

Manual patching requires human intervention to identify, download, and deploy patches. It often involves a more hands-on approach, allowing for careful testing and customization. It is suited for smaller environments or situations where a higher level of control is desired.

Organizations often use a combination of automated and manual patching, depending on their specific needs, the criticality of systems, and the complexity of their IT infrastructure.

In the cloud environment, patch management has evolved with innovative approaches to enhance efficiency and security. You must be familiar with the prominent approaches, as listed here:

- **IaC**: Embracing IaC allows organizations to define and manage infrastructure in a declarative manner. This means that patching can be automated through code, enabling the rapid deployment of updated configurations across cloud resources. IaC facilitates version-controlled, repeatable processes, ensuring consistency and reducing the risk of misconfigurations during patch application.

- **Immutable architecture**: Adopting an immutable architecture approach involves treating infrastructure components as unchangeable artifacts. Instead of updating existing instances, new instances with the necessary patches are deployed and the old instances are decommissioned. Immutable architecture enhances security by reducing the attack surface and minimizing the time during which vulnerable systems are active.

- **Software Composition Analysis (SCA)**: SCA tools analyze the software components and dependencies within an application, identifying vulnerabilities in third-party libraries or open source packages. Integrating SCA into the patch management process allows organizations to proactively address vulnerabilities in the software supply chain, enhancing overall security posture.

IaC Strategy

IaC is a transformative approach that involves managing and provisioning infrastructure through machine-readable scripts and configuration files. A well-defined IaC strategy enhances agility, consistency, and collaboration in the development and deployment of IT infrastructure. This approach aligns with modern DevOps practices, enabling teams to deliver infrastructure changes more rapidly and reliably while maintaining a high level of control and governance.

Here's a comprehensive IaC strategy:

- **Define** infrastructure components (servers, networks, and databases) in code using declarative or imperative scripts. Popular languages for IaC include YAML, JSON, and domain-specific languages such as **HashiCorp Configuration Language (HCL)**.

- Utilize **version control** systems (e.g., Git) to track changes, manage versions, and collaborate on infrastructure code. This ensures a history of modifications, facilitates collaboration among team members, and enables rollback to previous states.

- Implement **modular code** structures to enhance reusability. Break down infrastructure code into manageable modules that can be reused across projects, promoting consistency and efficiency.

- Maintain **documentation** within the code repository. Document the purpose, configuration, and usage of infrastructure components directly in the code to ensure that documentation stays synchronized with the code base.

- Implement **automated testing** for IaC scripts to validate configurations and catch errors early in the development process. This includes syntax checking, linting, and functional testing to ensure the correctness of the infrastructure code.

- **Integrate IaC into CI/CD pipelines** for automated testing and deployment. This allows for continuous integration of code changes, automated testing, and seamless deployment of infrastructure updates.

- Embrace the concept of **immutable infrastructure**, where components are treated as disposable and replaced with new instances rather than being modified. This enhances consistency, reduces drift, and ensures that infrastructure is always in a known, tested state.

- Embed **security practices into the IaC process** by implementing security policies and best practices directly into the code. This includes defining firewall rules, access controls, and encryption settings as part of the infrastructure code.

- Integrate **monitoring and compliance checks** into the IaC pipeline to ensure that deployed infrastructure adheres to security and compliance standards. Automated checks help maintain a secure and compliant environment.

- Design IaC scripts to be **scalable and flexible** to accommodate changes in infrastructure requirements. Consider parameterization and dynamic configuration options to support different environments and scaling needs.

Note

Declarative IaC involves expressing the desired state of the infrastructure without specifying the step-by-step process of achieving that state. It describes the end state, focusing on what the infrastructure should look like. Users declare the desired configuration, and the IaC tool determines the steps to achieve it. It promotes simplicity and abstraction, allowing for a more concise representation of infrastructure.

Idempotent IaC ensures that applying the same configuration multiple times produces the same result as applying it once. Applying IaC repeatedly does not cause unintended side effects or alter the system if the configuration is already in the desired state. It enables safer and more predictable deployments by preventing unnecessary changes or disruptions. It supports a consistent and reliable deployment process, reducing the risk of configuration drift.

Availability of Clustered Hosts

Clustered hosts refers to a group of interconnected computing nodes or servers that work together collaboratively to provide enhanced performance, availability, and scalability. Clustering involves the interconnection of multiple servers to operate as a unified system, sharing resources and distributing workloads efficiently. It also facilitates load balancing, fault tolerance, and efficient resource utilization.

Storage clusters, typically managed by CSPs, are configurations in which multiple storage devices or nodes are grouped together to provide a unified and scalable storage infrastructure. These clusters are designed to enhance storage performance, availability, and reliability by distributing data across multiple nodes. There are two main types of storage clusters:

- In **tightly coupled storage clusters**, multiple storage nodes operate in close proximity within a single physical location, forming an integrated and high-performance storage system. These clusters are characterized by low-latency communication and high throughput between nodes. The nodes are interconnected through dedicated, high-speed connections, allowing for seamless data access and efficient resource utilization. Tightly coupled clusters are often managed as a single administrative domain, simplifying configuration and maintenance tasks. Scalability is achieved by horizontally expanding the cluster with additional nodes, enhancing both storage capacity and performance.

- **Loosely coupled storage clusters**, on the other hand, involve storage nodes that are distributed across different locations, potentially spanning multiple data centers. These clusters rely on standard network protocols for communication, introducing higher latency compared to tightly coupled clusters. The geographical distribution provides benefits such as improved disaster recovery capabilities and fault tolerance. However, managing and configuring loosely coupled clusters may involve complexities associated with distributed systems and network considerations. Despite the potential for increased network latency, loosely coupled clusters offer flexibility in terms of resource placement and independence.

Distributed Resource Scheduling (DRS)

DRS is a pivotal feature in clustered environments, emphasizing the optimization of resource utilization and workload distribution across multiple hosts. Its continuous monitoring of host resource utilization allows for dynamic load balancing by intelligently migrating VMs between hosts, preventing resource bottlenecks, and maximizing overall performance. DRS operates with automation and intelligent algorithms, responding in real time to resource demand, availability, and configured rules. This proactive approach minimizes manual intervention, ensuring swift adaptation to changing workload patterns. Administrators can define policies, including affinity or anti-affinity rules, guiding DRS behavior to meet specific organizational requirements. DRS enables live VM migration without service disruption, facilitating optimal resource utilization for maintenance, load balancing, or addressing hardware failures seamlessly. DRS also considers power efficiency, consolidating VMs on a subset of hosts and powering down unused hosts during periods of low demand. This not only contributes to energy savings but also aligns with cost-effective practices and environmental sustainability goals.

Dynamic Optimization

Dynamic optimization is a feature in clustered environments that focuses on **real-time adaptability** and efficiency in managing virtualized resources. Key aspects of dynamic optimization are that it constantly monitors the performance and resource utilization of VMs within a cluster, and it responds in real time to changing workloads by dynamically adjusting resource allocations to meet demands.

Maintenance Mode

Maintenance mode is a crucial feature in clustered environments, providing a mechanism for performing updates, maintenance tasks, and hardware repairs without disrupting the overall operation of the cluster. Maintenance mode allows administrators to gracefully take a host or node offline for maintenance activities. This can include tasks such as applying updates, installing patches, or conducting hardware repairs. The host is temporarily marked as unavailable for VM workloads.

High Availability (HA)

HA is integral to clustered environments, ensuring continuous service availability by mitigating the impact of hardware failures or other disruptions. HA achieves fault tolerance through redundancy, employing multiple hosts capable of hosting VMs. Automated failover mechanisms detect and respond to failures swiftly, redirecting workloads to healthy hosts to minimize downtime. Continuous monitoring of host and VM health, along with VM restart policies, ensures efficient resource utilization and prioritizes critical workloads. HA systems integrate tightly with cluster management tools, coordinating workload redistribution during failover events while maintaining consistent cluster states through quorum and voting mechanisms. Proactive measures, such as health checks and live migrations, enhance system stability, scalability, and flexibility. Overall, HA is a cornerstone of disaster recovery planning, ensuring uninterrupted service delivery in dynamic and evolving clustered environments.

Availability of Guest OSs

CSPs offer a range of features and services to enhance the availability of guest OSs within their virtualized environments:

- CSPs provide **load-balancing** services that distribute incoming traffic across multiple instances of VMs running a guest OS. **Auto-scaling** capabilities automatically adjust the number of VM instances based on demand, ensuring optimal performance and availability.

- Most of the CSPs offer the concept of **availability zones**, which are physically separate data centers within a region. Deploying VMs across different availability zones enhances fault tolerance and resilience.

- **Live migration** features enable CSPs to move running VMs between physical hosts without downtime. Snapshotting allows users to capture the current state of a VM, facilitating quick backups and recovery.

- CSPs offer **managed backup and restore services**, allowing users to schedule and automate backups of VMs and their associated guest OS.

- CSPs often integrate **global CDNs** to distribute content and applications closer to end users. This enhances the performance and availability of guest OS-based applications by reducing latency and optimizing content delivery across geographically dispersed locations.

- CSPs may provide **automated patch management** services for guest OS instances, ensuring that the OS is up to date with the latest security patches and updates.

- CSPs typically define **Service-Level Agreements (SLAs)** that specify the level of availability and performance users can expect. These SLAs often include guarantees related to uptime, response times, and resolution of issues, providing users with assurances regarding the availability of their guest OS instances.

- CSPs provide **monitoring and analytics tools** that offer insights into the performance and health of guest OS instances. Real-time monitoring enables proactive responses to potential issues, contributing to the overall availability of applications and services.

Performance and Capacity Monitoring in Cloud Environments

Effective performance and capacity monitoring in the cloud enables organizations to maintain optimal system health, proactively address issues, and plan for future growth. Cloud administrators leverage a combination of automated tools, alerts, and analytics to ensure the reliable and efficient operation of cloud-based infrastructure:

- **Network monitoring**: Cloud administrators employ network monitoring tools to track bandwidth utilization, latency, and overall network health. Monitoring helps identify potential bottlenecks, ensures optimal data transfer, and aids in troubleshooting connectivity issues.

- **Compute monitoring**: Performance monitoring for compute resources involves tracking the utilization, processing power, and efficiency of VMs or containers. This includes monitoring CPU usage, disk I/O, and other metrics to optimize resource allocation and identify potential performance issues.

- **Storage monitoring**: Storage monitoring focuses on the performance and capacity of cloud storage solutions. Metrics such as **Input/Output Operations Per Second** (**IOPS**), latency, and available storage space are monitored to ensure efficient data storage, retrieval, and overall system responsiveness.

- **Memory monitoring**: Memory monitoring involves tracking the utilization and availability of **Random Access Memory** (**RAM**) in cloud instances. Monitoring memory usage helps prevent resource exhaustion, optimize application performance, and identify memory leaks or inefficiencies.

- **Response time monitoring**: Monitoring response times is crucial for assessing the overall performance and user experience of cloud-hosted applications. This includes tracking the time it takes for applications to respond to user requests, ensuring optimal responsiveness, and identifying areas for improvement.

In cloud environments, automated alerting systems are employed to notify administrators when monitored metrics exceed predefined thresholds, enabling proactive resolution of potential performance issues. Capacity planning involves forecasting resource needs using historical data and current trends, ensuring ample resources to prevent service interruptions.

Hardware Monitoring in Cloud Environments

Hardware monitoring in cloud environments encompasses the surveillance of various components to ensure optimal performance and prevent potential issues. Key aspects of hardware monitoring include the following:

- **Disk monitoring**: Continuous tracking of disk performance, usage, and health. Monitoring tools assess factors such as disk I/O operations, available storage space, and potential disk failures, ensuring reliable storage operations.

- **CPU monitoring**: Real-time monitoring of CPU utilization, processing speed, and overall performance. This helps identify resource bottlenecks, optimize workload distribution, and ensure efficient utilization of computational power.

- **Fan speed monitoring**: Monitoring the speed of system fans is crucial to prevent overheating. Deviations in fan speed can indicate potential cooling issues, and monitoring tools trigger alerts to address temperature regulation and prevent hardware damage.

- **Temperature monitoring**: Continuous monitoring of hardware temperatures, including CPUs, GPUs, and other critical components. Elevated temperatures can lead to performance degradation or hardware failures, making temperature monitoring essential for preemptive interventions.

Configuration of Host and Guest OS Backup and Restore Functions

Ensuring robust backup and restore functions for both host and guest OSs is integral to maintaining data integrity and system resilience in cloud environments. Cloud administrators should configure scheduled backups for the host OS, capturing system configurations, application settings, and critical data, utilizing disk imaging or file-level backups for full system recovery in case of host OS failures. For each VM or container running a guest OS, tailored backup configurations are essential, using backup agents or integrated solutions within VM management tools. Backup frequency and retention policies are established based on data criticality, balancing historical recovery points with storage considerations. Configurable storage options include dedicated backup storage volumes, cloud-based object storage, or external backup solutions, considering factors such as data durability, accessibility, and cost. Encryption mechanisms enhance data security during transit and storage, ensuring confidentiality. Automated scheduling minimizes manual intervention, ensuring consistent data protection, while periodic testing validates backup restoration integrity. Monitoring and alerting systems are configured for backup processes, providing prompt notifications in case of failures.

Summary

Throughout this chapter, you gained insights into the robust security provided by HSMs and TPMs, ensuring a high level of data protection. Operational and maintenance aspects were thoroughly explored, encompassing the intricate details of the IaC strategy, patch management processes, and techniques for hardening both host and guest OSs. The significance of network security controls was highlighted, emphasizing their critical role in safeguarding data integrity. Various network configurations, including VLAN, DNS, DHCP, and TLS, were examined in detail, offering a comprehensive understanding of how these elements contribute to a secure and efficient network infrastructure. The chapter also delved into the importance of management tools, shedding light on their crucial role in overseeing and optimizing cloud environments. Techniques for hardening these tools were discussed, ensuring their resilience against potential security threats. Overall, you've acquired a comprehensive understanding of key elements, strategies, and best practices in building and maintaining a secure and efficient cloud infrastructure.

The next chapter sheds light on operational controls and measures in cloud environments. You will go through the standards governing these operations in detail.

Exam Readiness Drill – Chapter Review Questions

Apart from a solid understanding of key concepts, being able to think quickly under time pressure is a skill that will help you ace your certification exam. That is why working on these skills early on in your learning journey is key.

Chapter review questions are designed to improve your test-taking skills progressively with each chapter you learn and review your understanding of key concepts in the chapter at the same time. You'll find these at the end of each chapter.

> **How to Access These Materials**
>
> To learn how to access these resources, head over to the chapter titled *Chapter 25, Accessing the Online Resources.*

To open the Chapter Review Questions for this chapter, perform the following steps:

1. Click the link – `https://packt.link/CCSPE1_CH17`.

    Alternatively, you can scan the following **QR code** (*Figure 17.1*):

Figure 17.1 – QR code that opens Chapter Review Questions for logged-in users

2.   Once you log in, you'll see a page similar to the one shown in *Figure 17.2*:

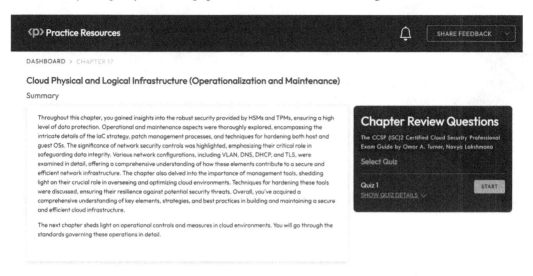

Figure 17.2 – Chapter Review Questions for Chapter 17

3.   Once ready, start the following practice drills, re-attempting the quiz multiple times.

Exam Readiness Drill

For the first three attempts, don't worry about the time limit.

ATTEMPT 1

The first time, aim for at least **40%**. Look at the answers you got wrong and read the relevant sections in the chapter again to fix your learning gaps.

ATTEMPT 2

The second time, aim for at least **60%**. Look at the answers you got wrong and read the relevant sections in the chapter again to fix any remaining learning gaps.

ATTEMPT 3

The third time, aim for at least **75%**. Once you score 75% or more, you start working on your timing.

> **Tip**
>
> You may take more than **three** attempts to reach 75%. That's okay. Just review the relevant sections in the chapter till you get there.

Working On Timing

Target: Your aim is to keep the score the same while trying to answer these questions as quickly as possible. Here's an example of how your next attempts should look like:

Attempt	Score	Time Taken
Attempt 5	77%	21 mins 30 seconds
Attempt 6	78%	18 mins 34 seconds
Attempt 7	76%	14 mins 44 seconds

Table 17.1 – Sample timing practice drills on the online platform

> **Note**
>
> The time limits shown in the above table are just examples. Set your own time limits with each attempt based on the time limit of the quiz on the website.

With each new attempt, your score should stay above **75%** while your "time taken" to complete should "decrease". Repeat as many attempts as you want till you feel confident dealing with the time pressure.

18
International Operational Controls and Standards

In the previous chapter, you covered the construction and implementation of cloud infrastructure, including hardware-specific security measures, operational aspects such as access controls and patch management, and topics such as **Infrastructure as Code (IaC)** strategy and performance monitoring. Building upon this, in this chapter, you will focus on operational controls and measures in cloud environments, underscoring the importance of standards governing these operations. You will learn how these standards ensure the effective, secure, and reliable functioning of services. Additionally, you will explore the crucial roles played by established frameworks such as **Information Technology Infrastructure Library (ITIL)** and **International Organization for Standardization/International Electrotechnical Commission (ISO/IEC)** 20000-1 in orchestrating seamless operations. Understanding these operational controls and standards is not only essential for success in the CCSP exam but is also vital for ensuring the smooth functioning of cloud services.

By the end of this chapter, you will be able to confidently answer questions on:

- Operational controls and standards
- Change management
- Continuity management
- Information security management
- Continual service improvement management
- Incident management and problem management
- Release management and deployment management
- Configuration management
- Service-level management, availability management, and capacity management

Operational Controls and Standards

Operational controls and standards in IT environments, whether traditional or cloud-based, encompass a set of measures and practices designed to ensure the effective, secure, and reliable operation of services. Operational controls and standards in the cloud share similarities with traditional IT environments, allowing established frameworks such as ITIL and ISO/IEC 20000-1 to be applied in both contexts.

ITIL

ITIL is a comprehensive framework of best practices for **Information Technology Service Management (ITSM)**. ITIL provides a structured approach to IT service delivery and support, aiming to align IT services with the needs of the business.

ITIL consists of a series of five core volumes, each focusing on a specific stage in the IT service life cycle:

- **Service Strategy**: This stage involves defining the overall strategy for IT services, ensuring they align with business objectives and customer needs.

- **Service Design**: This encompasses the planning and design of IT services to meet the requirements identified in the service strategy.

- **Service Transition**: This phase deals with the smooth transition of new or modified services into the live environment, minimizing disruptions.

- **Service Operation**: This is concerned with the day-to-day management of IT services to maintain their functionality and meet service-level agreements.

- **Continual Service Improvement (CSI)**: This emphasizes the ongoing evaluation and enhancement of IT services and processes, ensuring continuous optimization.

ISO/IEC 20000-1

ISO/IEC 20000-1 is a prominent international standard that defines the requirements for an **Information Technology Service Management System (ITSMS)**. This comprehensive framework is instrumental for organizations aiming to establish, implement, and continually enhance effective IT service management practices.

It focuses on creating a robust **Service Management System (SMS)** with policies, processes, and procedures for overseeing the entire IT services lifecycle. Applicable to internal and external IT service providers, it aligns with service delivery processes such as service level management, reporting, continuity, availability management, and budgeting. Notably, it integrates well with ITIL practices. Organizations can pursue certification to showcase compliance, confirming the establishment of an efficient ITSMS. Adhering to ISO/IEC 20000-1 enhances IT service delivery quality, customer satisfaction, and alignment with business objectives.

Change Management

Change management in IT refers to the structured process of planning, implementing, and controlling changes to the IT environment within an organization. It is a crucial aspect of IT service management that aims to ensure that changes are introduced in a way that minimizes risk, maintains system reliability, and aligns with business objectives. The primary goal of IT change management is to implement changes with minimal disruption to services and to prevent negative impacts on the organization's IT infrastructure.

ISO 27001:2013 integrates change management requirements into **Information Security Management Systems (ISMSs)**. ISO/IEC 20000-1:2018 encompasses change management requirements within the framework of **IT Service Management (ITSM)**. Similarly, ISO 9001:2015 addresses change management within the scope of **Quality Management Systems (QMSs)**.

The change management process typically involves several key stages:

1. **Request for Change (RFC)**: The process begins with the submission of an RFC, which could be initiated due to various reasons such as improvements, fixes, or compliance requirements. Change requests in the ITIL change management process typically go to the **Change Advisory Board (CAB)** or the **Change Control Board (CCB)**. These boards consist of key stakeholders, including representatives from IT, business units, and other relevant areas. An RFC should include an overview and purpose of the proposed change, details such as scope and objectives, an impact analysis covering business and technical aspects, a comprehensive implementation plan with testing and validation criteria, documentation, a communication plan, dependencies, cost considerations, a post-implementation review plan, and a rollback plan.

2. **Change Evaluation**: The proposed change initiates a formal process where the proposed change is thoroughly evaluated, including an assessment of its potential impact, risks, and benefits. This evaluation helps to determine whether the change should be approved, rejected, or requires further assessment based on factors such as alignment with organizational objectives, risk assessment, and potential benefits.

3. **Approval**: Once evaluated, the RFC goes through an approval process. This step involves obtaining the necessary authorizations and ensuring that all stakeholders are aware of the impending change. Approval indicates that the change is considered acceptable and can proceed to the implementation phase.

4. **Implementation**: The approved change is then scheduled and implemented. During this phase, it is crucial to follow predefined procedures, test the changes thoroughly, and have a backup plan in case unexpected issues arise.

5. **Review and Closure**: After implementation, the changes are reviewed to ensure they meet the desired outcomes. The change management process is then officially closed, and documentation is updated accordingly.

In change management, changes are categorized into three main types:

- **Standard Changes**: These are routine, low-risk changes that follow a pre-established and well-documented process. Standard changes are typically repetitive and have a known and accepted path for implementation, making them straightforward to execute without requiring extensive approval.

- **Normal Changes**: These changes are more significant than standard changes and might involve modifications to existing systems, applications, or infrastructure. Normal changes follow a structured process that includes thorough assessment, planning, and approval stages. They are typically implemented during scheduled change windows.

- **Emergency Changes**: Urgent and critical changes needed to address unforeseen issues or incidents fall under emergency changes. These changes aim to resolve immediate problems, and the process for emergency changes is expedited to ensure a swift response. However, despite the urgency, documentation and post-implementation review are required to maintain control and accountability.

Continuity Management

Continuity management, often referred to as **Business Continuity Management (BCM)**, is a holistic and strategic approach that organizations adopt to ensure the ongoing availability of critical business functions and processes in the face of disruptive events. The primary objective is to minimize the impact of incidents, such as natural disasters, cyber-attacks, and other unforeseen disruptions, on an organization's operations. ISO 22301, titled "Security and resilience – Business continuity management systems – Requirements," presents a framework for organizations to establish, implement, maintain, and enhance a **Business Continuity Management System (BCMS)**.

The continuity management process involves risk assessments to identify potential threats and vulnerabilities, followed by the development and implementation of comprehensive continuity plans. These plans outline strategies for maintaining essential services, managing resources, and facilitating timely recovery after a disruptive event. **Business Impact Analysis (BIA)** is conducted to assess the potential consequences of disruptions and prioritize critical business functions for continuity measures.

Continuity management is closely aligned with disaster recovery planning, which focuses on the restoration of IT infrastructure and data systems. Regular testing, training, and rehearsals of continuity plans ensure that organizations are well prepared to respond effectively during actual incidents.

Information Security Management

Information security management is a critical aspect of overall business governance that focuses on safeguarding an organization's information assets. It involves the development, implementation, and maintenance of policies, processes, and controls to ensure the confidentiality, integrity, and availability of sensitive information. Information security management aligns with various international standards and frameworks, such as ISO/IEC 27001, to establish a robust ISMS.

The process encompasses risk assessments, threat modeling, and the establishment of security controls to mitigate identified risks. Access controls, encryption, and regular security awareness training for employees are key components of an effective information security management strategy. Incident response plans and continuous monitoring mechanisms are implemented to detect and respond to potential security incidents promptly.

Information security management is a dynamic and evolving discipline that adapts to the ever-changing landscape of cyber threats and technological advancements. By prioritizing information security, organizations can protect sensitive data, maintain the trust of stakeholders, and comply with regulatory requirements.

Continual Service Improvement Management

Continual Service Improvement (**CSI**) is a vital phase within the ITSM life cycle that focuses on enhancing the efficiency, effectiveness, and overall quality of IT services. CSI management is a fundamental component of the ITSM framework, focusing on the ongoing enhancement of IT services to meet evolving business needs. The process follows a cyclical approach, employing the **Plan-Do-Check-Act** (**PDCA**) model to systematically identify, implement, and evaluate improvements.

In the "Plan" phase, potential areas for enhancement are identified, and improvement plans are developed. The subsequent "Do" phase involves implementing these plans, while the "Check" phase assesses the results through regular reviews and performance metrics. The "Act" phase integrates lessons learned, adjusting strategies and processes to optimize service delivery.

CSI management relies on **Key Performance Indicators** (**KPIs**) and metrics to measure the effectiveness of IT services against established objectives. Regular reviews, audits, and feedback mechanisms ensure that improvements align with organizational goals, fostering a culture of continual improvement. By embracing CSI management, organizations can adapt to changing requirements, boost customer satisfaction, and maintain a responsive and proactive IT service environment.

ISO/IEC 20000-1, titled "Information technology – Service management – Part 1: Service management system requirements," provides guidance on the establishment, implementation, maintenance, and continual improvement of a service management system. Within this standard, there is a strong emphasis on CSI, which involves regularly reviewing and enhancing service delivery processes, practices, and outcomes to better meet customer requirements and organizational objectives.

Incident Management

An incident is any event that causes, or may cause, an interruption to or reduction in the quality of a service. The primary goal of **incident management** is to restore normal service operations as quickly as possible, minimizing the impact on business operations. ISO/IEC 27035-2, titled "Information technology – Security techniques – Information security incident management – Part 2: Guidelines to plan and prepare for incident response," provides guidance on planning and preparing for incident response.

The process begins with the logging and categorization of incidents, often identified through user reports or automated monitoring systems. Incidents are then prioritized based on their impact and urgency, with high-priority issues receiving immediate attention. The Incident management team works to diagnose, resolve, and restore services to normal operation as swiftly as possible. Incident management aims not only to minimize service downtime but also to ensure transparent communication, efficient incident resolution, and continuous improvement.

Problem Management

Problem management focuses on proactively identifying and eliminating the root causes of recurring incidents to prevent future disruptions. While incident management addresses immediate issues to restore services, problem management takes a more strategic approach by investigating and resolving the underlying problems that lead to incidents. Problem management principles are frequently included in wider standards pertaining to IT service management or information security management, such as ISO/IEC 20000-1 and ISO/IEC 27001.

The problem management process commences with the identification of recurring incidents or patterns, categorizing and logging them based on their nature and potential impact on services. Rigorous **Root Cause Analysis (RCA)** is then conducted to delve into technical, process, or organizational factors contributing to the problem. Problems are prioritized by their impact and urgency, with high-priority issues, especially those affecting critical business functions, receiving immediate attention. The resolution plan outlines actions to eliminate the problem and prevent recurrence, with coordination with change management ensuring controlled implementation of changes. Linked closely with continual service improvement, problem management leverages lessons learned to enhance existing processes and avert similar issues in the future, ultimately preventing recurring incidents. This proactive approach results in more stable and reliable IT services, increasing service availability, minimizing disruption, and enhancing overall user satisfaction.

An Instance of Problem Management

Consider an organization where employees frequently fall victim to phishing emails that compromise corporate email accounts and sensitive company information. Each time a phishing attack occurs, the incident management team responds by notifying affected users, resetting compromised passwords, and conducting investigations to identify the source of the attack. However, these attacks persist, leading to unauthorized access to confidential data, financial fraud, and reputational damage to the company.

The problem management team steps in to investigate the recurring phishing attacks and uncover their root cause. Through analysis, they discover that the attacks are primarily facilitated by employees' lack of awareness and training on recognizing phishing emails. Additionally, they identify gaps in the company's email security protocols, such as ineffective spam filters and an absence of multi-factor authentication.

To address these underlying issues, the problem management team collaborates with the IT security team to implement a proactive phishing prevention strategy. This strategy includes the following:

- Comprehensive employee training and awareness programs to educate staff on identifying and reporting phishing emails.

- Implementation of advanced email security solutions, such as email authentication protocols (SPF, DKIM, DMARC), to detect and block phishing attempts.

- Deployment of multi-factor authentication for accessing corporate email accounts to prevent unauthorized access even if login credentials are compromised.

- Regular phishing simulation exercises to assess employees' awareness levels and identify areas for improvement.

As a result of these proactive measures, the frequency and success rate of phishing attacks decrease significantly. Employees become more vigilant and skilled at identifying phishing attempts, leading to fewer compromised email accounts and reduced risk of data breaches.

Release Management

Release management, covered in ISO/IEC 20000-1, is a critical component of the ITSM life cycle that focuses on planning, coordinating, and controlling the release of software and hardware changes into the live environment. The primary goal is to ensure a smooth and efficient transition of changes from development to production while minimizing disruptions to existing services. Release management encompasses activities such as planning, scheduling, risk assessment, testing, and communication with stakeholders to guarantee that the release meets quality standards and aligns with business objectives. By establishing well-defined release processes and collaborating across teams, release management contributes to the reliability and stability of IT services, fostering a controlled and predictable environment for implementing changes.

Deployment Management

Deployment management, also covered in ISO/IEC 20000-1, complements release management by concentrating on the execution phase of releasing software into the live environment. It involves the planning, coordination, and oversight of the installation and implementation of changes, ensuring that they are effectively deployed and configured. Deployment management works in tandem with release management to guarantee that the transition of changes is seamless and that the new or updated services operate as intended.

This process includes verification steps, rollback plans, and post-deployment reviews to address any issues promptly. By focusing on the practical aspects of implementing changes, deployment management plays a crucial role in realizing the objectives outlined in the release plan, contributing to the overall success of the IT service delivery.

Configuration Management

Configuration management focuses on establishing and maintaining accurate and up-to-date information about the **configuration items** (**CIs**) in an organization's IT infrastructure. CIs include hardware, software, documentation, and other components that contribute to delivering IT services. The primary goal of configuration management is to provide a comprehensive understanding of the configuration items, their relationships, and the configurations of the services they support. This information is crucial for various ITSM processes, including change management, incident management, and problem management. ISO/IEC 12207, titled *Systems and software engineering – Software life cycle processes*, provides guidelines for configuration management throughout the software development life cycle.

Configuration management begins by identifying and documenting all CIs and establishing a **Configuration Management Database** (**CMDB**) for storage. Each CI is uniquely identified, with relationships between CIs depicting dependencies within the IT infrastructure. The process extends to change control, where alterations are evaluated for their impact on existing configurations, ensuring the CMDB reflects the current IT environment in a controlled manner. Version control is pivotal, particularly for software; maintaining different versions and tracking changes guarantees correct deployment and reversible modifications.

Regular audits and reviews uphold the accuracy and completeness of the CMDB, identifying discrepancies, correcting inaccuracies, and enhancing overall configuration data quality. This meticulous configuration management approach significantly contributes to minimizing service disruptions by providing a clear understanding of the IT environment and enabling precise impact assessments during changes or incidents. Furthermore, it supports various ITSM processes by ensuring accurate and reliable information is readily available to IT teams and stakeholders, enhancing overall operational efficiency.

Service-Level Management

Service-Level Management (**SLM**) in the cloud is a critical process that ensures cloud services meet or exceed customer expectations. With the dynamic and scalable nature of cloud computing, SLM involves defining, negotiating, and managing **Service-Level Agreements** (**SLAs**) between cloud service providers and customers. These SLAs articulate the agreed-upon levels of service, including performance, availability, and response times. SLM is characterized by its ability to adapt to changing workloads and conditions, leveraging auto-scaling features to meet service levels efficiently. Real-time monitoring, analytics, and reporting play a crucial role in tracking and assessing the performance of cloud services, providing transparency and accountability.

Continuous improvement practices in SLM help organizations optimize their cloud services, ensuring they align with evolving business needs and contribute to overall customer satisfaction. ISO/IEC 20001 addresses SLM, focusing on defining and managing service levels outlined in SLAs, while also addressing deviations from agreed-upon standards. It ensures alignment of IT services with customer needs, monitors service performance, and facilitates continuous improvement efforts for enhanced service delivery.

Availability Management

Availability management in cloud computing is a critical aspect that revolves around ensuring the uninterrupted availability of IT services hosted in cloud environments. The first key consideration involves establishing clear and comprehensive SLAs with cloud service providers. These agreements should explicitly define the expected levels of service availability, performance metrics, and response times, providing a baseline for operational expectations. Within the ISO 20000 family, availability management is responsible for ensuring that services delivered by an organization meet agreed levels of availability.

An Instance of Availability Management

Within an SLA, it is imperative to distinctly articulate the expectations and commitments between the service provider and customers, particularly concerning scheduled and unscheduled maintenance. The following guidelines encompass key considerations for each category:

- Scheduled maintenance:

  - Clearly outline the regular maintenance windows or schedules during which the service provider may conduct updates, upgrades, or other maintenance activities.

  - Communicate the frequency, duration, and timing of scheduled maintenance with transparency, aiding customers in planning for potential service interruptions.

  - Incorporate a notification clause detailing how and when the service provider will apprise customers of upcoming scheduled maintenance, usually with advanced notice to mitigate the impact on users.

- Unscheduled maintenance (outages or downtime):

  - Establish a defined process for managing unscheduled maintenance, encompassing procedures to address unexpected outages, disruptions, and emergency maintenance.

  - Specify the expected response time for resolving issues leading to unscheduled downtime, thereby establishing accountability and defining customer expectations.

  - Include a notification clause addressing unscheduled maintenance, outlining how and when the service provider will communicate with customers during unforeseen service interruptions.

To enhance availability in cloud environments, organizations employ key strategies such as redundancy and failover mechanisms. Leveraging the global reach of cloud providers enables the deployment of resources across multiple regions, such as Availability Zones in Azure and AWS, ensuring redundancy and effective failover. This approach minimizes the impact of regional outages, contributing to a more resilient architecture. Dynamic resource allocation through auto-scaling adapts to varying workloads, optimizing resource utilization without compromising service availability. For instance, consider a scenario where an e-commerce website experiences a sudden surge in traffic during a holiday sale. With proper availability management strategies in place, such as auto-scaling to accommodate increased demand and redundant systems to mitigate potential failures, the website can maintain uninterrupted service for customers.

Integral to availability management in the cloud is robust data management. Implementing data replication across multiple data centers or regions ensures high data availability. Regular backups prevent data loss and facilitate swift recovery in unforeseen incidents. Real-time monitoring and alerting, alongside disaster recovery planning tailored to cloud environments, are essential components of an effective availability management strategy. Cloud monitoring tools provide insights into performance, enabling proactive responses, while automated alerts facilitate quick intervention, helping to maintain overall service availability.

Capacity Management

Capacity management, as covered in ISO/IEC 20000-1, in the cloud is a strategic process that revolves around optimizing the use of resources to meet varying workloads and ensure efficient service delivery. In the cloud environment, capacity can be dynamically scaled up or down based on demand, allowing organizations to adapt to changing business requirements. Capacity management encompasses planning, monitoring, and optimizing resources such as computing power, storage, and network bandwidth to ensure they align with the current and future needs of the organization.

In cloud computing, auto-scaling features play a crucial role in dynamically adjusting resource allocations in response to workload fluctuations. This ensures optimal resource utilization without manual intervention. Effective utilization of monitoring tools and analytics is crucial in tracking cloud resource performance, identifying bottlenecks, assessing utilization, and making informed decisions, enabling organizations to proactively adjust resource allocations and prevent performance issues, bottlenecks, and unnecessary costs. Additionally, consider a multi-cloud strategy, which involves distributing workloads across different cloud providers for increased resilience and reduced vendor lock-in. But be mindful of the complexities and costs introduced by multi-cloud environments. Alignment with SLAs is essential in the capacity management strategy, ensuring compliance with performance, availability, and reliability requirements. Collaboration between IT, finance, and business stakeholders is fundamental in this dynamic and continuous process, enabling organizations to achieve efficient resource utilization, cost savings, and agility in responding to evolving business needs in the cloud.

Summary

In this chapter, you gained insights into the vital role of operational controls and standards, discovering how they ensure effective, secure, and reliable IT service operations in both traditional and cloud environments.

The chapter emphasized the significance of frameworks such as ITIL and ISO/IEC 20000-1, providing a structured approach to IT service delivery and support.

You studied change management focusing on the structured process of planning, implementing, and controlling changes within the IT environment while minimizing risks. Continuity management emerged as a strategic approach to ensure the ongoing availability of critical business functions, particularly in the face of disruptive events.

Information security management, a critical aspect of business governance, focuses on safeguarding information assets. Continual service improvement management, a fundamental phase in the IT service management life cycle, and its role in enhancing the efficiency, effectiveness, and quality of IT services were also discussed.

Incident management was portrayed as a process dedicated to the swift resolution of disruptions to IT services, with the primary goal of restoring normal service operations promptly. Problem management, taking a strategic approach, aims to proactively identify and eliminate root causes of recurring incidents, preventing future disruptions.

You also understood release management's critical role in planning, coordinating, and controlling the release of software and hardware changes into the live environment. The chapter then shed light on deployment management's complementary role in overseeing the execution phase of releasing software into the live environment. You must ensure changes are effectively deployed and configured. Finally, you studied capacity management as a strategic process for optimizing the use of resources to meet varying workloads and ensuring efficient service delivery, incorporating auto-scaling features and collaboration between stakeholders.

In the next chapter, you will learn about forensic data collection methodologies, evidence management, and other key concepts for the collection, acquisition, and preservation of digital evidence in the cloud.

Exam Readiness Drill – Chapter Review Questions

Apart from a solid understanding of key concepts, being able to think quickly under time pressure is a skill that will help you ace your certification exam. That is why working on these skills early on in your learning journey is key.

Chapter review questions are designed to improve your test-taking skills progressively with each chapter you learn and review your understanding of key concepts in the chapter at the same time. You'll find these at the end of each chapter.

> **How to Access These Materials**
>
> To learn how to access these resources, head over to the chapter titled *Chapter 25, Accessing the Online Resources.*

To open the Chapter Review Questions for this chapter, perform the following steps:

1. Click the link – `https://packt.link/CCSPE1_CH18`.

    Alternatively, you can scan the following **QR code** (*Figure 18.1*):

Figure 18.1 – QR code that opens Chapter Review Questions for logged-in users

2. Once you log in, you'll see a page similar to the one shown in *Figure 18.2*:

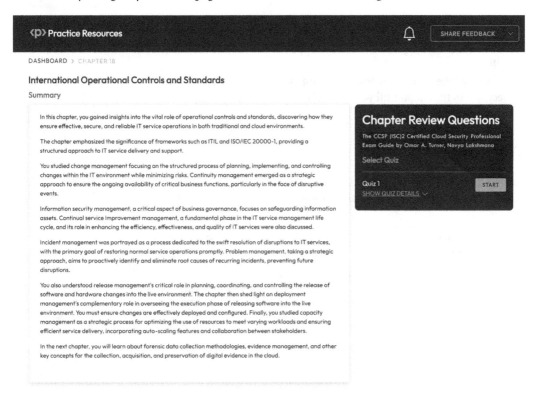

Figure 18.2 – Chapter Review Questions for Chapter 18

3. Once ready, start the following practice drills, re-attempting the quiz multiple times.

Exam Readiness Drill

For the first three attempts, don't worry about the time limit.

ATTEMPT 1

The first time, aim for at least **40%**. Look at the answers you got wrong and read the relevant sections in the chapter again to fix your learning gaps.

ATTEMPT 2

The second time, aim for at least **60%**. Look at the answers you got wrong and read the relevant sections in the chapter again to fix any remaining learning gaps.

ATTEMPT 3

The third time, aim for at least **75%**. Once you score 75% or more, you start working on your timing.

Tip

You may take more than **three** attempts to reach 75%. That's okay. Just review the relevant sections in the chapter till you get there.

Working On Timing

Target: Your aim is to keep the score the same while trying to answer these questions as quickly as possible. Here's an example of how your next attempts should look like:

Attempt	Score	Time Taken
Attempt 5	77%	21 mins 30 seconds
Attempt 6	78%	18 mins 34 seconds
Attempt 7	76%	14 mins 44 seconds

Table 18.1 – Sample timing practice drills on the online platform

---

Note

The time limits shown in the above table are just examples. Set your own time limits with each attempt based on the time limit of the quiz on the website.

With each new attempt, your score should stay above **75%** while your "time taken" to complete should "decrease". Repeat as many attempts as you want till you feel confident dealing with the time pressure.

19
Digital Forensics

This chapter explores digital forensics in detail, as it is crucial for cybercrime investigation as well as the CCSP exam. It gives prominence to the digital forensics lifecycle, covering identification, collection planning, preservation, analysis, and presentation stages. The collection methods are discussed to provide you with insights into forensically sound practices for evidence gathering. Moreover, you'll study evidence management, focusing on chain of custody documentation and legal considerations. The chapter aims to offer you an understanding of digital forensics, prioritizing key elements that are essential for effective investigative procedures.

By the end of this chapter, you will be able to confidently answer questions on:

- Digital forensics
- Forensic data collection methodologies
- Evidence management
- Collecting, acquiring, and preserving digital evidence

Digital forensics

Digital forensics (or computer forensics) is like being a detective but for the digital world. It involves gathering, examining, and saving electronic evidence to investigate cybercrimes, incidents, or digital conflicts. It encompasses the application of forensic techniques to digital media, devices, and networks with the goal of uncovering, preserving, and presenting evidence in a legally sound manner. Digital forensics plays a crucial role in understanding and mitigating cybersecurity incidents and is employed in various contexts, including law enforcement, corporate investigations, cybersecurity, and legal proceedings.

You will now investigate some standards and guidelines for managing the electronic discovery process, digital evidence collection, and legal considerations regarding these in the cloud:

- **ISO/IEC 27050:2016 – Information Technology – Security Techniques – Electronic discovery**: This standard focuses on **electronic discovery (e-discovery)**, which involves the identification, collection, and management of **electronically stored information (ESI)** during legal proceedings. E-discovery is particularly important in the context of litigation, regulatory compliance, and other legal processes where electronic information is involved. ISO/IEC 27050 aims to provide a framework for organizations to manage e-discovery processes effectively and securely.

- **ISO/IEC 27037:2012 – Guidelines for Identification, Collection, Acquisition, and Preservation of Digital Evidence**: This standard provides guidelines for the identification, collection, acquisition, and preservation of digital evidence. It is particularly relevant in the context of digital forensics, where the investigation of cyber incidents and crimes involves the analysis of digital evidence. Adhering to these guidelines helps maintain the integrity and reliability of digital evidence throughout the investigation process.

- **Cloud Security Alliance (CSA) Security Guidance Domain 3: Legal Issues: Contracts and Electronic Discovery**: The guidance delves into various aspects of contractual relationships in cloud environments, emphasizing elements such as defining responsibilities, specifying service levels, addressing data ownership and transfer, establishing liability frameworks, and ensuring compliance with relevant legal and regulatory requirements. The guidance explores legal requirements related to e-discovery in cloud environments, providing insights into crafting effective data retention policies, ensuring data preservation during legal proceedings, maintaining evidence integrity, and fostering collaboration between cloud customers and providers during legal processes.

The standards mentioned above provide a foundation for managing electronic discovery and digital evidence in the cloud. However, effectively implementing these standards requires a clear understanding of the specific stages involved in handling digital evidence throughout its lifecycle. ISO/IEC 27037 offers a structured approach by defining key stages that ensure the integrity and admissibility of digital evidence in legal proceedings.

This connection emphasizes the importance of both the standards (providing best practices) and the lifecycle stages (providing a structured workflow) for a comprehensive approach to digital evidence management in the cloud. Here are the key stages of the **digital evidence lifecycle** as per ISO/IEC 27037:

- **Identification**: This stage involves recognizing and identifying potential sources of electronic evidence that may be relevant to an investigation or legal matter. Identification may include determining the locations of relevant data and understanding its potential significance.

- **Collection planning**: Before collecting electronic evidence, a collection plan is established. This plan outlines the scope of the collection, the methods to be used, and the personnel involved. It considers legal and privacy considerations, ensuring that the collection process adheres to applicable laws and regulations.

- **Collection**: Actual collection involves gathering digital evidence in a forensically sound manner. This stage includes capturing data from various sources, such as cloud storage, virtual machines, containers, and application logs. The goal is to preserve the original state of the evidence to maintain its integrity.

- **Preservation**: Preservation ensures that collected electronic evidence remains unchanged and unaltered throughout its lifecycle. This stage involves using appropriate techniques and tools to create a secure and unmodifiable copy of the data. Preservation measures help maintain the evidentiary value of the digital evidence.

- **Recording**: Documentation is a crucial aspect of the digital evidence lifecycle. During the recording stage, detailed documentation is created, including information about the evidence, its source, the collection process, and any relevant metadata. This documentation contributes to establishing a clear chain of custody.

- **Transportation**: If digital evidence needs to be moved from one location to another, proper transportation procedures are followed. This includes secure packaging, tracking, and other measures to prevent tampering or loss during transit.

- **Analysis**: Analysis involves examining the collected evidence to extract relevant information, draw conclusions, and support the objectives of the investigation or legal proceedings.

- **Presentation**: The presentation stage involves presenting the findings of the analysis in a clear and understandable manner. This may include preparing reports, creating visual representations, and providing expert testimony in legal proceedings.

> **Note**
>
> Digital forensics and e-discovery are related fields within the broader scope of information security and legal processes, but they serve different purposes and involve distinct processes. Digital forensics focuses on investigating and analyzing digital evidence for crimes, while e-discovery involves identifying and managing ESI for legal purposes, such as litigation and compliance.

Forensic data collection methodologies

Forensic data collection involves securing digital evidence from electronic devices while preserving its integrity for legal or investigative purposes. This critical step utilizes specialized tools and techniques to ensure the evidence remains unaltered and can be used in court or for further analysis. Here are the steps to secure digital evidence efficiently:

- Use **forensically sound methods and tools** for evidence collection to avoid contamination or alteration, such as using write-blocking mechanisms to prevent any alterations to the original data during the collection process. In situations where data recovery is required, employ methods that are forensically sound and do not compromise the integrity of the recovered data.

- **Verify the integrity** of the collected data by comparing hash values before and after the collection process.

- Maintain **detailed logs** throughout the entire evidence collection process. Log every action taken, including identification of potential evidence, tools, and methods used, timestamps, and individuals involved. **Document metadata** and other contextual information that provides additional insights into the origin, usage, and history of the digital artifacts.

- Securely **label and package** collected items, employing measures against physical and environmental threats, including the use of tamper-evident seals, to preserve evidence integrity during transportation.

- Ensure that evidence is collected in a manner that complies with local laws and regulations to maintain its **admissibility** in legal proceedings. Also, follow **ethical guidelines** and consider the impact on individuals' **privacy rights**.

- Document factors that may affect the collection process in the cloud, such as data replication across regions, and rapid creation and destruction of resources in the cloud.

- **Working on copies** in digital forensics is imperative to safeguard the integrity of original evidence. This practice ensures that the untouched, unaltered source remains preserved, addressing legal requirements for an unbroken chain of custody and enhancing the credibility and admissibility of evidence in judicial proceedings.

- In digital forensics, **prioritizing the collection of the most volatile data**, such as information stored in RAM, register contents, routing tables, process tables, live network connects, and temporary files, is essential. Swift identification and retrieval of such data at the outset of an investigation ensure the preservation of real-time system state, minimizing the risk of alteration and enhancing the accuracy of forensic analysis.

- **Communication** with CSPs and key stakeholders, including legal teams and law enforcement, is indispensable in digital forensics. Collaborating with CSPs is vital for obtaining access to cloud-hosted data and navigating technical nuances. Effective communication with legal teams ensures adherence to legal requirements and helps address any regulatory concerns. Coordination with law enforcement agencies is essential for compliance with investigative procedures and sharing critical information. This collaborative communication framework enables the exchange of guidance, requirements, and expertise, ultimately enhancing the accuracy, legality, and efficiency of the digital forensic process.

> **Note**
>
> **The Clarifying Lawful Overseas Use of Data (CLOUD) Act** is US legislation that empowers law enforcement agencies to access data stored by American technology companies, even if the data is stored on servers located outside the US. It facilitates international data-sharing agreements and allows the US government to enter into bilateral agreements with other countries for streamlined access to electronic evidence. The CLOUD Act aims to address legal challenges related to cross-border data access and enhances cooperation between nations in criminal investigations.

Evidence management

Maintaining a secure and unbroken chain of custody is critical in legal proceedings to ensure that the evidence is credible and has not been tampered with, contaminated, or mishandled. The chain of custody is the record-keeping process that tracks the handling, possession, control, transfer, and examination of evidence, both physical and digital, during an investigation or legal proceeding. Any break in the chain of custody can potentially compromise the admissibility of the evidence in court.

Key components of a chain of custody typically include:

- **Collection**: Detailed documentation of when, where, and by whom the evidence was collected.
- **Transportation**: Record of how the evidence was transported, including who transported it and the means of transportation.
- **Storage**: Information about where and how the evidence was stored, along with any changes in possession.
- **Analysis**: Documentation of any examinations, testing, or analyses performed on the evidence, including the names of analysts and the dates of examination.
- **Courtroom presentation**: The chain of custody documentation is often presented in court to establish the reliability and authenticity of the evidence.

A **legal hold**, also known as a litigation hold or preservation order, is a directive to an organization to retain and safeguard all relevant documents and data that may be pertinent to a legal proceeding. This is typically issued when litigation, investigations, or regulatory compliance matters are anticipated or have commenced. The purpose of a legal hold is to ensure the preservation of potentially relevant information, preventing its destruction, alteration, or deletion, and allowing its later retrieval for legal review or disclosure. Organizations must suspend routine data disposal processes and take proactive measures to comply with the legal hold to avoid spoliation of evidence and potential legal consequences.

In cloud environments, it is imperative to explicitly include provisions related to legal holds in contractual agreements. These clauses should outline responsibilities for both the cloud service provider and the customer, specifying the procedures for implementing and lifting legal holds. Addressing legal holds in contracts ensures that relevant data is properly preserved during legal proceedings, safeguarding against potential data loss or alterations that could impact the outcome of litigation or regulatory investigations.

The primary attributes of digital evidence include:

- **Relevance**: Evidence should be directly related to the incident under investigation. Relevance ensures that the information collected is valuable and contributes meaningfully to understanding the case.

- **Authenticity**: The authenticity of evidence verifies its origin and integrity. It must be established that the evidence has not been tampered with and accurately reflects the state of the system or data at the time of collection.

- **Accuracy**: Accurate evidence, reflecting the true state of affairs, involves precise data collection, analysis, and reporting to prevent incorrect conclusions that could jeopardize the integrity of the investigative process. Truthfulness is paramount, demanding an honest and objective presentation of facts to ensure that information accurately represents the reality of the situation.

- **Completeness**: Complete evidence, encompassing the entirety of information, offers a comprehensive view of the incident. This involves thorough documentation, covering all relevant aspects of the investigation and ensuring that no crucial information is omitted, even if it negatively impacts the case. Upholding ethical standards and legal obligations, it is imperative not to hide evidence, as doing so is illegal and undermines the integrity of the forensic process. Transparency in presenting the entirety of the evidence, regardless of its implications, is essential for upholding justice and maintaining the credibility of the investigative process.

- **Admissibility**: Admissible evidence conforms to legal requirements, ensuring that it can be presented in a court of law. When collecting digital evidence, it is imperative to adhere to legal means to ensure the admissibility of the evidence in a court of law. Compliance with legal procedures, chain of custody documentation, and authentication are critical factors for admissibility. It is essential to note that hearsay, which is information based on what someone else has said rather than on firsthand knowledge, indirect evidence, or altered evidence might not be admissible in court.

- **Convincing**: Convincing evidence is presented in a clear and persuasive manner. It involves effectively communicating findings, interpretations, and conclusions, making it comprehensible to stakeholders, legal professionals, and other non-technical audiences. The ability to convey the significance of the evidence strengthens its impact and persuasiveness in the context of an investigation or legal proceedings.

- **Reliability**: Reliable evidence is evidence that is consistently accurate and trustworthy. It involves using validated tools, following standard procedures, and ensuring that the data's integrity is maintained throughout the forensic process.

Collecting, acquiring, and preserving digital evidence

Technical readiness for evidence collection in digital forensics involves ensuring that systems, tools, and processes are well prepared to gather, analyze, and preserve digital evidence in a forensically sound manner. Here are the key components of technical readiness:

- **Forensic tools and software**: Ensure that forensic tools are up to date and compatible with the latest technologies and operating systems. Regularly test the functionality of forensic software to verify its effectiveness in various scenarios.

- **Logging and monitoring systems**: Set up robust logging and monitoring systems to capture relevant events and activities across the digital environment. Integrate logging systems with **Security Information and Event Management (SIEM)** tools for centralized analysis. Enable logging for all systems, applications, and network devices, prioritizing the critical ones. Define clear log retention policies outlining the duration that logs should be retained. Regularly review logs to identify unusual or suspicious activities. Implement real-time monitoring tools for immediate threat detection. Ensure logs include timestamps, user actions, and system changes. Enforce secure log storage practices, including encryption and access controls.

- **Network and system baselines**: Create baselines for normal network and system behavior to facilitate the identification of anomalies during investigations. Regularly update baselines to reflect changes in the digital environment. Document normal network traffic, system performance, and user behavior as baselines. Set up alerts for deviations from established baselines.

- **File Integrity Monitoring (FIM)**: Deploy FIM tools to detect unauthorized changes to critical files. Set up alerts for any deviations from established file integrity baselines. In a cloud consisting of immutable container images, serverless architecture, or **Function as a Service (FaaS)**, similar requirements to FIM can be addressed through alternative approaches. For example, the integrity of container images can be ensured by implementing CI/CD pipelines with built-in security checks. This includes scanning container images to ensure that only authorized changes are deployed. Additionally, ensuring the integrity of serverless functions can involve implementing code signing and deployment verification mechanisms to validate the authenticity and integrity of deployed functions.

- **Backup and recovery systems**: Establish regular backup procedures for critical data and system configurations. Periodically test the restoration process to ensure backups can be successfully recovered. Encrypt backup files to protect sensitive information. Leverage backups not only for data recovery but also for comparing the current state with the previous good state, allowing forensic investigators to identify and analyze differences.

- **Storage security**: Implement encryption for stored forensic data to maintain confidentiality. Apply strict access controls to prevent unauthorized access to stored evidence.

- **Chain of custody processes**: Develop and maintain a robust chain of custody process, documenting every stage of evidence handling. Ensure forensic practitioners and incident responders are trained in proper chain of custody procedures.

- **Incident response plans**: Integrate evidence collection processes with overall incident response plans. Regularly conduct simulated incidents to test the effectiveness of evidence collection procedures.

- **Data and records retention policies**: Establish clear data and records retention policies, outlining the duration of retention for different types of data. Ensure that data retention policies comply with relevant legal and regulatory requirements.

- **Collaboration with Cloud Service Providers (CSPs)**: Ensure contracts and agreements with CSPs for seamless collaboration during evidence collection. Ensure that arrangements are in place for secure and lawful access to cloud-stored data.

Applying Technical Readiness in Cloud Forensics

While the core principles of technical readiness remain the same, cloud environments introduce unique challenges. Forensic tools must handle data dispersion and dynamic resource allocation. Robust logging and collaboration with cloud providers become even more critical to ensure successful evidence collection and maintain the chain of custody in the cloud. Digital forensics in the cloud presents several challenges (as listed below) that distinguish it from traditional digital forensics conducted on on-premises systems:

- **Data dispersion**: Cloud environments often involve the storage and processing of data across multiple geographically distributed locations. Investigators must navigate the dispersion of data, considering issues related to jurisdiction, legal requirements, and the complexities of collecting evidence spread across diverse cloud servers.

- **Dynamic resource allocation**: Cloud services dynamically allocate resources based on demand, leading to a constantly changing infrastructure. Traditional forensic methods designed for static environments may struggle to keep pace with the dynamic nature of cloud services, requiring forensic tools and techniques that can adapt to on-the-fly changes.

- **Shared infrastructure and multi-tenancy**: Cloud providers often use shared infrastructure, leading to multi-tenancy, where multiple users share the same physical resources. Isolating and attributing specific activities to individual users/customers becomes challenging, requiring forensic investigators to develop strategies for accurate attribution and ensuring data privacy compliance. One customer's investigation has the potential to unintentionally impact another customer's data security.

- **Ephemeral nature of evidence**: Cloud data and instances can be ephemeral, with temporary storage and short-lived virtual machines. Investigators must act swiftly to collect evidence before it is automatically deleted, emphasizing the need for real-time monitoring and rapid response capabilities.

- **Limited visibility and control**: Cloud customers have limited visibility and control over the underlying infrastructure managed by CSPs. This limited control poses challenges for forensic practitioners who may struggle to access crucial data, logs, or metadata. To address this, it becomes essential for cloud customers to establish clear contracts and agreements with their service providers. These contracts should outline the procedures for collaboration and information sharing, ensuring that forensic investigators can obtain the necessary data without compromising the security or privacy of the cloud infrastructure. Establishing such contractual frameworks is paramount to navigating challenges related to data access and collaboration with CSPs.

- **Encryption and security measures**: Decrypting and accessing data for forensic purposes may require specific credentials, keys, or cooperation from the CSP, introducing additional complexity and considerations. This is especially true in server-side encryption, where the CSP takes responsibility for encrypting the data at rest, utilizing encryption keys managed by the provider.

- **Lack of standardization**: The lack of standardized forensic procedures for cloud environments. Digital forensic practitioners need to adapt to a rapidly evolving landscape, staying informed about emerging standards, best practices, and legal considerations specific to cloud forensics.

- **Legal and jurisdictional challenges**: Legal and jurisdictional issues become more complex in the cloud, involving multiple regions and potentially conflicting regulations. Forensic investigators must navigate diverse legal frameworks, comply with international laws, and work with legal teams to ensure the admissibility of evidence across jurisdictions.

- **Vendor-specific tools and formats**: CSPs often use proprietary tools and data formats. Forensic tools must support a variety of vendor-specific formats, requiring adaptability and interoperability to ensure effective evidence collection and analysis.

Best practices for evidence preservation

Effective evidence preservation is crucial in digital forensics to maintain the integrity, authenticity, and admissibility of collected evidence. Here are key best practices for evidence preservation:

- **Chain of custody documentation**: Maintain a detailed and well-documented chain of custody for all pieces of evidence. Train personnel involved in evidence handling on proper chain of custody procedures.

- **Forensic imaging**: Capture forensic images of storage devices to preserve the state of the evidence. Use strong cryptographic hash functions to create checksums and verify the integrity of forensic images.

- **Secure storage**: Store evidence in encrypted containers or filesystems to protect against unauthorized access. Implement strict access controls to limit who can view or manipulate stored evidence.

- **Backup procedures**: Create multiple copies of evidence to safeguard against accidental loss or corruption. Store backup copies in different physical locations to mitigate the risk of physical damage.

- **Documentation**: Document the context, collection methodology, and any changes made during evidence processing. Include precise timestamps for each stage of evidence preservation.

- **Adherence to legal and regulatory requirements**: Be aware of and comply with relevant legal and regulatory requirements regarding evidence preservation. Align evidence preservation practices with established data retention policies.

- **Tamper-evident seals**: Use tamper-evident seals on evidence containers to detect any unauthorized access or tampering. Document the placement and condition of seals in the chain of custody records.

- **Digital signatures**: Apply digital signatures to digital evidence to verify its authenticity. Employ trusted and validated forensic tools to sign digital evidence.

- **Environment control**: Limit physical access to the evidence storage environment to authorized personnel only. Maintain suitable environmental conditions (temperature, humidity) to prevent damage to physical evidence.

- **Regular audits**: Regularly review audit trails and access logs for the evidence storage system. Conduct periodic audits of evidence storage procedures to ensure compliance with best practices.

- **Collaboration with legal teams**: Collaborate with legal teams to ensure that evidence preservation practices align with legal requirements. Seek legal advice to prevent spoliation by preserving evidence in a manner that maintains its authenticity and reliability.

- **Preventing interference**: Consider utilizing a Faraday cage for physical evidence containing electronic components to block electromagnetic signals and prevent remote tampering. Implement shielding techniques for electronic devices to protect against radiofrequency interference during evidence collection. When dealing with wireless devices, maintain radio silence to prevent external interference during evidence preservation.

Summary

This chapter provided an overview of digital forensics. You understood its multidisciplinary nature and critical role in investigating cybercrimes. Standards such as ISO/IEC 27050:2016 and ISO/IEC 27037:2012 were explored, shedding light on e-discovery and guidelines for digital evidence collection.

You went through the digital evidence lifecycle stages and understood the importance of technical readiness, forensic tools, logging systems, baselines, and collaboration with CSPs. Challenges in cloud forensics were discussed, emphasizing the need for adaptability to dynamic environments, legal considerations, and collaboration with diverse stakeholders.

You discovered evidence preservation best practices, chain of custody, forensic imaging, secure storage, and collaboration with legal teams. The chapter concluded by addressing interference prevention through Faraday cages and highlighting the unique challenges presented by cloud forensics, reinforcing the significance of adaptability and interoperability in that field.

In the next chapter, you'll learn about best practices for establishing resilient communication channels and procedures essential for managing various impacts, covering common communication methods with vendors, customers, regulators, partners, and other stakeholders.

Exam Readiness Drill – Chapter Review Questions

Apart from a solid understanding of key concepts, being able to think quickly under time pressure is a skill that will help you ace your certification exam. That is why working on these skills early on in your learning journey is key.

Chapter review questions are designed to improve your test-taking skills progressively with each chapter you learn and review your understanding of key concepts in the chapter at the same time. You'll find these at the end of each chapter.

> **How to Access These Materials**
>
> To learn how to access these resources, head over to the chapter titled *Chapter 25, Accessing the Online Resources*.

To open the Chapter Review Questions for this chapter, perform the following steps:

1. Click the link – `https://packt.link/CCSPE1_CH19`.

    Alternatively, you can scan the following **QR code** (*Figure 19.1*):

Figure 19.1 – QR code that opens Chapter Review Questions for logged-in users

2. Once you log in, you'll see a page similar to the one shown in *Figure 19.2*:

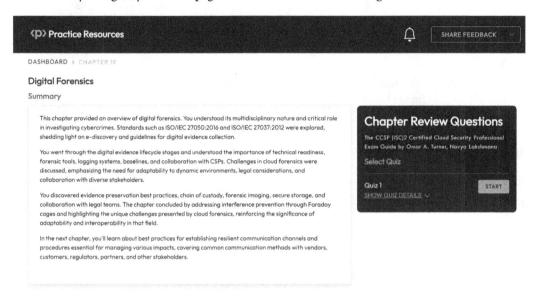

Figure 19.2 – Chapter Review Questions for Chapter 19

3. Once ready, start the following practice drills, re-attempting the quiz multiple times.

Exam Readiness Drill

For the first three attempts, don't worry about the time limit.

ATTEMPT 1

The first time, aim for at least **40%**. Look at the answers you got wrong and read the relevant sections in the chapter again to fix your learning gaps.

ATTEMPT 2

The second time, aim for at least **60%**. Look at the answers you got wrong and read the relevant sections in the chapter again to fix any remaining learning gaps.

ATTEMPT 3

The third time, aim for at least **75%**. Once you score 75% or more, you start working on your timing.

> **Tip**
> You may take more than **three** attempts to reach 75%. That's okay. Just review the relevant sections in the chapter till you get there.

Working On Timing

Target: Your aim is to keep the score the same while trying to answer these questions as quickly as possible. Here's an example of how your next attempts should look like:

Attempt	Score	Time Taken
Attempt 5	77%	21 mins 30 seconds
Attempt 6	78%	18 mins 34 seconds
Attempt 7	76%	14 mins 44 seconds

Table 19.1 – Sample timing practice drills on the online platform

> **Note**
> The time limits shown in the above table are just examples. Set your own time limits with each attempt based on the time limit of the quiz on the website.

With each new attempt, your score should stay above **75%** while your "time taken" to complete should "decrease". Repeat as many attempts as you want till you feel confident dealing with the time pressure.

Managing Communications

In this chapter, you'll learn the essential principles of communication with key stakeholders in cloud services, addressing both cloud providers and customers. Emphasizing transparency and efficiency, this chapter covers identifying target audiences, defining clear messages, determining optimal timing, stating communication purposes, and selecting appropriate channels. You will explore tailored communication strategies and focus on the importance of proactive engagement, clear expectations, and ongoing collaboration to ensure success and compliance in the dynamic landscape of cloud services.

By the end of this chapter, you will be able to confidently answer questions on managing communications with vendors, customers, partners, regulators, and other stakeholders.

Managing communication with relevant parties

Establishing robust lines of communication is imperative for both **Cloud Service Providers** (**CSPs**) and customers. In the realm of IT operations, where collaboration with various stakeholders is crucial, effective communication plays a pivotal role. It emphasizes the importance of providing timely, accurate, concise, and relevant information to ensure mutual understanding and smooth interaction with relevant stakeholders. This commitment to transparent and efficient communication is foundational in navigating the complexities of cloud services and fostering a cooperative environment.

Key principles and practices for achieving effective communication are as follows:

- **Identify the target audience or stakeholders**: This includes vendors, customers, regulators, partners, and any other individuals impacted by or involved in the communication – for example, "All customers of the organization."

- **Clearly title your message**: This could be related to cloud security policies, updates, incidents, training programs, or any other relevant topic. Be specific and avoid unnecessary complexity – for example, "Notification of upcoming changes to the organization's cloud service affecting customer experience."

- **Determine the timing of the communication**: This includes both regular updates and communication during specific events or incidents. Timeliness is crucial, especially in security-related matters. An example is "Two weeks before the scheduled changes."

- **Clearly articulate the purpose or objective of the communication, tailoring the message according to the audience**: Explain why the information is important, how it aligns with organizational goals, and the impact it may have on stakeholders. Providing context helps individuals understand the significance. An example is "The organization is enhancing security measures to protect sensitive data and comply with industry regulations. Using **Multi-Factor Authentication (MFA)** adds an extra layer of protection against unauthorized access."

- **Inform recipients of what changes to expect and whether there's action that they need to take**: Provide details on the specific changes. If there are any actions required from the recipients, clearly outline them in the message to ensure understanding and compliance. An example is "When you log in to the cloud services, you will be prompted to enroll for MFA." Additionally, specify the steps recipients need to take to enable MFA on their accounts, such as following the prompts to set up MFA.

- **Outline the method and channels of communication**: Choose appropriate communication channels, such as email, meetings, internal messaging platforms, or training sessions. Consider the preferences and accessibility of your target audience. An example is "Email and in-app notification."

Vendors

Managing communication with vendors is pivotal for fostering strong partnerships and efficient collaboration in the supply chain. Clear and transparent communication channels are essential for conveying expectations, addressing concerns, and aligning strategies. Regular updates on procurement processes, delivery schedules, and changes in requirements are crucial components of maintaining a collaborative vendor relationship.

In emergency preparedness, organizations should establish and communicate clear emergency paths and call lists to vendors, ensuring rapid dissemination of critical information during disasters. This aligns with business continuity and disaster recovery plans, facilitating swift communication.

Maintaining a comprehensive vendor inventory list with prioritization is vital in the realm of effective vendor management. This list should encompass key details about each vendor, including their capabilities, performance metrics, and the criticality of their contributions to the supply chain. Moreover, it is essential to take note of any single points of failure within the vendor ecosystem. Incorporating this knowledge into communication strategies ensures a proactive approach to managing vendor relationships and enhancing overall supply chain stability.

Establishing clear customer support channels with vendors is also crucial for effective issue resolution and maintaining a positive working relationship. Prioritizing vendors allows organizations to allocate resources effectively, focus communication efforts, and strategically manage relationships, contributing to the resilience and efficiency of the supply chain, even in the face of security incidents.

Customers

Managing communication with customers in the context of cloud services is fundamental for business success. Organizations must prioritize transparent and proactive engagement to build trust, loyalty, and satisfaction. This involves timely updates on product releases, service improvements, and swift resolution of customer inquiries. In the cloud landscape, ensuring seamless and secure solutions requires consistent communication on service updates, disruptions, and changes. Whether through traditional channels or modern platforms, addressing customer inquiries promptly enhances their overall experience. Celebrating successes, implementing security measures, and sharing innovative features through effective communication reinforce the commitment to excellence.

In the procurement process, contract and **Service-Level Agreement (SLA)** negotiations are pivotal. Clear and comprehensive communication during these negotiations establishes the foundation for a transparent partnership. Both parties must articulate expectations, service levels, and specific requirements, addressing potential issues such as data security and compliance. Maintaining up-to-date documentation, including user guides, is crucial for clarity and transparency. User guides empower customers to maximize service utilization, reducing ambiguity and enhancing their experience. Clear communication about the shared responsibility model is essential, ensuring customers understand their role in securing data and applications. Irrespective of whether it's IaaS, PaaS, or SaaS, customers hold responsibility, particularly in ensuring security measures for data, applications, and **Identity and Access Management (IAM)**.

Fostering clear communication during negotiations, emphasizing documentation, user guides, shared responsibility, and customer involvement in security aspects establishes a robust framework aligned with customer expectations. This proactive approach not only prevents misunderstandings but also forms the basis for issue resolution during the service engagement, contributing to a lasting and successful customer relationship in the dynamic cloud services landscape.

Partners

Managing communication with partners involves clear paths that delineate the flow of information, ensuring effective dialogue throughout the partnership life cycle. Thorough onboarding processes are crucial, encompassing partner vetting procedures that align with stringent security best practices. Partnerships often involve the intricate use of federated services, granting access levels akin to organizational employees. This complexity necessitates meticulous onboarding, including the provision of clear security expectations to partners operating independently.

Moreover, this proactive communication approach goes beyond just onboarding. Continuous and transparent communication becomes a linchpin for effective collaboration. For instance, in scenarios involving logistics partners, ongoing updates and coordination are vital, ensuring smooth delivery schedules and addressing any logistical challenges promptly.

The importance of robust offboarding procedures cannot be overstated. When concluding a partnership, a well-documented offboarding policy ensures the efficient termination of partner access to all enterprise systems. This not only aligns with overarching security policies but also safeguards the integrity of systems and infrastructure, mitigating the potential risks associated with terminated partnerships.

In essence, effective communication with partners is a multifaceted strategy that encompasses clear communication paths, meticulous onboarding and ongoing communication, and robust offboarding procedures.

> **Note**
>
> The organization retains ultimate responsibility for the security and integrity of customer data, regardless of whether any issues, outages, or breaches are caused by the CSP or other factors.

Regulators

Effectively managing communication with regulators is crucial for maintaining compliance and fostering positive regulatory relationships. The organization's credibility relies on timely reporting, transparent disclosure of business practices, and strict adherence to regulatory standards. Proactively engaging with regulators, discussing industry changes and compliance initiatives, and seeking guidance ensures alignment with regulatory requirements. In the dynamic realm of cloud services, addressing location-specific requirements necessitates a nuanced approach, considering the unique challenges introduced by cloud infrastructure. Organizations must promptly inform regulators of any non-compliance, demonstrating a commitment to regulatory adherence and maintaining trust. In cases of non-compliance, providing insights into root causes, implemented measures, workarounds, and estimated resolution timelines is essential. Early communication with regulators allows a proactive and collaborative approach, promoting transparency and commitment to issue resolution.

Legal and compliance teams play a pivotal role in understanding and addressing the communication requirements outlined by various regulators. This awareness, driven by data type, geographical considerations, and services provided, enhances compliance efforts. Recognition of specific communication needs, such as those outlined by GDPR in Europe or HIPAA in the United States, ensures tailored and effective compliance strategies, reinforcing the organization's commitment to regulatory adherence.

Other stakeholders

Depending on the type of company and the services provided, there may be additional stakeholders unique to the industry. In addition to the main stakeholders, it's crucial to consider other key players, such as investors, security researchers, public relations professionals, local communities, and employees. For public communication, involving PR professionals ensures a strategic and positive image projection. Engaging with security researchers establishes a valuable channel for exchanging information, fostering a collaborative approach to cybersecurity.

Transparent communication regarding corporate social responsibility initiatives, organizational changes, and community outreach projects becomes even more paramount. This not only builds trust and credibility among a diverse range of stakeholders but also reinforces the organization's commitment to social and environmental responsibility.

Internally, effective communication is the cornerstone of ensuring that employees are well informed, aligned with organizational goals, and feel valued within the company. Utilizing transparent communication channels creates a sense of inclusivity and shared purpose, contributing to a cohesive and motivated workforce. Regular updates on company objectives, milestones, and any changes in strategy foster a culture of transparency and trust within the organization.

Maintaining open channels for feedback from all stakeholders, including employees, local communities, and the public, is crucial for a well-rounded communication approach.

Here's an example of how CSPs should communicate with various stakeholders in the event of a security incident:

- Vendors:

  - Explain potential impacts on procurement processes, delivery schedules, and requirements, if any.

  - Provide details on how the incident affects vendor operations and their products/services.

  - Collaborate with vendors to understand steps they're taking and assess the impact on the CSP's data and services, if the security incident is related to their products/services.

  - Discuss any actions the CSP should take to mitigate the incident's effects.

- Cloud customer services:

  - Notify customers promptly and provide comprehensive information about the incident's nature and potential impacts on their data or services.

  - Advise customers on necessary actions, such as enabling MFA for enhanced security.

  - Ensure transparency regarding adherence to SLAs and contracts during incident resolution.

- Partners:

  - Clarify procedures for partners accessing systems during troubleshooting/restoration.

  - Collaborate with partners on response actions and inform them of any steps they should take to safeguard systems.

- Regulators:

  - Determine whether any standards or regulations have been violated, such as data breach laws.

  - Keep regulators informed about incident details and actions taken or planned to address the situation.

- Other stakeholders:

  - Engage PR professionals in strategic communication to maintain a positive public image.

  - Collaborate with security researchers to exchange information and enhance cybersecurity measures.

  - Internally, ensure employees are well informed about the incident and their role in the response efforts.

Overall, transparent, timely, and proactive communication is crucial to effectively manage security incidents and maintain stakeholder trust.

Summary

In this chapter, you explored the fundamental principles of effective communication with diverse stakeholders in cloud services. Emphasizing transparency, clarity, and proactive engagement, the chapter provided insights into identifying target audiences, crafting clear messages, determining optimal communication timing, and selecting suitable channels. The tailored communication strategies outlined for vendors, customers, partners, regulators, and additional stakeholders underscore the importance of clear expectations and collaborative efforts to navigate the dynamic landscape of cloud services successfully.

By mastering these principles, you are now equipped to confidently navigate and manage communications with vendors, customers, partners, regulators, and other stakeholders in the intricate realm of cloud services. The chapter's focus on proactive engagement, transparent communication, and strategic collaboration empowers you to foster strong relationships, ensure compliance, and contribute to the overall success and resilience of your organization within the dynamic cloud services environment.

In the next chapter, you'll learn the best practices for establishing the primary requirements of a **Security Operations Center (SOC)**. You'll also look at a wide range of tools related to monitoring and logging that are necessary for effective SOC management.

Exam Readiness Drill – Chapter Review Questions

Apart from a solid understanding of key concepts, being able to think quickly under time pressure is a skill that will help you ace your certification exam. That is why working on these skills early on in your learning journey is key.

Chapter review questions are designed to improve your test-taking skills progressively with each chapter you learn and review your understanding of key concepts in the chapter at the same time. You'll find these at the end of each chapter.

> **How to Access These Materials**
>
> To learn how to access these resources, head over to the chapter titled *Chapter 25, Accessing the Online Resources*.

To open the Chapter Review Questions for this chapter, perform the following steps:

1. Click the link – `https://packt.link/CCSPE1_CH20`.

    Alternatively, you can scan the following **QR code** (*Figure 20.1*):

Figure 20.1 – QR code that opens Chapter Review Questions for logged-in users

2. Once you log in, you'll see a page similar to the one shown in *Figure 20.2*:

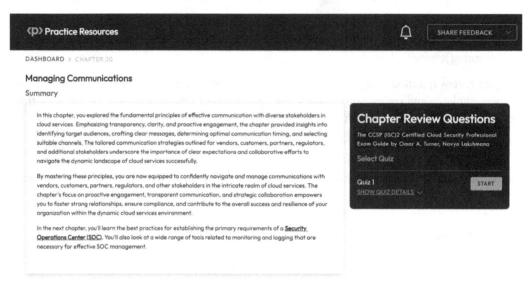

Figure 20.2 – Chapter Review Questions for Chapter 20

3. Once ready, start the following practice drills, re-attempting the quiz multiple times.

Exam Readiness Drill

For the first three attempts, don't worry about the time limit.

ATTEMPT 1

The first time, aim for at least **40%**. Look at the answers you got wrong and read the relevant sections in the chapter again to fix your learning gaps.

ATTEMPT 2

The second time, aim for at least **60%**. Look at the answers you got wrong and read the relevant sections in the chapter again to fix any remaining learning gaps.

ATTEMPT 3

The third time, aim for at least **75%**. Once you score 75% or more, you start working on your timing.

> **Tip**
>
> You may take more than **three** attempts to reach 75%. That's okay. Just review the relevant sections in the chapter till you get there.

Working On Timing

Target: Your aim is to keep the score the same while trying to answer these questions as quickly as possible. Here's an example of how your next attempts should look like:

Attempt	Score	Time Taken
Attempt 5	77%	21 mins 30 seconds
Attempt 6	78%	18 mins 34 seconds
Attempt 7	76%	14 mins 44 seconds

Table 20.1 – Sample timing practice drills on the online platform

> **Note**
>
> The time limits shown in the above table are just examples. Set your own time limits with each attempt based on the time limit of the quiz on the website.

With each new attempt, your score should stay above **75%** while your "time taken" to complete should "decrease". Repeat as many attempts as you want till you feel confident dealing with the time pressure.

21
Security Operations Center Management

In this chapter, you will delve into crucial elements of cybersecurity operations, focusing on the **Security Operations Center** (**SOC**), vulnerability assessments, and incident management. You will explore the SOC's pivotal role in continuous monitoring, incident response, and threat intelligence emphasizing its significance in maintaining organizational security. You will also learn Intelligent monitoring practices, log capturing, and analysis through **Security Information and Event Management** (**SIEM**) are highlighted. Further, you will study the incident management process, from preparation to continuous improvement, with considerations for cloud customers. And finally, the chapter will shed light on vulnerability assessments covering scan types, accuracy evaluation, and the importance of continuous monitoring in cloud environments.

By the end of this chapter, you will be able to confidently answer questions on the following:

- SOC
- Monitoring of security controls
- Log capture and analysis
- Incident management
- Vulnerability assessments

SOC

The SOC acts as a centralized hub within an organization, continuously monitoring and analyzing security events to detect, respond to, and mitigate potential cybersecurity threats. It plays a crucial role in maintaining the overall security posture and resilience of the organization's digital assets. A SOC operates 24/7 to provide continuous monitoring of an organization's IT environment, networks, and systems. The SOC operates at tactical (mid-term) and operational (day-to-day) levels, effectively addressing immediate threats and maintaining ongoing security for the organization. This dual focus ensures both timely responses to current incidents and a proactive stance against future risks. A SOC serves as the initial point of contact for reporting potential security incidents within an organization.

In the shared responsibility model of cloud security, both the cloud service providers and customers maintain their SOCs, based on the service model (IaaS, PaaS, and SaaS). In this collaborative approach, the parties work in tandem, aligning efforts to create a comprehensive security posture.

Some of the primary functions of an SOC are the following:

- **Monitoring and Detection**: Continuously monitor the organization's IT infrastructure, networks, and systems for potential security incidents. Detect and analyze anomalies or suspicious activities that may indicate a security threat. Ensure adherence to policies, standards, and regulatory requirements by actively monitoring for any signs of non-compliance.

- **Incident Response**: Review alerts or events to see if they are incidents, then evaluate incidents to determine their severity and relevance. Prioritize incidents based on the level of threat and potential impact on the organization. Collaborate with the Incident Response team to assess security incidents. Share pertinent details, provide context on threats, and work jointly on strategies for containment, eradication, recovery, and post-incident analysis.

- **Threat Intelligence**: Gather, analyze, and apply threat intelligence to enhance the SOC's understanding of potential risks and emerging threats. Stay informed about the evolving cybersecurity landscape.

- **Vulnerability Management**: Identify and assess vulnerabilities in the organization's systems and applications. Work toward mitigating or remediating vulnerabilities to reduce the risk of exploitation.

- **Continuous Monitoring and Reporting**: Establish robust practices for ongoing surveillance to swiftly detect and respond to security incidents. Monitor for indicators of compromise and unauthorized access in real time, ensuring a proactive approach to maintaining a secure environment. Additionally, the SOC monitors security controls to ensure their effectiveness in mitigating threats. Regularly generate comprehensive metrics-based reports to document monitoring findings, enhancing visibility and aiding in the continuous improvement of security measures.

> **Note**
>
> ISO 18788 is an international standard that outlines requirements for establishing, implementing, operating, monitoring, reviewing, maintaining, and improving the management of security operations. The standard emphasizes the importance of a systematic, risk-based approach to security operations management.

Monitoring of Security Controls

Monitoring is the process of observing and overseeing activities, events, or systems to ensure they are functioning as intended. It involves regularly checking for deviations from expected behavior and identifying issues or potential risks.

Continuous monitoring takes monitoring a step further by implementing real-time or near-real-time monitoring practices. It involves ongoing and automated observation of systems, networks, and data to promptly detect any anomalies, security events, or potential threats. It aims to provide immediate awareness and response capabilities, enhancing the organization's overall cybersecurity posture.

NIST Special Publication 800-37, the **Risk Management Framework** (**RMF**) emphasizes the importance of continuous monitoring as a key component of the risk management process. Continuous monitoring involves the ongoing oversight and assessment of security controls, risk posture, and the effectiveness of security measures throughout the system's lifecycle. Rather than being a one-time event, continuous monitoring is integrated into the risk management process to provide real-time awareness of security-related activities.

Monitoring security controls extends beyond evaluating their technical performance; it encompasses a detailed examination of the data they produce. This process involves analyzing information from diverse security measures, including firewalls, **Intrusion Detection Systems** (**IDSs**), **Intrusion Prevention Systems** (**IPSs**), honeypots, network security groups, and **Artificial Intelligence** (**AI**). The goal is to identify irregular patterns, potential threats, or anomalies that could signal security incidents within the organization.

Firewalls are network security devices designed to monitor and control incoming and outgoing network traffic based on predetermined security rules. Monitoring firewall logs helps ensure that only authorized traffic is allowed and that any attempts at unauthorized access are detected. Unusual patterns or anomalies in traffic are investigated.

IDSs analyze network or system activities to identify potential security incidents and IPSs proactively block or prevent recognized threats. The SOC team assesses IDS/IPS alerts, investigates potential incidents, and takes necessary actions to prevent or mitigate security threats. Reviewing the actions detected by an IDS is essential for the SOC to ensure effective detection. The actions taken by an IPS are verified to eliminate false positives and examine genuine activities that are blocked, allowing the SOC to delve deeper into true positive cases. This validation process is crucial for maintaining the accuracy and efficiency of security measures.

Azure **Network Security Groups** (**NSGs**), **AWS Security Groups**, or **GCP Firewall Rules** define security rules that control inbound and outbound traffic to **Network Interfaces** (**NICs**), VMs, and subnets. Monitoring NSG configurations ensures that traffic is allowed or denied according to the organization's security policies. Any changes or misconfigurations are promptly identified and corrected.

Honeypots are decoy systems designed to attract and detect unauthorized access or attacks. Monitoring honeypot activity provides insights into potential threats. If there is any interaction with the honeypot, it can indicate malicious intent, triggering an investigation by the SOC team.

AI is a transformative force in enhancing continuous monitoring capabilities, bringing automation, efficiency, and advanced threat detection to the cybersecurity landscape. AI algorithms can rapidly analyze immense volumes of data in real time, enabling the swift identification of patterns and anomalies and empowering security teams with the ability to respond promptly to emerging cyber threats. AI leverages behavioral analysis to establish a baseline of normal activities within a network. By flagging deviations or unusual behaviors, it aids in the proactive identification of potential security incidents. **Machine Learning** (**ML**) further enhances continuous monitoring by allowing algorithms to learn from historical data, adapt to changes in network behavior, and ensure more accurate anomaly detection.

Predictive analysis is another facet of AI's role in continuous monitoring, wherein it anticipates potential threats based on historical trends and known patterns. This proactive approach enables security teams to take preventive measures before security incidents unfold. AI systems also integrate with threat intelligence feeds, providing real-time context for security alerts and aiding in prioritizing responses based on threat severity and relevance.

User and Entity Behavior Analytics (**UEBA**) focuses on monitoring and analyzing user and entity behavior within a network. This helps in identifying suspicious activities, compromised accounts, and insider threats by understanding normal behavior and highlighting deviations. UEBA combines different methods, including AI, ML, data mining, pattern recognition, advanced analytics, statistical analysis, and rule-based algorithms, to thoroughly detect anomalies and potential threats.

Effective log collection and analysis are crucial for monitoring an organization's infrastructure, networks, and systems, as well as its security controls.

Log Capture and Analysis

NIST SP 800-92, titled "Guide to Computer Security Log Management," provides valuable insights into managing logs effectively within an organization. A log is a record of events or activities that occur within the system. Each entry in a log contains information related to a specific event, providing a chronological and detailed account of actions, transactions, and incidents. Logs are generated by various components of an information system, including security tools, operating systems, applications, and networking equipment.

Logs are collected for several reasons:

- Logs capture and record security-related events, helping organizations monitor for suspicious activities, potential threats, and unauthorized access.

- Logs provide a detailed timeline of events during a security incident, aiding in the investigation, analysis, and mitigation of breaches or attacks. Logs support the correlation of events and the analysis of root causes, enabling organizations to understand the relationships between different incidents and identify underlying issues.

- Many regulations and standards mandate the collection and retention of logs to demonstrate compliance with security and privacy requirements.

- Logs provide insights into the performance, health, and usage of systems and applications, facilitating proactive maintenance and troubleshooting.

- Logs serve as a valuable resource for digital forensics, helping investigators reconstruct events and understand the sequence of actions during an incident.

- Logs provide an audit trail for accountability, allowing organizations to track user activities, system changes, and policy violations.

- Logs are essential for diagnosing and resolving issues, as they contain information about errors, warnings, and system events.

- By analyzing logs, organizations can plan for future resource needs, ensuring that systems have adequate capacity to handle increasing workloads.

- Logs contribute to user behavior analytics, helping organizations detect abnormal patterns and potential insider threats.

- Logs establish accountability by recording actions and changes within the organization. Nonrepudiation ensures that users cannot deny their actions.

Log capture and analysis are essential aspects of the broader subject of log management.

Log Management

Log management involves the collection, analysis, retention, and protection of log data generated by an organization's systems, networks, applications, and security devices. Logs are records of events and activities within an IT environment, providing valuable insights into system behavior, user actions, and potential security incidents. Effective log management is crucial for maintaining the security, compliance, and operational efficiency of an organization. Logs may often contain sensitive data, such as usernames, hostnames, IP addresses, and details of security incidents. Unauthorized access to logs can lead to data breaches and compromise the confidentiality of critical information.

Here are some log management recommendations:

- Develop comprehensive **policies and procedures** for log management, covering the entire lifecycle of log data, including generation, transmission, storage, analysis, and disposal.

- **Prioritize log management** activities based on the organization's perceived reduction of risk. Allocate resources effectively to ensure that critical log management functions are performed.

- Address the **preservation of original logs** to ensure their integrity for evidentiary purposes. This may involve sending copies of logs to centralized devices and using tools to analyze and interpret network traffic.

- Ensure all systems and devices generating logs are synchronized to a reliable time source using the **Network Time Protocol (NTP)** or similar mechanisms. **Time synchronization** mechanisms such as the NTP can align the clocks of different devices to a common reference time, typically provided by highly accurate time servers connected to atomic clocks. Consistent timestamps across logs enable accurate correlation of events, facilitating efficient troubleshooting, forensic analysis, compliance auditing, and ensuring the integrity and reliability of log data.

- Develop a robust **log management infrastructure** that includes the necessary hardware, software, networks, and storage media. This infrastructure should support the secure generation, transmission, storage, analysis, and disposal of log data.

- Select **robust hardware components**, including servers, storage devices, and network equipment, capable of handling diverse log data volumes, ensuring redundancy and scalability to meet evolving organizational requirements. Ensure redundancy and scalability to accommodate the organization's growing needs. Note that hardware-related considerations are the responsibility of cloud service providers, as customers do not have direct access to or control over the underlying infrastructure.

- Choose log management **software with secure capabilities** for generation, transmission, storage, analysis, and disposal, featuring real-time monitoring, correlation, and reporting, ensuring compatibility with diverse log sources and formats.

- Design a **network architecture** that facilitates the efficient and secure transmission of log data from various sources to the central log management system. Implement secure protocols for transmitting logs to prevent unauthorized access or tampering during transit.

- Choose appropriate **storage solutions** that align with the organization's data retention policies and legal requirements. Implement secure and redundant storage to prevent data loss. Consider options for both online and offline/archive storage based on the organization's needs.

- Implement **security measures** such as access controls, encryption for data in transit and at rest, and regular security audits to protect the log management infrastructure. Safeguard against unauthorized access to log data, which may contain sensitive information. Regularly conduct testing procedures to verify the effective collection of logs from various sources.

- Ensure the log management infrastructure encompasses **advanced analysis capabilities**, facilitating correlation across diverse sources, detecting patterns and anomalies, and providing meaningful insights. Design with **scalability** to accommodate growth, incorporating flexibility for emerging technologies and changes in log sources, formats, and volumes.

- Align the log management infrastructure with relevant **compliance requirements** and regulations. This may include industry-specific standards or legal mandates governing the storage and protection of log data.

- Implement mechanisms for **secure deletion of log data** in accordance with data retention policies and legal requirements. Secure deletion ensures that sensitive information is permanently and irreversibly removed from the log storage, reducing the risk of unauthorized access or data exposure.

Security Information and Event Management (SIEM)

SIEM is a comprehensive solution that combines **Security Information Management** (**SIM**) and **Security Event Management** (**SEM**) to provide a centralized and integrated approach to managing an organization's security information and events. SIEM systems collect, aggregate, correlate, and analyze log data from various sources throughout an organization's technology infrastructure. The goal is to identify patterns, detect anomalies, and provide insights into potential security threats and incidents.

SIEM involves a multi-step process to effectively manage log data for cybersecurity purposes. Here's a breakdown of the key stages in the SIEM lifecycle:

1. **Collect**: SIEM systems collect log data from various sources across an organization's IT infrastructure, including network devices, servers, applications, and security appliances. Logs may include information about user activities, system events, network traffic, and security-related events.

2. **Normalization**: During the normalization phase, the collected log data is processed to ensure a consistent format across different sources. Normalization helps standardize log entries, making it easier to correlate events and compare data from diverse systems.

3. **Aggregation**: Aggregation involves consolidating log data from multiple sources into a centralized repository. Aggregated logs provide a comprehensive view of an organization's IT environment and streamline the analysis process.

4. **Correlation**: Correlation involves analyzing log entries to identify patterns, relationships, or anomalies. SIEM systems correlate events from different sources to distinguish between normal and potentially malicious activities.

5. **Analysis**: Log data is analyzed using various techniques, including rule-based analysis, statistical analysis, and ML. The goal is to detect security threats, unauthorized access, or abnormal behavior within the IT infrastructure.

6. **Alerting**: SIEM systems generate alerts and notifications based on predefined rules and thresholds. Alerts notify security teams of potential security incidents or deviations from normal behavior, allowing for rapid response.

7. **Archiving**: Log data is securely archived for an extended period to support historical analysis and forensic investigations. Archiving ensures that organizations can access past log entries for compliance, audits, or incident response purposes.

8. **Reporting and Dashboards**: SIEM platforms generate detailed reports summarizing security events and provide visual dashboards for real-time monitoring, offering insights into security metrics and overall posture.

9. **Integration**: SIEM systems often integrate with other security tools, such as firewalls, antivirus solutions, and IDSs. Integration enhances the overall security ecosystem, allowing for a more comprehensive and coordinated response to threats.

Incident Management

Incident management refers to the process of identifying, managing, and resolving security incidents within an organization. It involves the systematic approach of detecting, responding to, mitigating, and recovering from security events that could potentially impact the confidentiality, integrity, or availability of information or systems. The goal of incident management is to minimize the impact of security incidents, prevent their recurrence, and ensure the organization's overall cybersecurity resilience.

The Computer Security Incident Handling Guide, NIST SP 800-61, offers a structured approach to incident management, delineating the following essential steps in the process:

1. Preparation

    I.   Craft a comprehensive policy that delineates the organization's strategic approach to incident response, defining its scope, objectives, and guiding principles.

    II. Assemble a dedicated team with clearly defined roles and responsibilities, ensuring diverse expertise across technical, legal, and communication domains.

    III. Implement training programs for the **Incident Response Team** (**IRT**) to enhance technical skills, communication, and incident handling procedures.

    IV. Extend awareness initiatives to all relevant staff, fostering a proactive security culture throughout the organization.

    V.   Create detailed incident response plans for various incident scenarios, outlining specific actions to be taken during each phase of incident response, from detection to recovery.

    VI. Simulate real-world incidents through tabletop exercises and drills to rigorously test the effectiveness of response plans and enhance the team's adaptability to diverse situations.

VII. Build partnerships and communication channels with external entities, including law enforcement, regulatory bodies, and industry peers, to facilitate collaborative incident resolution.

VIII. Enforce industry-standard security best practices to secure networks, systems, and applications. Regularly update and patch systems, conduct security assessments, and employ IDSs to monitor and respond to potential threats.

2. Detection and Analysis

I. Systematically log relevant events across networks, systems, and applications, conducting regular reviews to proactively identify potential security incidents and deviations from the norm.

II. Employ advanced monitoring solutions to continuously scan for unusual activities, automatically generating alerts that promptly notify the IRT of potential anomalies and security threats.

III. Initiate a detailed and comprehensive investigation upon receiving alerts, deploying skilled personnel to assess the nature and severity of potential incidents, while confirming their legitimacy through careful correlation of information from diverse sources.

IV. Evaluate and prioritize incidents based on their potential impact on critical assets or services, considering the dimensions of confidentiality, integrity, and availability, and determine the urgency of the response required to mitigate potential harm or prevent further damage.

V. Implement a clearly defined escalation process, involving higher levels of expertise or management as incidents progress or intensify, ensuring timely communication and coordination within the IRT and with relevant stakeholders.

VI. Collect and preserve data related to incidents for further analysis and potential forensic examination, creating a comprehensive dataset to facilitate a thorough understanding of the incident's scope and characteristics.

VII. Engage in meticulous root cause analysis to identify the underlying factors contributing to the incident, aiming to understand the intricacies of the event and inform future prevention strategies, with an additional effort to attribute the incident to a specific threat actor or source if feasible.

3. Containment, Eradication, and Recovery

I. Employ immediate and targeted containment strategies to halt the spread of the incident, effectively isolating affected systems and preventing further damage to the organization's infrastructure.

II. Undertake a comprehensive eradication effort, focusing on identifying and eliminating the root causes of the incident to ensure a sustainable resolution and prevent any potential recurrence.

III.   Execute a meticulous recovery process, restoring affected systems and data to their pre-incident state, with a particular emphasis on minimizing downtime and ensuring a seamless return to normal business operations.

IV.   Validate the security of restored systems before they are brought back to normal operation, conducting thorough assessments to ensure that vulnerabilities have been effectively addressed and mitigated.

V.   Document all actions taken throughout the containment, eradication, and recovery phases, creating a detailed record of the incident response process for future reference, analysis, and continuous improvement.

4. Post-Incident Activity

I.   Initiate a thorough post-incident review and analysis, delving into the intricacies of the event to gain comprehensive insights into the incident's dynamics, including its detection, response, and resolution.

II.   Based on the lessons learned from the post-incident analysis, systematically update and enhance incident response plans, ensuring they reflect the most current and effective strategies for addressing potential future incidents.

III. Facilitate the sharing of valuable information with relevant parties, including internal stakeholders, industry peers, and, where appropriate, law enforcement or regulatory agencies, fostering a collaborative and informed approach to cybersecurity.

IV.   Methodically preserve evidence related to the incident, adhering to best practices in digital forensics, to support potential legal actions, investigations, or compliance requirements that may arise because of the security incident.

V.   Conduct targeted awareness and training sessions for both the IRT and broader organizational staff, integrating the lessons learned from the incident into training materials to fortify the overall security posture.

VI.   Embrace a culture of continuous improvement by actively seeking feedback and experiences from the incident handling processes, allowing for iterative enhancements and refinements to the organization's overall incident response capabilities.

Cloud customers should be aware of the following aspects related to incident response:

- Understand the **Cloud Service Provider's** (**CSP's**) process for notifying customers about security incidents. Be aware of the communication channels through which incident notifications will be sent.

- Inquire about the expected response timeframes from the CSP when a security incident occurs. Understand the time it may take for the CSP to acknowledge, investigate, and resolve incidents.

- Review the **Service Level Agreement** (**SLA**) to check whether incident response commitments, including response and resolution times, are explicitly defined. Ensure that the SLA aligns with the customer's expectations for incident handling.

- Understand the escalation procedures in place if the incident response process does not meet the agreed-upon timeframes. Know who to contact and how to escalate concerns regarding incident response.

- Evaluate the CSP's transparency in providing incident reports and updates. Check whether the CSP offers a detailed post-incident report, including the root cause analysis and steps taken for remediation.

- Clarify the customer's role and responsibilities in the incident response process. Understand any specific actions or information the CSP may require from the customer during an incident.

- Inquire about the CSP's continuous improvement practices in incident response. Check if lessons learned from previous incidents contribute to enhancing security measures.

- Be aware of any legal and compliance considerations related to incident response, both for the CSP and the customer. Understand how data breaches and incidents are handled in accordance with applicable laws and regulations.

> **Note**
>
> An **attack vector** refers to the specific pathway or method used by an attacker to gain unauthorized access to a computer system, network, or application in order to compromise its security. Attack vectors exploit vulnerabilities or weaknesses in the target system, and they can take various forms, including phishing, malware, exploits, social engineering, drive-by downloads, **Man-in-the-Middle** (**MitM**) attacks, brute-force attacks, zero-day attacks, watering-hole attacks, and USB-drop attacks.

Vulnerability Assessments

Vulnerability assessment is a systematic process of identifying, evaluating, and prioritizing potential vulnerabilities in computer systems, networks, applications, and organizational processes. The primary goal is to proactively discover weaknesses that could be exploited by attackers to compromise the confidentiality, integrity, or availability of information assets.

The vulnerability assessment process begins with **Scope Definition**, aiming to clearly outline the assessment's boundaries by identifying assets, networks, and applications within the defined scope, and establishing overarching goals and objectives.

After scope definition, the process transitions to **Asset Inventory**, striving to create a comprehensive list of assets within the designated scope by identifying and documenting hardware, software, network devices, and other relevant assets for a thorough assessment. Hardware-related assessments are internally conducted by CSPs across all service models (IaaS, PaaS, SaaS); customers depend on the provider to ensure hardware security and integrity due to the lack of direct infrastructure access.

Once the asset inventory is complete, the next stage is **Vulnerability Identification**, which aims to pinpoint potential vulnerabilities within the identified assets. Activities involve leveraging automated vulnerability scanning tools, conducting manual testing, and incorporating threat intelligence to uncover weaknesses, misconfigurations, and potential entry points.

After identifying vulnerabilities, the process progresses to **Vulnerability Assessment**, where the objective is to assess the severity and potential impact of the identified vulnerabilities. Activities include assigning risk scores based on severity and prioritizing vulnerabilities based on their risk level and potential impact on the organization.

Following the vulnerability assessment, the focus shifts to **Risk Analysis**, involving the analysis of identified vulnerabilities in the context of the organization's risk tolerance. Activities include evaluating the potential impact of vulnerabilities on business operations, data confidentiality, integrity, and availability, considering risk factors such as likelihood and potential exploits.

With a comprehensive understanding of vulnerabilities and associated risks, the next step is **Reporting**. The objective here is to communicate findings (including software flaws, missing patches, open ports, services that should not be running, weak passwords, misconfigurations, outdated software versions, insecure network configurations, unnecessary user privileges, and security policy violations) and risks to stakeholders.

Subsequently, the process moves to **Remediation Planning**, aiming to develop a plan to address and mitigate identified vulnerabilities. Activities involve working with relevant teams to create a remediation plan, prioritizing and scheduling patching, configuration changes, or other measures to reduce or eliminate vulnerabilities.

With the remediation plan in place, the focus shifts to the **Implementation of Remediation**, where the objective is to apply security measures to address vulnerabilities. Activities involve implementing the remediation plan, including patching systems, updating configurations, and applying security controls.

Following remediation, the process enters the **Verification and Validation** stage, where the objective is to verify the effectiveness of applied remediation. Activities include conducting follow-up assessments to confirm that vulnerabilities have been successfully remediated and validating that the risk posture has been improved.

Finally, to maintain ongoing security, the process includes **Continuous Monitoring**, aiming to establish continuous monitoring practices to detect new vulnerabilities and changes in the threat landscape. Activities involve implementing regular vulnerability scans, analyzing threat intelligence, and conducting periodic assessments.

When considering vulnerability assessments in a cloud environment, customers should consider several factors related to patching, frequency, responsibilities, and the SLA, including the following:

- When will the CSP apply patches to address identified vulnerabilities?
- Who is responsible for vulnerability assessments of what is in the cloud?
  - Responsibility is shared between the customer and the CSP, following the shared responsibility model in IaaS, SaaS, and PaaS.
- How frequently does the CSP release and apply security patches?
- What security responsibilities does the customer bear in terms of vulnerability remediation?
- Are specific terms related to vulnerability response outlined in the SLA?
- How does the CSP communicate about identified vulnerabilities and remediation efforts?
- Can customers perform vulnerability scans or tests on their cloud infrastructure?
  - You can often find out whether and how vulnerability assessments/tests can be conducted on cloud platforms through various online channels, such as CSP websites or documentation, which include resources such as terms of service, terms of use, security policies, and compliance documentation.
- What is the incident response plan for critical vulnerabilities?

Vulnerability scans can be categorized based on whether credentials (such as usernames and passwords) are required to conduct the scan:

- Authenticated Vulnerability Scans
  - Credentials required: Authenticated scans involve providing valid login credentials to the system or network being scanned.
  - Provide a more comprehensive assessment as they have access to more detailed information.
  - Can identify vulnerabilities that are only visible to authenticated users.
  - Offer a more accurate representation of the security posture.
- Unauthenticated Vulnerability Scans
  - No credentials required: Unauthenticated scans do not rely on specific login credentials and assess vulnerabilities from an external perspective.
  - Faster to execute as they don't require authentication steps.
  - Useful for external assessments, simulating attacks from an outsider's perspective, and identifying vulnerabilities that are visible to potential external attackers.

Vulnerability scans can be categorized based on their level of intrusiveness, which refers to the impact the scan has on the target system or network:

- Non-Intrusive Vulnerability Scans:

  - Low impact: Non-intrusive scans aim to identify vulnerabilities without actively engaging with the target systems.

  - Minimize disruption to normal system operations.

  - Suitable for environments where any potential impact needs to be avoided.

  - Provides an overview of potential vulnerabilities without actively probing the system.

- Intrusive Vulnerability Scans:

  - High impact: Intrusive scans actively interact with the target systems, probing for vulnerabilities and potential weaknesses.

  - Offer a more thorough assessment by actively testing vulnerabilities.

  - Can identify deeper or more complex security issues.

  - Provide a detailed understanding of how systems respond to specific probing.

> **Note**
>
> **Cloud Security Posture Management (CSPM)** tools and platforms are designed to assist organizations in securing and ensuring compliance within their cloud environments. These solutions typically offer features such as continuous monitoring, risk assessment, and remediation of security misconfigurations across various cloud services and resources. By identifying misconfigurations, compliance violations, and security risks, CSPM enables organizations to proactively mitigate these issues and maintain a robust security posture. Additionally, CSPM tools integrate seamlessly with DevOps processes to ensure security throughout the software development lifecycle.

Vulnerability scans can be categorized based on the specific area they target within an IT environment. Here are some common types:

- Network Vulnerability Scans:

  - Focus on identifying vulnerabilities in network devices, such as routers, switches, and firewalls. CSPs oversee this area since hardware access is restricted to them, not customers.

  - Aim to discover weaknesses that could be exploited to compromise the overall network security.

- Application Vulnerability Scans:

  - Concentrate on finding vulnerabilities in software applications.

  - Look for coding errors, misconfigurations, and other issues that may be exploited by attackers.

- Web Application Vulnerability Scans:

  - Specifically designed to identify vulnerabilities in web applications.

  - Target common issues such as SQL injection, **Cross-Site Scripting** (**XSS**), and other web-related vulnerabilities.

- Host-Based Vulnerability Scans:

  - Assess the security of individual hosts or devices (such as servers and workstations).

  - Look for vulnerabilities in the operating system, installed applications, and configurations.

- Database Vulnerability Scans:

  - Focus on database systems to find security weaknesses.

  - Identify issues such as misconfigurations, weak access controls, or outdated software.

- Wireless Network Vulnerability Scans:

  - Assess the security of wireless networks, including Wi-Fi routers and access points. CSPs manage this since hardware access is restricted to them.

  - Look for vulnerabilities that could lead to unauthorized access or data interception.

In the context of vulnerability scans, the terms **true positive**, **false positive**, **true negative**, and **false negative** refer to the accuracy of the scan results in identifying vulnerabilities. Here are examples of each of these terms:

- **True Positive** (**TP**): The vulnerability scan correctly identifies a real vulnerability that exists. Example: The scan correctly identifies an outdated and vulnerable version of web server software running on a server.

- **False Positive** (**FP**): The vulnerability scan incorrectly flags a non-existent vulnerability. Example: The scan mistakenly identifies a service running on a host as vulnerable to a specific exploit, but further investigation reveals it's a false alarm.

- **True Negative** (**TN**): The vulnerability scan correctly determines the absence of a vulnerability, and there is indeed no vulnerability. Example: The scan correctly reports that a particular server is not susceptible to a known security flaw because the necessary software is not installed.

- **False Negative** (**FN**): The vulnerability scan fails to detect an actual vulnerability that exists. Example: The scan overlooks an unpatched and exploitable vulnerability in a critical system, providing a false sense of security.

These two assets work together to effectively manage vulnerabilities:

- **Common Vulnerabilities and Exposures** (**CVE**): A standardized system for uniquely identifying and naming cybersecurity vulnerabilities, expressed in the format CVE-YYYY-NNNNN.

- **Common Vulnerability Scoring System** (**CVSS**): A framework that assesses the severity and characteristics of vulnerabilities, providing a standardized method for prioritizing them based on impact and exploitability, with scores ranging from 0 to 10.

Summary

This chapter guided you through the world of cybersecurity operations. You learned about the essential steps involved and explored various types of vulnerability management. The discussion covered the shared responsibility model in cloud security, highlighting the collaborative efforts required from both Cloud Service Providers (CSPs) and their customers. You discovered the power of intelligent monitoring for proactive threat detection, utilizing tools like firewalls and artificial intelligence.

The SIEM lifecycle equipped you with a comprehensive framework for managing security events. Additionally, incident management practices empowered you with a systematic approach to handling security incidents. You explored the incident handling process outlined in NIST SP 800-61, covering preparation, detection and analysis, containment, eradication, recovery, post-incident activities, and continuous improvement.

The emphasis shifted to the significance of a Security Operations Center (SOC) and its pivotal role as a centralized hub. The chapter explained how the SOC actively monitored, analyzed, and mitigated cybersecurity threats around the clock (24/7). It provided continuous oversight of your organization's IT environment, networks, and systems. The SOC also served as the initial point of contact for reporting potential security incidents within your organization.

Vulnerability assessments were outlined to provide you with a nuanced understanding of how organizations proactively identify and address potential vulnerabilities. This included authenticated and unauthenticated scans, as well as non-intrusive and intrusive approaches.

In the next chapter, you will learn about compliance with legal and contractual requirements, as well as policies, standards, guidelines, baselines, and procedures.

Exam Readiness Drill – Chapter Review Questions

Apart from a solid understanding of key concepts, being able to think quickly under time pressure is a skill that will help you ace your certification exam. That is why working on these skills early on in your learning journey is key.

Chapter review questions are designed to improve your test-taking skills progressively with each chapter you learn and review your understanding of key concepts in the chapter at the same time. You'll find these at the end of each chapter.

> **How to Access These Materials**
>
> To learn how to access these resources, head over to the chapter titled *Chapter 25, Accessing the Online Resources*.

To open the Chapter Review Questions for this chapter, perform the following steps:

1. Click the link – `https://packt.link/CCSPE1_CH21`.

    Alternatively, you can scan the following **QR code** (*Figure 21.1*):

Figure 21.1 – QR code that opens Chapter Review Questions for logged-in users

2. Once you log in, you'll see a page similar to the one shown in *Figure 21.2*:

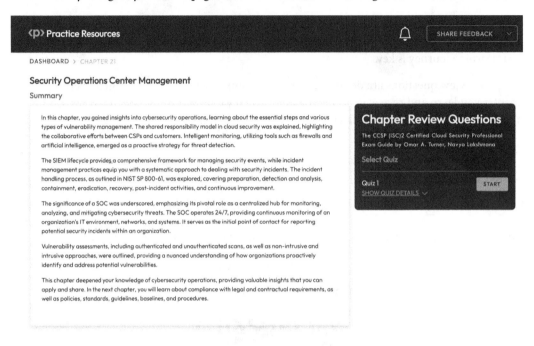

Figure 21.2 – Chapter Review Questions for Chapter 21

3. Once ready, start the following practice drills, re-attempting the quiz multiple times.

Exam Readiness Drill

For the first three attempts, don't worry about the time limit.

ATTEMPT 1

The first time, aim for at least **40%**. Look at the answers you got wrong and read the relevant sections in the chapter again to fix your learning gaps.

ATTEMPT 2

The second time, aim for at least **60%**. Look at the answers you got wrong and read the relevant sections in the chapter again to fix any remaining learning gaps.

ATTEMPT 3

The third time, aim for at least **75%**. Once you score 75% or more, you start working on your timing.

> **Tip**
>
> You may take more than **three** attempts to reach 75%. That's okay. Just review the relevant sections in the chapter till you get there.

Working On Timing

Target: Your aim is to keep the score the same while trying to answer these questions as quickly as possible. Here's an example of how your next attempts should look like:

Attempt	Score	Time Taken
Attempt 5	77%	21 mins 30 seconds
Attempt 6	78%	18 mins 34 seconds
Attempt 7	76%	14 mins 44 seconds

Table 21.1 – Sample timing practice drills on the online platform

> **Note**
>
> The time limits shown in the above table are just examples. Set your own time limits with each attempt based on the time limit of the quiz on the website.

With each new attempt, your score should stay above **75%** while your "time taken" to complete should "decrease". Repeat as many attempts as you want till you feel confident dealing with the time pressure.

22
Legal Challenges and the Cloud

In this chapter, you'll gain an understanding of the legal requirements and unique risks associated with the cloud environment. You'll explore how conflicting international legislation impacts cloud operations and grasp the important legal frameworks and guidelines governing the industry. Additionally, you'll learn about eDiscovery considerations, integrating standards such as the **International Organization for Standardization (ISO)/International Electrotechnical Commission (IEC)** 27050 and guidance from the **Cloud Security Alliance (CSA)**, along with forensic requirements. You'll differentiate between two crucial roles: **data owner/controller** and **data custodian/processor**. Furthermore, you'll learn about the impact of the cloud on risk management and study topics such as outsourcing and designing contracts for cloud services.

By the end of this chapter, you will be able to confidently answer questions on the following:

- Legal requirements and unique risks within the cloud environment
- Implications of cloud to enterprise risk management
- Outsourcing and cloud contract design

Legal Requirements and Unique Risks Within the Cloud

The adoption of cloud computing has revolutionized the way organizations manage and process data, offering scalability, flexibility, and cost efficiency. However, the dynamic nature of the cloud environment brings forth a multitude of legal requirements and unique risks that businesses must navigate to ensure compliance, security, and operational resilience.

Conflicting International Legislation

In traditional on-premise environments, the physical location of data is often clear, making it easier to navigate legal requirements within a specific jurisdiction. However, in the cloud, data can be dispersed across various regions, leading to challenges in determining the applicable legal framework. **Jurisdictional boundaries** become more complex as data travels, resides, and is processed in different countries, each with its own set of data protection, privacy, and security laws.

The absence of a **unified, overarching international law** for cloud computing complicates compliance efforts. Organizations must grapple with a patchwork of regulations, treaties, and standards that may vary significantly from one country to another.

Divergent legal requirements across jurisdictions can expose businesses to risks such as fines, penalties, or even the loss of business opportunities. Failure to comply with specific regulations can result in severe consequences, impacting on the company's reputation and financial standing.

Adopting multiple cloud service providers for high availability and redundancy introduces an additional layer of complexity. Each provider may have different data management practices and operate under distinct legal jurisdictions, requiring careful consideration and management of legal risks.

Breach notification rules, which stipulate how organizations must inform authorities and affected individuals in the event of a data breach, can vary widely. Staying compliant with these rules becomes more challenging when dealing with data dispersed across multiple cloud providers and jurisdictions.

Copyright and intellectual property laws may differ globally, impacting how organizations handle, store, and transmit digital assets in the cloud. Organizations need to be vigilant about potential infringements and ensure compliance with the laws of each jurisdiction where they operate.

Cloud services often involve the transfer of data across borders. International import/export laws may come into play, especially when dealing with sensitive or regulated data. Organizations must align their cloud practices with these laws to avoid legal complications.

Collaborate with legal experts and compliance teams with knowledge of international laws to navigate and interpret the diverse legal landscape. Establish clear data governance policies that consider the implications of data dispersion and address compliance with various international laws. Regularly monitor changes in laws and regulations, adapting data management practices accordingly. Implement a dynamic compliance strategy that evolves with the legal landscape. Engage in open communication with **cloud service providers (CSPs)** to understand their data handling practices, security measures, and commitment to compliance with relevant laws. You should request and review CSPs' data handling policies before engaging with them. Nevertheless, once a service is hosted on the cloud, the responsibility for data ultimately falls on the customers as the data owners.

Example Scenario

There are data protection conflicts between the **General Data Protection Regulation** (**GDPR**) and the US Patriot Act. A multinational corporation, ABC Corp, headquartered in the **European Union** (**EU**), decides to leverage cloud services provided by a US-based CSP. Here are the rules they must adhere to in these regions:

- **EU data protection laws (GDPR)**: Strict regulations on the processing and transfer of personal data, requiring explicit consent and adherence to stringent data protection principles

- **US Patriot Act and national security laws**: Grants US authorities access to data for national security reasons, potentially conflicting with GDPR principles and raising concerns about the protection of EU citizens' personal data

There are different breach notification rules in different jurisdictions. Rules from some of the prime jurisdictions are listed here:

- **EU GDPR**: Organizations in the EU must report data breaches to authorities within 72 hours and, if it poses a high risk, notify affected individuals promptly

- **California Consumer Privacy Act (CCPA)**: California businesses must notify residents of a data breach without unreasonable delay and no later than 45 days after discovering the incident if it poses a risk to individuals' personal information

- **Health Insurance Portability and Accountability Act, US (HIPAA)**: Covered entities and business associates under HIPAA must notify affected individuals and the US Department of Health and Human Services of breaches involving unsecured protected health information, without unreasonable delay and no later than 60 days after discovery

Evaluation of Legal Risks Specific to Cloud Computing

When choosing a CSP, cloud customers need to carefully evaluate legal risks to ensure compliance, data security, and protection of their interests. Here are the key considerations:

- Assess the CSP's **data protection and privacy practices** to ensure alignment with applicable regulations, such as GDPR, or industry-specific requirements. Evaluate how the provider handles data storage, processing, and transfer, and clarify data ownership and control.

- Review the CSP's **security measures**, **certifications** (e.g., ISO/IEC 27001), and **compliance** with industry standards. Ensure that the provider has robust security protocols in place to protect against unauthorized access, data breaches, and other security threats.

- Thoroughly examine the CSP's **contracts** and **service-level agreements** (**SLAs**). Clarify responsibilities, liabilities, and dispute resolution mechanisms. Ensure that the terms align with legal requirements and the organization's specific needs.

- Understand the **geographical locations** of the CSP's data centers and assess the legal implications of data residency, considering potential conflicts with data protection laws and sovereignty issues in different jurisdictions.

- Evaluate the CSP's policies on **data portability** and consider **exit strategies**. Ensure that data can be easily migrated or retrieved in a usable format if the relationship with the provider ends, minimizing risks associated with vendor lock-in.

- Clarify ownership and usage rights for **intellectual property**, including data and applications hosted in the cloud. Ensure that the CSP's terms do not infringe the organization's intellectual property rights.

- Assess the CSP's **incident response capabilities**, including their ability to detect and respond to security incidents promptly. Understand the provider's breach notification processes and timelines, ensuring compliance with legal requirements.

- Verify that the CSP complies with relevant industry regulations and standards. Determine the extent to which the provider allows and facilitates **audits** of their security and compliance practices, providing transparency and assurance.

- Understand the terms and conditions related to **service termination** and **data destruction**. Ensure that the CSP has proper processes in place to delete data securely and in compliance with legal requirements when the service is terminated.

- Assess the CSP's **business continuity** and **disaster recovery** capabilities. Verify that their plans align with legal obligations and industry best practices, minimizing legal risks associated with service interruptions.

- Prioritize CSPs that demonstrate **transparent communication** regarding their legal and security practices. Providers that openly share information about their policies and procedures contribute to building trust with customers.

- Evaluate the CSP's **shared responsibility model** to clarify roles and responsibilities. While the customer holds ultimate data responsibility, understanding who is accountable for specific tasks is crucial.

- Assess whether any data transactions occur beyond the cloud customer's control. If so, check how the CSP ensures compliance with relevant regulations and determine whether they can provide evidence of these practices in a court of law upon request.

Legal Framework and Guidelines

Legal frameworks and guidelines play a crucial role in providing the regulatory structure and standards that govern various aspects of activities, industries, and technologies, including cloud computing. Organizations operating in the cloud environment should be aware of and comply with the applicable legal frameworks and guidelines to ensure data protection, privacy, and overall legal compliance. The specifics of compliance may vary depending on factors such as the industry, the geographic location, and the nature of the data being processed.

eDiscovery and Forensics Requirements

The **shared responsibility model** presents difficulties in maintaining a consistent and documented chain of custody, as different entities, including the CSP and the customer, may bear responsibility for distinct aspects of data management. This division of responsibilities can potentially result in gaps or uncertainties in the chain of custody.

The preparedness of a CSP for eDiscovery becomes challenging when proper log retention and comprehensive logging are deficient. Insufficient log retention periods or incomplete logging by the CSP can hinder forensic investigations, potentially jeopardizing the chain of custody and impeding the reconstruction of incident timelines. Organizations relying on the CSP's logging infrastructure need to thoroughly assess its capabilities, negotiate clear terms in service agreements regarding logging practices and data retention, and establish effective communication and coordination for incident response readiness. In cases where native CSP logging proves inadequate, organizations may need to implement supplementary logging strategies and regularly test and validate the overall effectiveness of the eDiscovery process. Considering alternative data sources is also crucial for a comprehensive eDiscovery approach.

A distinctive risk in cloud eDiscovery is the difficulty associated with physically searching and seizing cloud resources, such as storage or hard drives. In contrast to traditional environments where direct examination is possible through physical access to servers and storage devices, the cloud introduces complexities that can make this process challenging, if not impossible. This can involve leveraging specialized eDiscovery tools and techniques designed for cloud environments, such as data preservation and collection mechanisms provided by CSPs. Additionally, legal agreements are established between the cloud customer and the CSP to facilitate eDiscovery processes and ensure compliance with legal requirements.

Taking into account eDiscovery as a security requirement during the selection and contract negotiation phases with a cloud vendor is a prudent strategy. It is imperative to proactively identify and communicate any risks to the CSP and advocate for the necessary due care and due diligence measures.

NISTIR 8006 NIST Cloud Computing Forensic Science Challenges

The NISTIR 8006 report titled *Cloud Computing Forensic Science Challenges* focuses on researching challenges related to digital forensics in the cloud environment. The primary objectives include identifying these challenges, analyzing their root causes, and categorizing them to aid in the development of standards and technology research to mitigate such challenges. The document outlines the methodology used in collecting and aggregating forensic science challenges, including literature searches, stakeholder input, and group discussions.

> **Note**
>
> *Chapter 19, Digital Forensics,* discusses additional resources and guidelines pertinent to eDiscovery and digital forensics, including standards such as ISO/IEC 27050 and the CSA Security Guidance.

Understanding the Implications of Cloud to Enterprise Risk Management

The shift from traditional on-premises environments to cloud computing has profound implications for **enterprise risk management** (ERM). Cloud technologies offer flexibility, scalability, and cost efficiencies, but they introduce new risk considerations that organizations must carefully navigate. By addressing the unique challenges posed by data security, shared responsibility models, multitenancy, compliance, business continuity, vendor relationships, and financial considerations, organizations can effectively manage risks and capitalize on the benefits offered by cloud technologies.

Assessing Provider Risk Management Programs

Assessing CSP's risk management program is essential to guarantee the security, compliance, and reliability of the services they offer. The evaluation involves a comprehensive approach to various aspects, starting with the examination of the CSP's **security policies and procedures**. This includes a thorough assessment of their documentation, covering critical **security measures** such as data encryption, access controls (such as multi-factor authentication, identity and access management, and the principle of least privilege), and incident response.

The way the CSP handles **data encryption** is paramount. Understanding their approach to encrypting data in transit and at rest ensures alignment with industry best practices, including robust encryption algorithms and key management. This includes both the encryption of customer data using CSP-managed encryption keys and the capability to accommodate customer-managed encryption keys. **Physical security** measures at the CSP's data centers are another vital aspect of the assessment. This involves evaluating access controls, surveillance systems, and environmental controls to safeguard the physical infrastructure.

Incident response and disaster recovery plans form a critical part of a CSP's risk management strategy. A thorough review of these plans ensures the existence of clear processes for handling security incidents and recovering data in case of disasters. **Network security infrastructure**, encompassing firewalls, intrusion detection/prevention systems, and network segmentation, is pivotal to safeguarding data. A detailed assessment ensures the robustness of these measures.

Vendor management is essential, especially if the CSP relies on third-party vendors. The assessment should focus on how the CSP manages and monitors these relationships to ensure alignment with security standards. **Monitoring and auditing** capabilities are indicative of a CSP's ability to detect and respond to security threats. This includes assessing log management, intrusion detection, and the frequency of security audits.

SLAs play a crucial role in setting expectations for uptime, data availability, and the CSP's responsibilities during service disruptions. A thorough review ensures alignment with business requirements. Data ownership and portability terms should be clarified, and the CSP's support for data portability assessed. This ensures that data can be easily retrieved when needed.

Considering the **financial stability** of the CSP is important, as a financially stable provider is more likely to invest in robust security measures and infrastructure. Transparency and communication practices are indicative of a CSP's commitment to building trust. Evaluating how the CSP communicates security practices, incidents, and updates contributes to the overall assessment.

Seeking references from existing customers provides real-world insights into the CSP's services and their effectiveness in managing risks. Legal and contractual aspects, including terms related to liability, indemnification, and dispute resolution, are critical considerations in the overall risk management assessment.

Ensuring compliance with industry standards and regulations is paramount when assessing a CSP's risk management programs. Certifications such as ISO 27001 and HIPAA, along with evidence of compliance certifications, serve as crucial benchmarks in evaluating the CSP's commitment to meeting established standards. Furthermore, examining any legal or regulatory actions against the CSP provides valuable insights into their regulatory standing.

Incorporating a range of certifications, including **Service Organization Control 2 (SOC 2)**, ISO 27001, **Federal Risk and Authorization Management Program (FedRAMP)**, and CSA **Security, Trust, Assurance, and Risk (STAR)**, is integral to the comprehensive evaluation of a CSP's security and risk management practices. These certifications offer third-party validation, attesting to the provider's dedication to maintaining robust security measures. It is imperative to regularly check for **recertification** and updates to ensure ongoing compliance and assurance.

One key dimension to consider is the CSP's risk profile and risk appetite. Understanding the provider's risk profile involves examining their historical and current exposure to cybersecurity threats, past incidents, and vulnerabilities. This assessment provides insights into the CSP's vulnerability landscape, helping to gauge their preparedness and resilience against potential risks.

Equally important is evaluating the CSP's risk appetite, which defines the level of risk the provider is willing to accept in pursuit of its business objectives. This involves understanding the trade-offs between risk and business performance. A CSP with a clear and well-defined risk appetite is better positioned to align its risk management programs with its strategic goals, providing transparency to clients regarding the inherent risks associated with the services.

To assess the security posture of a CSP, you need to be familiar with important frameworks and certifications. The following provide different levels of assurance and focus on various aspects of cloud security:

- **SOC 2**: Verify whether the CSP has undergone a SOC 2 audit, specifically focusing on security, availability, processing integrity, confidentiality, and privacy (if applicable). Request the SOC 2 Type II report, which provides an independent assessment of the effectiveness of the CSP's controls over a specified period.

- **ISO 27001**: Confirm the CSP's ISO 27001 certification, which is a globally recognized standard for **information security management systems** (**ISMSs**). Review the scope of their certification to ensure it aligns with the services you are utilizing.

- **ISO 27017**: Verify whether the CSP has obtained ISO 27017 certification, which specifically addresses information security controls for cloud services. Review the certification to ensure compliance with best practices for cloud security.

- **ISO 27018**: Confirm whether the CSP adheres to ISO 27018 standards, focusing on the protection of Personally Identifiable Information (PII) in cloud environments.

- **FedRAMP**: If the CSP serves US government entities, check for FedRAMP authorization. US FedRAMP facilitates a cost-effective, risk-based approach for federal government adoption of cloud services, empowering agencies to utilize modern cloud technologies while emphasizing security and federal information protection.

- **CSA STAR Level 1 – Self-Assessment**: Evaluate whether the CSP has completed a self-assessment using the CSA STAR Level 1 criteria. This involves the provider completing the **Consensus Assessments Initiative Questionnaire** (**CAIQ**), a set of industry-accepted questions about their security posture.

- **CSA STAR Level 2 – Third-Party Certification**: Check whether the CSP has obtained third-party certifications based on the CSA STAR Level 2 criteria. This involves an independent assessment of the provider's security controls against the **Cloud Controls Matrix** (**CCM**).

- **CSA STAR Level 3 – Continuous Monitoring**: Assess whether the CSP has implemented continuous monitoring practices, as outlined in the CSA STAR Level 3 criteria. This involves ongoing assessment and reporting of security controls' effectiveness.

Difference Between Data Owner/Controller and Data Custodian/Processor

The terms **data owner/controller** and **data custodian/processor** are distinctions made in the context of data governance and management. They refer to different roles and responsibilities in handling and processing data. *Table 22.1* provides a breakdown of their key differences:

Data owner/controller	Data custodian/processor
Responsibilities: • **Decision-making authority:** The data owner or controller is responsible for making decisions about the data. This includes determining the purpose of data collection, how it should be used, and who should have access to it. • **Compliance:** Ensures that data processing activities comply with relevant regulations and legal requirements.	**Responsibilities:** • **Data handling:** The data custodian or processor is responsible for the technical aspects of data handling, storage, and processing • **Execution of operations:** Implements the decisions made by the data owner/controller regarding data usage
Rights: • **Access control:** Controls access to the data, deciding who can view, modify, or delete it • Has the authority to modify or update the data as necessary	**Rights:** • **Access control (technical):** Implements access controls according to the specifications provided by the data owner/controller • **Data processing:** Processes data according to the defined purposes and procedures
Accountability: Holds legal responsibility for the data and may be subject to legal consequences if data processing activities violate regulations.	**Accountability:** While the data custodian is responsible for handling data securely, they may not bear legal liability for how the data is used unless it violates technical protocols
Examples: In an organization, the **Chief Information Officer (CIO)** or a specific data governance team might act as data owners/controllers. The customer holds the ultimate responsibility for data and serves as the data owner/controller.	**Examples:** In an organization, IT administrators or database administrators may act as data custodians/processors, while cloud service providers' roles depend on the service model (IaaS, PaaS, or SaaS)

Table 22.1 – Roles and responsibilities of data owner/controller and data custodian/processor

The data owner/controller governs and dictates the rules for data, while the data custodian/processor handles the technical aspects of implementing those rules. The roles often work in collaboration to ensure effective and secure data management.

Regulatory Transparency Requirements

Cloud security professionals should be well versed in regulatory transparency requirements imposed on data controllers to ensure compliance and robust data protection practices. Here are key regulatory transparency requirements, focusing on breach notification, **Sarbanes-Oxley (SOX)**, and the GDPR that are particularly relevant to cloud security professionals:

Breach Notification

Breach notification requirements are pivotal components of data protection regulations, necessitating organizations to promptly and transparently inform individuals and relevant authorities in the event of a security incident. The definition of a data breach encompasses unauthorized access, acquisition, or disclosure of personal data, including both accidental and intentional breaches.

Timeframes for notification are a crucial aspect, often mandating organizations to notify affected parties within a specified period. This emphasis on prompt disclosure underscores the importance of immediate action when a breach occurs. Regulations detail the content that must be included in notifications, encompassing information about the breach's nature, compromised data types, and recommended actions for affected individuals.

Communication channels are specified, dictating how organizations should convey breach notifications. This could be by email, or public announcements, depending on the circumstances. Thresholds may also be defined, triggering the obligation to notify based on factors such as the number of affected individuals or the types of compromised information.

Exemptions or delays may be allowed under certain circumstances, such as when law enforcement requests a delay for investigative purposes. Understanding the consequences of non-compliance is crucial, as penalties can range from fines to legal action and reputational damage. Cloud security professionals need to be vigilant about variations in breach notification requirements internationally, particularly when breaches involve individuals in multiple jurisdictions.

Documentation and record-keeping play a vital role, with regulations often mandating the detailed documentation of breach incidents and the steps taken for resolution. Coordination with regulatory authorities may also be required, necessitating cloud security professionals to engage with these entities and provide necessary information for investigations.

Post-breach communication is an ongoing obligation, with regulations often requiring organizations to keep affected individuals informed about investigation progress, remediation efforts, and additional steps they should take to protect themselves. By understanding these nuances and intricacies of breach notification requirements, cloud security professionals can proactively develop and implement robust incident response plans that align with regulatory expectations, ensuring compliance and fostering transparency in the face of security incidents.

SOX

SOX, enacted in 2002, is a landmark US federal law primarily aimed at enhancing transparency, accountability, and the reliability of financial reporting within publicly traded companies. It was enacted in response to corporate scandals that eroded investor confidence.

SOX criminalizes the alteration, destruction, mutilation, or concealment of records or documents with the intent to obstruct federal investigations. This reinforces the importance of maintaining accurate and unaltered financial records, contributing to transparency in financial reporting.

SOX enhances transparency by requiring CEOs and CFOs to certify the accuracy of financial reports and internal controls. It holds top executives personally accountable for the information presented in financial statements, promoting a culture of accuracy and transparency.

SOX mandates companies to establish and maintain effective internal controls over financial reporting. This ensures transparency in documenting and assessing internal control processes, assuring the reliability of financial reporting.

SOX emphasizes the independence of audit committees overseeing financial reporting. This independence enhances transparency by ensuring that the audit process is objective and free from undue influence.

SOX includes provisions to protect whistleblowers who report violations of security laws, accounting irregularities, or fraudulent activities. This promotes transparency by encouraging individuals within organizations to come forward with information about potential misconduct.

SOX requires timely and transparent disclosure of material changes in financial conditions or operations. This ensures that investors and stakeholders are promptly informed about significant events that may impact the company's financial status.

GDPR

GDPR aims to protect the privacy and personal data of individuals within the EU and **European Economic Area** (**EEA**). Transparency is a fundamental principle of GDPR, and several provisions within the regulation emphasize the need for organizations to be transparent in their data processing activities. Here are key aspects of the GDPR's transparency requirements:

- **Lawfulness, fairness, and transparency**: GDPR's Article 5 outlines the principles of data processing, including that personal data must be processed lawfully, fairly, and in a transparent manner. Organizations are required to be clear and open about how they collect, process, and use personal data.

- **Information to be provided to data subjects**: GDPR specifies the information organizations must provide to individuals (data subjects) regarding the processing of their personal data. This includes details such as the identity of the data controller, purposes of processing, recipients of data, and the right to access and rectify personal information.

- **Privacy notices and consent**: Organizations must provide clear and easily accessible privacy notices to individuals before collecting their personal data. These notices should outline the purposes of processing and any other relevant information. Consent, when required, must be obtained in a transparent manner, and individuals should be informed of their right to withdraw consent at any time.

- **Right of access**: Individuals can confirm from the data controller whether their personal data is being processed and can get access to that data, along with detailed information about its processing, recipients, and storage duration. Additionally, individuals have the right to receive a copy of their personal data undergoing processing. This emphasizes the transparency of data processing activities, allowing individuals to be aware of and verify the lawfulness of the processing.

- **Right to rectification and erasure**: Individuals have the right to rectify inaccuracies in their personal data and request its erasure under certain circumstances. This reinforces transparency, as organizations are obliged to correct or delete inaccurate or outdated information.

- **Data Protection Impact Assessments (DPIAs)**: In cases where processing operations are likely to result in a high risk to individuals' rights and freedoms, organizations are required to conduct a DPIA. The DPIA process enhances transparency by assessing and mitigating risks associated with data processing activities.

- **Data breach notification**: Organizations must notify the relevant supervisory authority and, in certain cases, data subjects about data breaches without undue delay. This requirement promotes transparency by ensuring that individuals are informed about security incidents that may impact their personal data.

- s**Accountability**: Organizations are accountable for compliance with GDPR and must be able to demonstrate such compliance. This involves maintaining documentation, conducting risk assessments, and implementing measures that emphasize transparency in data processing practices.

- **Data protection officer**: Organizations that engage in large-scale or systematic processing of personal data may be required to appoint a **data protection officer** (**DPO**). The DPO plays a role in ensuring transparency and advising on compliance with data protection regulations.

Note

Safeguards are proactive measures implemented beforehand to prevent risks, including access controls and encryption. In contrast, **countermeasures** are reactive strategies activated in response to identified threats or incidents, such as incident response plans and disaster recovery. Safeguards build a secure foundation, while countermeasures provide swift responses to mitigate impacts and prevent recurrences, collectively forming a comprehensive risk management approach.

Risk Treatment

Risk treatment is a key phase in the risk management process where organizations take specific actions to address identified risks. The primary goal is to modify the risk's likelihood or impact, ensuring it aligns with the organization's risk appetite and objectives. Let's take a look at some risk treatment strategies.

Risk Avoidance

Risk avoidance is a risk management strategy that involves the deliberate decision not to engage in certain activities or ventures that carry significant risks. Organizations opt for risk avoidance when they recognize the potential negative consequences of a particular risk and choose to sidestep or withdraw from those activities to prevent exposure. This strategy aligns with the principle that preventing a risk from materializing is often more effective and efficient than dealing with its consequences. Risk avoidance acknowledges that some risks can be eliminated entirely by not engaging in specific actions or ventures.

Risk avoidance should be considered under the following circumstances:

- **Unacceptable impact**: When the potential impact of a specific risk is deemed unacceptable, and the organization wishes to prevent exposure to that impact. This approach is especially justified when the cost of impact is considerably high and available remediation measures are insufficient to reduce it significantly.

- **Unpredictable or high-impact risks**: For risks that are highly unpredictable, have severe potential consequences, or involve uncertainties that make effective mitigation challenging.

Example: A pharmaceutical company is considering the launch of a new drug with high research and development costs. However, during a thorough market analysis, they identified a potential legal and regulatory risk associated with the drug's side effects. The company decides to avoid this risk altogether.

Risk Mitigation

Risk mitigation is a risk management strategy focused on reducing the likelihood or impact of identified risks. It involves implementing preventive measures and contingency plans to minimize the potential harm and enhance the organization's resilience in the face of uncertainty.

It's crucial to understand that risk can never be entirely eliminated. Instead, organizations aim to reduce risk to an acceptable level based on their risk tolerance, business objectives, and available resources. Residual risk refers to the level of risk that remains after implementing risk mitigation measures.

Example: A software development company mitigates the risk of project delays by cross-training team members on critical tasks and maintaining flexible project timelines.

Organizations often use risk transfer/share alongside mitigation strategies. While mitigation reduces the overall risk, transferring the residual financial risk through mechanisms such as insurance provides an additional layer of protection.

Risk Transfer/Sharing

Risk transfer, or **risk sharing**, is a risk management strategy where an organization shifts the financial burden or responsibility of potential losses to another party. This is often achieved through mechanisms such as insurance policies, contractual agreements, or partnerships. Insurance policies are commonly used for risk transfer. Insurance policies provide financial coverage against specific risks, allowing organizations to share the burden with insurers. Formal contractual agreements with suppliers, partners, or service providers may include clauses that allocate certain risks and responsibilities, sharing the potential impacts.

Example: An e-commerce company transfers the risk of potential financial losses due to supply chain disruptions by securing a comprehensive insurance policy. If unforeseen events affect the supply chain, the insurance provider bears the financial impact, enabling the company to share the risk externally.

Risk Acceptance

Risk acceptance is a risk management strategy wherein an organization consciously acknowledges and tolerates a certain level of risk without implementing specific measures to mitigate or transfer it. This strategy is chosen when the potential impact is deemed acceptable, and resources are directed towards managing more critical aspects of the business.

Organizations opt for risk acceptance when the cost of mitigating a particular risk exceeds the potential impact, and the risk is deemed manageable within acceptable limits. However, it's important to note that risk acceptance might not be a viable option when there are regulatory mandates or obligations associated with specific activities. In situations where compliance is mandatory, organizations may need to explore alternative risk management strategies, balancing regulatory requirements with the need to mitigate potential impacts.

Example: A technology company accepts the minor risk of occasional software bugs in non-critical applications. Recognizing that the impact is minimal and the cost of extensive mitigation measures outweighs potential benefits, the company decides to allocate resources to more significant areas of concern.

Different Risk Frameworks

Several risk frameworks are used across industries to guide organizations in identifying, assessing, and managing risks.

ISO 31000:2018

ISO 31000:2018 is a globally recognized standard that provides comprehensive guidance on risk management. Applicable to organizations of any size, sector, or industry, this standard is designed to help entities establish effective risk management processes tailored to their specific contexts.

At its core, ISO 31000:2018 is built upon fundamental principles that include the integration of risk management into organizational governance, a structured and comprehensive approach, customization to the organization's context, and a commitment to continual improvement.

The standard offers a structured framework for risk management, emphasizing alignment with organizational objectives and governance structures. The risk management process is outlined, comprising key steps such as establishing the context, risk assessment, risk treatment, monitoring and review, and communication and consultation. This iterative process enables organizations to adapt to evolving internal and external factors.

ISO 31000:2018 stresses the importance of integrating risk management into various organizational processes, including governance, strategy development, planning, and performance management. Clear criteria for assessing and evaluating risks, including risk tolerance and appetite, are highlighted. This enables organizations to make informed decisions about which risks to accept, treat, or avoid. The standard recognizes the dynamic nature of risk criteria, allowing for adjustments as the organization's risk landscape evolves.

Effective communication and consultation with stakeholders are emphasized as essential elements of the risk management process. Transparency in reporting and sharing information related to risks fosters a collaborative approach and ensures that relevant insights contribute to the decision-making process.

ISO 31000:2018 encourages organizations to establish systematic processes for monitoring and reviewing the effectiveness of their risk management activities. This ongoing evaluation supports continual improvement and enables organizations to respond proactively to changes in their risk environment.

EU Agency for Cybersecurity (ENISA)

ENISA is an EU agency headquartered in Athens, Greece. Established in 2004, its primary mission is to enhance the overall cybersecurity posture of the EU. ENISA collaborates with EU member states, institutions, businesses, and the public to develop and promote effective cybersecurity strategies, policies, and practices.

ENISA's contributions to risk management include the following:

- ENISA provides guidance and best practices on risk management, aiding organizations in identifying, assessing, and mitigating cybersecurity risks

- The agency contributes to the development of risk management frameworks and methodologies, promoting standardized approaches for EU member states

- ENISA engages in capacity-building initiatives related to risk management, offering training programs and workshops to enhance the skills of cybersecurity professionals

- ENISA fosters collaboration and information sharing among EU member states, facilitating a collective response to emerging cyber threats

- ENISA provides policy support to the European Commission and member states, assisting in the development of strategic initiatives that strengthen the EU's resilience against cyber risks

Security Risks in the Cloud Identified by ENISA

ENISA has identified several security risks associated with cloud computing. Some key areas of concern are as follows:

- Risks related to the management, storage, and processing of sensitive data in the cloud, including compliance with data protection regulations

- Concerns about unauthorized access, insufficient identity verification, and inadequate access controls in cloud environments

- Risks associated with the encryption of data in transit and at rest, ensuring that proper encryption measures are implemented

- Challenges related to incident response and forensic capabilities in the event of security incidents or breaches in cloud environments

- Risks associated with dependence on a single cloud service provider, limiting flexibility and making migration challenging

- Concerns about the security of the supply chain, including the security practices of cloud service providers and their subcontractors

- The need for clear understanding and communication of responsibilities between cloud service providers and their customers in a shared responsibility model

NIST Special Publication 800-37

National Institute of Standards and Technology (NIST) Special Publication 800-37, titled *Guide for Applying the Risk Management Framework to Federal Information Systems*, outlines the NIST **Risk Management Framework (RMF)**. This framework provides guidance for federal agencies to manage and strengthen the security of their information systems.

Here are the key components of NIST Special Publication 800-37:

- **RMF**: It describes the structured process for managing information security risk within federal agencies

- It identifies a series of steps for effective risk management: Prepare, Categorize, Select, Implement, Assess, Authorize, and Monitor

- It integrates security into the **system development life cycle (SDLC)** to ensure that security measures are addressed from the initiation to the decommissioning of information systems

- It emphasizes the importance of continuous monitoring to promptly detect and respond to evolving security threats and vulnerabilities

- It provides guidance on managing security risks for interconnected systems, considering the dynamic nature of modern information technology environments

- It acknowledges the need for flexibility to accommodate various risk environments and information system types

- It defines the roles and responsibilities of key stakeholders, including senior leaders, authorizing officials, and security professionals involved in the risk management process

- It aligns with other NIST publications, such as NIST Special Publication 800-53, which provides security controls and baselines

> **Note**
>
> NIST SP 800-146, titled *Cloud Computing Synopsis and Recommendations*, serves the purpose of demystifying the field of cloud computing in straightforward language and offering guidance for information technology decision-makers. The document includes clear definitions of cloud computing terminology, as well as an exploration of the advantages and risks associated with all service models. Recognizing that cloud computing is still evolving, with its full strengths and weaknesses yet to be fully researched and tested, the publication provides recommendations on the suitable utilization of cloud computing. It also outlines the current knowledge constraints, emphasizing areas for future analysis and research (see `https://csrc.nist.gov/pubs/sp/800/146/final`).

Metrics for Risk Management

Various metrics are used to measure and monitor risk in different contexts. It's important for organizations to tailor their risk metrics to align with their specific industry, objectives, and risk appetite. Regularly reviewing and updating these metrics ensures that risk management practices remain effective in the dynamic business environment. Here are some common metrics for risk management:

- **Risk exposure**: The potential financial loss or impact associated with a specific risk. Monetary value or percentage representing the potential loss.

- **Risk likelihood**: The probability that a specific risk event will occur. Expressed as a probability percentage (e.g., 10% likelihood).

- **Risk impact**: The magnitude of the consequences if a risk event occurs. Quantified in terms of financial loss, time delays, or other relevant units.

- **Risk severity**: A combination of likelihood and impact, indicating the overall severity of a risk. Often measured on a scale (e.g., low, medium, high).

- **Risk appetite**: The level of risk that an organization is willing to accept in pursuit of its objectives. Often qualitative, expressed in terms of low, medium, or high-risk tolerance.

- **Risk velocity**: The speed at which a risk event could occur and impact the organization. It offers a timeframe for potential risk occurrence.

- **Risk register**: A comprehensive list of identified risks, including their likelihood, impact, and mitigation strategies.

- **Residual risk**: The level of risk that remains after implementing risk mitigation measures. Measured in terms of likelihood and impact post-mitigation.

- **Incident response time**: The time taken to respond to and mitigate the impact of a risk event. Measured in hours, days, or other relevant time units.

- **Single-loss expectancy (SLE)**: The expected monetary loss for a single occurrence of a specific risk event. Calculated as the product of the asset value and the exposure factor (SLE = asset value * **exposure factor (EF)**).

- **EF**: The percentage of asset value that is expected to be lost in the event of a risk occurrence. Expressed as a percentage (e.g., 10%, 50%).

- **Annual loss expectancy (ALE)**: The expected monetary loss for a specific risk event over the course of a year. Calculated as the product of the SLE and the **annualized rate of occurrence (ARO)** (ALE = SLE * ARO).

- **ARO**: The estimated frequency of a specific risk event occurring within a year. Expressed as a frequency (e.g., 0.1 events per year).

- **Mean time to detect (MTTD)**: The average time it takes to detect that a risk event has occurred. Measured in hours, days, or other relevant time units.

- **Mean time to contain (MTTC)**: The average time it takes to contain the impact of a risk event once it has been detected. Measured in hours, days, or other relevant time units.

- **Mean time to recover (MTTR)**: The average time it takes to restore normal operations after a risk event has been contained. Measured in hours, days, or other relevant time units.

- **Mean time between failures (MTBF)**: The average time between the occurrence of one failure (risk event) and the next for a particular system or process. Measured in hours, days, or other relevant time units.

- **Number of security incidents**: The total count of security incidents that an organization experiences within a specific time period. A numerical value representing the number of security incidents.

- **Number of attempts stopped (or blocked)**: The count of unauthorized access attempts or malicious activities that were successfully prevented or blocked by security measures. A numerical value indicating the volume of stopped attempts.

- **Server uptime**: The percentage of time that servers are operational and available for use. Expressed as a percentage, calculated as (Total Uptime / Total Time) * 100.

- **Server downtime duration**: The total time a server is unavailable due to planned or unplanned downtime. Measured in hours, minutes, or percentages.

- **Patch compliance**: The percentage of servers/systems/devices that have the latest security patches and updates applied. Expressed as a percentage, calculated as (patched endpoints / total endpoints) * 100.

- **Time to deploy patches**: The average time taken to implement and deploy security patches or updates across the organization. Measured in hours, days, or other relevant time units.

Assessment of Risk Environment

Assessing the risk environment across different domains, such as service, vendor, infrastructure, and business, is critical for effective risk management. Here's an overview of how you might approach risk assessment in each of these areas:

Service Risk Assessment

Conducting a **service risk assessment** is integral to systematically evaluating and managing risks associated with the organization's services. The primary objective is to identify vulnerabilities and threats, focusing on key considerations such as service availability (assess the risk of service downtime and its impact on business operations), data security (evaluate the risks related to the confidentiality, integrity, and availability of sensitive data), compliance (assess the risks associated with non-compliance with relevant regulations and standards), and adherence to **service-level agreements** (evaluate the risk of failing to meet agreed-upon service levels). The metrics employed, including service uptime percentage, number of service incidents, and compliance score, provide quantifiable insights, enabling the organization to proactively address and mitigate risks, ensuring the resilience and reliability of its services. Regular service risk assessments contribute to maintaining operational excellence, customer satisfaction, and adherence to regulatory requirements.

Vendor Risk Assessment (Including CSPs)

Organizations conduct a comprehensive **vendor risk assessment** to evaluate and manage risks associated with third-party vendors and suppliers, particularly CSPs. The primary objective is to ensure the security, compliance, and business continuity of services provided by these external entities. Key considerations include evaluating the CSP's security practices, emphasizing their adherence to the shared responsibility model, scrutinizing compliance, and assessing alignment with relevant regulations and industry standards. Additionally, the assessment evaluates the CSP's preparedness for disruptions and data availability during business continuity events. Contractual obligations are examined, encompassing data ownership, access controls, and indemnification clauses.

Recognizing the critical link between dependency and risk, organizations acknowledge that greater dependency translates to greater risk. To address this, organizations implement risk mitigation strategies, develop redundancy plans and contingency measures, and closely monitor the performance and security practices of vendors with higher levels of criticality. Legal and reputational considerations further underscore the importance of diligently assessing and mitigating risks associated with highly dependent vendors. Neglecting such assessments may lead to increased legal liability in the event of a security breach or service interruption, potentially causing profound reputational damage. Metrics such as vendor risk score, number of vendor security incidents, compliance audit results, and a dependency index quantifying the level of organizational dependency on each vendor aid in quantifying and managing these risks effectively.

Infrastructure Risk Assessment

Conducting an **infrastructure risk assessment** is pivotal for organizations to systematically evaluate and manage risks associated with their IT infrastructure. The primary objective is to safeguard the reliability, security, and resilience of critical technology components. Key considerations include network security, where risks related to unauthorized access and data breaches are thoroughly assessed. Additionally, the evaluation extends to system vulnerabilities, scrutinizing the risks associated with unpatched systems and software, a critical aspect in maintaining a secure infrastructure. Disaster recovery is also a focal point, with a meticulous assessment of risks related to the organization's ability to recover from disasters. Moreover, physical security risks are evaluated to address concerns related to unauthorized access to the physical infrastructure. The CSP bears responsibility for physical and hardware aspects across all service models.

To quantify the effectiveness of risk management strategies, organizations employ several metrics. Server uptime percentage is a crucial measure of the operational continuity of servers. The number of security vulnerabilities is tracked to identify and address potential weaknesses in the infrastructure. **Disaster recovery time** (**RTO**) is monitored closely, providing insights into the organization's preparedness and responsiveness to disruptive events. These metrics collectively contribute to the organization's ability to proactively manage risks, enhance infrastructure resilience, and ensure the continued availability and security of critical IT components.

Business Risk Assessment

Conducting a business risk assessment is paramount for organizations aiming to systematically evaluate risks that could impact the overall success and objectives of the business. The primary objective is to identify and manage potential threats, ensuring the organization's resilience in the face of dynamic business environments. Key considerations encompass market risks, requiring an evaluation of potential impacts stemming from changes in the market. Financial risks are assessed to gauge stability, evaluate investments, and project potential impacts on revenue. Additionally, reputation risks are scrutinized, focusing on potential threats to the organization's public image and brand. Strategic risks are assessed to identify potential challenges and pitfalls associated with the organization's strategic initiatives.

To quantify the impact and likelihood of identified risks, organizations utilize specific metrics. **Revenue at risk** is a critical metric, providing insight into potential financial impacts on the organization's income streams. **Net income variability** is monitored to understand fluctuations and vulnerabilities in the financial structure. Customer satisfaction scores serve as a valuable metric to gauge the impact of risks on the organization's relationships with its customer base. These metrics collectively contribute to a comprehensive understanding of business risks, enabling organizations to proactively manage challenges, adapt to changes, and maintain a resilient and successful business operation.

Various Frameworks and Standards for the Assessment of the Risk Environment

Mitigating security risks requires a thorough assessment of your IT environment. Here's a breakdown of three key frameworks used for such assessments:

Common Criteria (ISO/IEC 15408-1) is used for evaluating and certifying the security features and capabilities of information technology products. It provides a framework for assessing the security aspects of products and systems, offering a common set of criteria and evaluation methods. This standard is widely recognized and utilized for ensuring the security of IT products in various industries.

CSA STAR is designed to improve transparency and provide assurance regarding the security practices of cloud service providers. STAR encompasses various levels, with the STAR Certification being one of them. It involves a rigorous third-party independent assessment of a cloud service provider's security posture. The CSA STAR framework helps organizations make informed decisions about selecting and using cloud services based on their security and risk management requirements.

EU Cybersecurity Certification Scheme on Cloud Services: The EU has introduced a cybersecurity certification framework to enhance the overall cybersecurity posture of products and services within the EU. The certification scheme includes specific provisions for cloud services. This framework aims to establish a common set of cybersecurity standards and requirements, providing assurance to users and consumers about the security of certified products and services.

Understanding Outsourcing and Cloud Contract Design

Outsourcing and **cloud contract design** are integral components of modern business strategies, involving the delegation of certain functions or services to external parties through contractual agreements. Outsourcing is a business practice where an organization contracts out certain tasks, functions, or processes to external service providers rather than handling them in-house. These tasks can range from customer support and IT services to manufacturing and back-office operations.

Business Requirements

When engaging with a CSP, organizations must articulate their business requirements clearly through various contractual documents to establish terms, conditions, and expectations for the services provided.

- **SLA**: An SLA delineates agreed-upon service levels, performance benchmarks, and support expectations from the CSP. It defines measurable metrics such as uptime, response times, and resolution times, along with delineating responsibilities, penalties for non-compliance, incentives for exceeding targets, and escalation procedures. Security professionals should actively participate in defining security requirements and ensuring that they are comprehensively addressed within the SLA.

- **Master service agreement (MSA)**: An MSA serves as a comprehensive contract outlining the general terms and conditions that govern the relationship between the organization and the CSP. It lays the groundwork for specific transactions detailed in subsequent **statements of work (SOWs)** or addenda. The MSA usually covers detailed security and privacy requirements, and security professionals should collaborate with legal teams to ensure the robust inclusion of security clauses within the MSA.

- **SOW**: An SOW supplements the MSA by providing detailed specifications for specific projects, services, or deliverables. It outlines project scope, timelines, resource requirements, and acceptance criteria, facilitating granular security requirements tailored to each project.

- **Memorandum of understanding (MOU)**: An MOU is a document outlining the terms and details of a collaborative agreement between two parties. While less formal than a legal contract, an MOU can provide a framework for cooperation, information sharing, or joint initiatives. An MOU might be utilized to establish a broader understanding and the objectives before formalizing specific contractual obligations.

- **Business partnership agreements (BPAs)**: BPAs formalize the terms and conditions of a partnership between businesses, specifying mutual goals, responsibilities, and benefits. In the context of engaging with a CSP, a BPA could be employed to outline the collaborative aspects, strategic goals, and shared responsibilities between the organization and the CSP, providing a comprehensive framework for the partnership.

- **Non-disclosure agreement (NDA)**: An NDA is a legal contract between two or more parties that outlines the confidential information they will share with each other and the restrictions on the use and disclosure of that information. The purpose of an NDA is to protect sensitive information and trade secrets, ensuring that the parties involved do not disclose or use the information for unauthorized purposes.

These documents collectively establish a structured framework that ensures clarity, transparency, and alignment of expectations between the organization and the CSP. Regular reviews and updates to these agreements are essential to accommodate evolving business needs, technological advancements, and regulatory requirements. Legal counsel, relevant stakeholders, and security professionals should collaborate closely in drafting and negotiating these contracts to accurately reflect the organization's interests, objectives, and detailed security and privacy requirements within the overall engagement.

Vendor Management

Managing relationships with CSPs involves unique considerations and risks. Effective vendor management with CSPs encompasses various aspects, including assessments, risk mitigation, and contractual safeguards.

Vendor Assessments

Conducting thorough vendor assessments is a crucial prerequisite before establishing any partnership. This comprehensive evaluation process encompasses a thorough examination of a vendor's capabilities, security measures, regulatory compliance, financial stability, and overall reputation. The assessment should extend beyond the initial stages and remain an ongoing practice, covering critical areas such as data security practices, reliability, and the ability to consistently meet SLAs.

When assessing CSPs, particular attention must be given to their security protocols and data protection measures. It is imperative to scrutinize the CSP's approach to data encryption, access controls, and overall security measures to ascertain the safeguarding of sensitive information. Ensuring compliance with industry-specific regulations and standards is equally vital; whether it be GDPR for data protection, HIPAA for healthcare information, or SOC 2 for service organizations, alignment with these benchmarks is integral to risk mitigation.

Historical performance evaluation of the CSP is a key aspect of the vendor assessment process. This involves a comprehensive analysis of the CSP's track record, including historical uptime, responsiveness to incidents, and overall reliability. The assessment of historical performance provides valuable insights into the CSP's ability to consistently deliver services at the required levels and respond effectively to potential disruptions or security incidents.

Vendor Lock-In Risks

Vendor lock-in, a scenario where a customer becomes heavily reliant on a specific vendor's products or services, can present substantial challenges, making it onerous or expensive to transition to an alternative provider. The risks associated with vendor lock-in include restricted flexibility, potential cost escalations, and diminished negotiation power. To mitigate these risks effectively, organizations must carefully scrutinize contracts to identify and address the nuances of vendor lock-in.

Ensuring data portability is a paramount consideration in averting the adverse effects of vendor lock-in. Organizations should insist on clear mechanisms within contracts that facilitate seamless data extraction and migration. This provision becomes crucial should the organization decide to switch providers, empowering them with the ability to retain control over their data and prevent hindrances during the transition.

Interoperability is another key element in avoiding vendor lock-in pitfalls, especially in the realm of CSPs. When selecting CSPs, prioritize vendors that offer robust support for interoperability standards and APIs. Prioritize solutions that offer flexibility and scalability, allowing you to adapt and expand your IT infrastructure as your organization grows and evolves.

Contractual flexibility plays a pivotal role in mitigating vendor lock-in risks. Including provisions in contracts that permit the scaling of services up or down provides organizations with the agility needed to adapt to changing business needs. Whether expanding operations or downsizing, contractual flexibility ensures that the organization retains control over its service levels without succumbing to the constraints of a rigid agreement.

Vendor Viability

Assessing a vendor's **viability** is a critical aspect of vendor management, requiring a comprehensive evaluation of their financial stability, business continuity plans, and overall health. The vendor's financial standing plays a pivotal role in determining its capability to sustain ongoing operations and fulfill contractual obligations. Regular monitoring and effective communication channels with vendors are essential to stay abreast of any changes in their business landscape that might impact the partnership.

Financial stability stands out as a cornerstone in assessing a vendor's viability, especially when dealing with CSPs. It involves a thorough examination of the CSP's financial health and stability to ensure its ability to provide continuous and reliable services. This evaluation helps mitigate the risk of engaging with a financially insecure vendor, reducing the likelihood of service disruptions or unmet contractual obligations.

Additionally, evaluating redundancy and failover capabilities is crucial for understanding a CSP's preparedness to handle unexpected disruptions. A robust assessment of the CSP's infrastructure redundancy measures and failover capabilities aids in mitigating the risks associated with service disruptions. Understanding how the CSP manages redundancy in their systems and implements failover mechanisms provides insights into their commitment to maintaining service continuity.

Regular monitoring of the vendor, including continuous assessment of their financial health and infrastructure capabilities, is fundamental. This ongoing diligence ensures that the organization remains informed about any changes in the vendor's business landscape that might impact the stability of services. Effective communication channels facilitate a proactive approach, allowing the organization to address any emerging concerns or changes in the vendor's viability promptly.

Escrow

Escrow agreements serve as crucial safeguards in vendor relationships, particularly when dealing with mission-critical software or services. These agreements act as a contingency plan, providing protection in the event of the vendor's failure or inability to fulfill their obligations. In such circumstances, an escrow agreement enables the customer to access vital components such as source code, documentation, or other essential materials necessary for a smooth transition to an alternative solution. This strategic approach effectively mitigates the risks associated with vendor insolvency or unexpected disruptions.

Establishing escrow agreements for critical data and services is imperative, particularly when engaging with CSPs. These agreements ensure that in the event of the CSP's failure or termination of services, the customer retains access to crucial data and services. This contingency plan provides a safety net, allowing the organization to continue its operations seamlessly even under adverse circumstances, safeguarding against potential data loss or service interruptions.

In cases where custom applications are hosted on the cloud, considering escrow agreements for source code access becomes pertinent. This additional layer of protection ensures that, if applicable, the organization has access to the source code of custom applications. This access is invaluable for maintaining and modifying the applications as needed, facilitating a smoother transition in the event of unforeseen circumstances or the need to switch providers.

Contract Management

Several key aspects should be carefully addressed in the contract to ensure a comprehensive and mutually beneficial partnership.

Right to Audit

Include clauses granting the customer the right to audit the CSP's processes, controls, and security measures. This ensures transparency and allows the customer to verify compliance with agreed-upon standards and regulations. CSPs usually have a lot of customers and making specific audits for each one can be very complicated and need a lot of resources for both the provider and the customer. In addressing this complexity, organizations can opt for a more pragmatic approach by incorporating standardized audits such as SOC 2 or ISO 27001, recognized industry benchmarks that offer a comprehensive evaluation of the CSP's controls and security practices.

Metrics and Definitions

Define clear metrics for service performance, availability, and other key indicators. Ensure that all terms and definitions are well defined to prevent misunderstandings and disputes.

Termination

Termination in a contract refers to the conclusion or end of the agreement between two parties. It's important to clearly define the conditions and procedures for termination to prevent misunderstandings and ensure a smooth transition if the contract needs to end. Specify any associated penalties or fees and the process for transitioning services to another provider. Determine whether there's a notice period required before termination. A notice period is the warning that one party must give to the other before ending the contract. This helps both parties prepare for the transition.

Check whether there's an early termination fee. Some contracts include an early termination fee, which is a charge imposed if one party ends the contract before its agreed-upon expiration date. This fee is meant to compensate for the potential losses or costs incurred due to early termination.

Litigation and Dispute Resolution

Litigation refers to the process of taking legal action, typically through the court system, to resolve disputes between parties. Outline the procedures for dispute resolution, whether through negotiation, arbitration, mediation, or litigation. Arbitration is a method of dispute resolution where parties involved in a conflict agree to submit their disputes to a neutral third party, known as an arbitrator or a panel of arbitrators. Engaging legal counsel is instrumental in crafting well-defined dispute resolution mechanisms that suit the nature of the contract and promote fair and effective conflict resolution. The contract should include the court or neutral third party's location and applicable country laws for litigation or arbitration.

Assurance and Service Levels

Assurance involves capturing the expectations of both parties and establishing mechanisms to ensure service levels, performance, and security are guaranteed. It is a proactive approach to instill confidence and trust in the delivery of services. To achieve assurance, clear definitions of the expectations from both providers and customers are essential. This includes articulating the standards, benchmarks, and quality parameters that need to be met. A fundamental tool for achieving assurance is the implementation of SLAs. SLAs set specific, measurable targets for service performance, availability, and other critical aspects.

Indemnification

Indemnification is a contractual provision wherein one party agrees to compensate the other for specific losses, damages, or liabilities incurred as a result of specified events or circumstances. Define the scope of indemnification, specifying the types of claims or liabilities for which the CSP will indemnify the customer. This may include breaches of data security, intellectual property infringement, or violations of applicable laws. Clearly articulate any limitations on indemnification, such as monetary caps or exclusions for certain types of damages. This ensures both parties have a clear understanding of the extent of indemnification.

Indemnification typically involves financial compensation. However, it can also include non-monetary forms of restitution, such as providing goods or services to rectify the situation or covering legal fees and expenses incurred as a result of the indemnified party's actions.

Compliance

Ensure that the contract explicitly addresses compliance with relevant laws, regulations, and industry standards. Specify which party is responsible for meeting specific compliance requirements.

Access to Cloud/Data

Clearly define the customer's rights to access their data stored in the cloud, especially in the event of contract termination. Address data ownership, retrieval procedures, and any associated costs.

Cyber Risk Insurance

Addressing **cyber risk insurance** in a contract agreement involves incorporating clauses that specifically outline the terms and conditions related to cybersecurity insurance coverage. It is designed to protect organizations from financial losses resulting from cybersecurity incidents or data breaches.

Cyber risk insurance should explicitly outline covered risks, encompassing data breaches, network security incidents, business interruption from cyber events, and related liabilities. You should specify policy limits, indicating the maximum amount the insurance provider pays for a covered incident, and detail any deductibles payable by the insured party. You should also clearly identify exclusions or limitations, delineating circumstances not covered. The insured party must conduct regular risk assessments, implement cybersecurity best practices, and mitigate potential risks, with compliance possibly being a condition for maintaining coverage. Define responsibilities in a cyber incident, encompassing cooperation with the insurer's investigation, providing necessary information, and taking actions to mitigate further damage. Additionally, address subrogation, clarifying whether the insurance provider retains or waives rights to pursue legal action against responsible third parties after compensating the insured. Understanding and incorporating these elements ensures a comprehensive and effective cyber risk insurance framework.

The coverage can vary between policies and providers, but it typically includes costs associated with the following:

- **Data breach response costs**: Expenses related to notifying affected individuals, providing credit monitoring services, and managing public relations in the aftermath of a data breach.

- **Forensic investigation costs**: The costs of hiring cybersecurity experts to investigate and determine the cause and extent of a cyber incident.

- **Legal costs**: Legal fees and costs associated with lawsuits, regulatory investigations, and compliance. This may include fines and penalties imposed by regulatory authorities and legal defense costs.

- **Business interruption losses**: Compensation for income lost during downtime caused by a cyber incident. This can include reimbursement for revenue losses, additional expenses incurred during the interruption, and extra costs to restore operations.

- **Ransomware payments and extortion expenses**: Some policies may cover ransom payments made to cybercriminals in the event of a ransomware attack. Costs associated with responding to extortion threats, which could include payments or other expenses incurred to prevent the release of sensitive information.

- **Recovery costs**: Cyber risk insurance often covers recovery costs, encompassing expenses for restoring compromised systems, data recovery, infrastructure rebuilding, and other measures essential for returning to normal operations post a cyber incident.

Supply-Chain Management

Supply-chain management involves implementing standards and best practices to ensure the security and resilience of the supply chain. Including the supply chain in BCDR plans is crucial for ensuring comprehensive preparedness. One notable standard is ISO/IEC 27036, which specifically addresses information security for supplier relationships. ISO/IEC 27036 focuses on providing guidance for establishing, implementing, maintaining, and continually improving an ISMS for supplier relationships. The standard emphasizes the importance of risk management in supplier relationships, helping organizations identify and mitigate cybersecurity risks associated with their supply chain.

ISO/IEC 27036 encourages a collaborative approach between organizations and their suppliers, fostering effective communication and coordination to enhance overall supply chain security. It provides a framework for selecting and implementing security controls that are relevant to supplier relationships. This includes measures to protect information exchanged, shared systems, and collaborative processes.

ISO/IEC 27036 outlines assurance mechanisms to ensure that suppliers meet the required security standards. This may include assessments, audits, and ongoing monitoring of supplier performance. The standard addresses legal and regulatory compliance within the context of supplier relationships, helping organizations and their suppliers adhere to applicable laws and regulations related to information security.

ISO/IEC 27036 includes guidance on developing incident response plans and procedures to effectively address and mitigate security incidents within the supply chain. It promotes a culture of continuous improvement, encouraging organizations to regularly assess and enhance their information security practices in collaboration with their suppliers.

Incorporating ISO/IEC 27036 into supply-chain management practices strengthens cybersecurity measures, promotes resilience, and contributes to the overall security posture of organizations and their extended supply-chain networks.

Summary

This chapter equipped you with a comprehensive understanding of the legal intricacies and risks surrounding cloud computing. You explored conflicting international legislation, eDiscovery standards, and risk management strategies, building a strong foundation for navigating the legal landscape of cloud environments. The chapter highlighted the differentiation of roles in data handling, guided you through regulatory transparency requirements, and presented diverse risk treatment approaches. It culminated with an in-depth exploration of outsourcing and cloud contract design, addressing your business requirements, vendor management considerations, and critical aspects of contract management. Finally, the inclusion of supply-chain management principles, guided by ISO/IEC 27036, ensured you have a holistic view of legal and risk considerations for effective cloud governance.

In the next chapter, you'll learn about privacy regulations and country-specific legislations concerning PII and **protected health information (PHI)**, exploring key jurisdictional variations in data privacy laws.

Exam Readiness Drill – Chapter Review Questions

Apart from a solid understanding of key concepts, being able to think quickly under time pressure is a skill that will help you ace your certification exam. That is why working on these skills early on in your learning journey is key.

Chapter review questions are designed to improve your test-taking skills progressively with each chapter you learn and review your understanding of key concepts in the chapter at the same time. You'll find these at the end of each chapter.

> **How to Access These Materials**
>
> To learn how to access these resources, head over to the chapter titled *Chapter 25, Accessing the Online Resources*.

To open the Chapter Review Questions for this chapter, perform the following steps:

1. Click the link – `https://packt.link/CCSPE1_CH22`.

    Alternatively, you can scan the following **QR code** (*Figure 22.1*):

Figure 22.1 – QR code that opens Chapter Review Questions for logged-in users

2. Once you log in, you'll see a page similar to the one shown in *Figure 22.2*:

Figure 22.2 – Chapter Review Questions for Chapter 22

3. Once ready, start the following practice drills, re-attempting the quiz multiple times.

Exam Readiness Drill

For the first three attempts, don't worry about the time limit.

ATTEMPT 1

The first time, aim for at least **40%**. Look at the answers you got wrong and read the relevant sections in the chapter again to fix your learning gaps.

ATTEMPT 2

The second time, aim for at least **60%**. Look at the answers you got wrong and read the relevant sections in the chapter again to fix any remaining learning gaps.

ATTEMPT 3

The third time, aim for at least **75%**. Once you score 75% or more, you start working on your timing.

> **Tip**
>
> You may take more than **three** attempts to reach 75%. That's okay. Just review the relevant sections in the chapter till you get there.

Working On Timing

Target: Your aim is to keep the score the same while trying to answer these questions as quickly as possible. Here's an example of how your next attempts should look like:

Attempt	Score	Time Taken
Attempt 5	77%	21 mins 30 seconds
Attempt 6	78%	18 mins 34 seconds
Attempt 7	76%	14 mins 44 seconds

Table 22.2 – Sample timing practice drills on the online platform

> **Note**
>
> The time limits shown in the above table are just examples. Set your own time limits with each attempt based on the time limit of the quiz on the website.

With each new attempt, your score should stay above **75%** while your "time taken" to complete should "decrease". Repeat as many attempts as you want till you feel confident dealing with the time pressure.

23
Privacy and the Cloud

In this chapter, you will delve into the essence of privacy, exploring its challenges and established standards and the pivotal role of jurisdiction. You will examine various types of private data and the significance of **Privacy Impact Assessments** (**PIAs**) in safeguarding diverse forms of sensitive information. Furthermore, you will consider the impact of country-specific legislation on privacy, highlighting the nuanced approaches to data protection across different regions.

By the end of this chapter, you will be able to confidently answer questions on the following:

- Privacy issues
- Difference between contractual and regulated private data
- Country-specific legislation related to private data
- Jurisdictional differences in data privacy
- Standard privacy requirements
- PIAs

Privacy Issues

Privacy refers to the right of individuals to control their personal information and keep it away from unauthorized access or disclosure. It involves the protection of sensitive data from being accessed, shared, or used without the consent of the individuals to whom that information pertains. Privacy is often regulated by laws and regulations to ensure that individuals have control over their personal data.

Private data, also known as personal data, is information that pertains to an identifiable individual. This can include a wide range of details, such as names, addresses, phone numbers, email addresses, financial records, health information, and social security numbers.

Privacy issues in the digital age encompass a range of concerns related to the handling of personal information. These include the rising threat of data breaches, where unauthorized access compromises individuals' private data; the proliferation of surveillance technologies that encroach on privacy rights; extensive data tracking and profiling by online platforms without transparent disclosure or explicit consent; the misuse of data-mining techniques; and the potential consequences of widespread location tracking. Lack of informed consent, government surveillance programs in the name of national security, and the collection of biometric data raise additional privacy challenges. The growth of **Internet of Things (IoT)** devices and the lack of transparency in data practices further contribute to the complex landscape of privacy issues, necessitating comprehensive measures to protect individuals' privacy rights.

> **Note**
> Privacy is concerned with the protection of personal information and the rights of individuals, while security is a broader concept focused on safeguarding information and systems from a range of threats. Privacy is a subset of security.

In the following section, you'll learn about bodies such as the OECD to understand how personal information should be managed.

The Organization for Economic Co-operation and Development (OECD)

The OECD is an international organization composed of member and non-member countries, focused on promoting policies that enhance economic and social well-being globally. The OECD provides a platform for countries to collaborate, share experiences, and develop best practices in various policy areas, including economic policy, education, innovation, and environmental sustainability.

Regarding privacy principles, the OECD has established guidelines known as the "OECD Privacy Guidelines" or "OECD Privacy Principles." These principles were initially developed in 1980 and have served as a foundational framework for privacy protection. The OECD Privacy Principles are as follows:

- **Collection Limitation Principle**: Personal data should be collected only for specified, explicit, and lawful purposes. The collection should be adequate, relevant, and not excessive for the intended purposes.

- **Data Quality Principle**: Personal data should be accurate, complete, and kept up to date to ensure that it is suitable for the purposes for which it is collected and used.

- **Purpose Specification Principle**: The purposes for which personal data is collected should be specified at the time of collection, and any further use should be limited to purposes compatible with the original ones.

- **Use Limitation Principle**: Personal data should not be disclosed, made available, or otherwise used for purposes other than those specified, except with the consent of the data subject or by the authority of the law.

- **Security Safeguards Principle**: Organizations responsible for personal data must take appropriate technical and organizational measures to safeguard the data against unauthorized access, disclosure, alteration, and destruction.

- **Openness Principle**: There should be transparency about data policies, practices, and the existence of personal data processing. Individuals should be informed about the purposes of data collection and have access to information about how their data is handled.

- **Individual Participation Principle**: Data subjects have the right to know whether an organization holds information about them, the nature of that information, and the purpose for which it is used. They should also have the right to challenge the accuracy of the data and have it corrected, amended, or deleted.

- **Accountability Principle**: Organizations are accountable for complying with the privacy principles. This includes implementing measures to ensure adherence to the principles and providing mechanisms for individuals to address privacy concerns.

These principles provide a comprehensive framework for the fair and responsible handling of personal data. While they were originally formulated in the context of data protection, they have influenced the development of privacy laws and regulations globally. The principles emphasize the importance of balancing the need for data use with the protection of individual privacy rights.

For your reference, here is the OECD website: `https://www.oecd.org/general/data-protection.htm`.

Difference between Contractual and Regulated Private Data

Contractual private data is governed by agreements or contracts between parties. It is primarily defined by the terms and conditions outlined in legal agreements and contracts. The protection and use of contractual private data are determined by the specific contractual obligations agreed upon by the parties involved. The **Payment Card Industry Data Security Standard** (**PCI-DSS**) is an example of contractual private data. It is a set of security standards designed to ensure that companies that accept, process, store, or transmit credit card information maintain a secure environment. PCI-DSS is contractual in nature because it is imposed by payment card issuers such as Mastercard, Visa, and Amex, through contracts with merchants.

Regulated private data is governed by external laws and regulations imposed by governmental or industry authorities. Compliance is mandatory, and the legal framework sets the standards for the protection and handling of such data. The protection of regulated private data is not negotiable and is defined by laws and regulations. Organizations handling this data must adhere to specific requirements to ensure privacy, security, and confidentiality. **Protected Health Information (PHI)** and **Personally Identifiable Information (PII)** are examples of regulated private data. PHI is primarily governed by the **Health Insurance Portability and Accountability Act (HIPAA)** in the healthcare sector, and PII is subject to various data protection laws, such as the **General Data Protection Regulation (GDPR)** in the European Union or state-level data breach notification laws.

Country-Specific Legislation Related to Private Data

Various countries have enacted legislation to regulate the protection and handling of private data, reflecting the increasing importance of privacy rights. These laws aim to safeguard personal information from unauthorized access and promote transparency in data processing. They address concerns raised by digital technologies and cross-border data flows. In the next section, you'll learn about some of the country-specific legislation related to private data.

European Union

You'll learn about EU-specific regulations such as the EU Data Protection Directive and GDPR, exploring their significance in shaping data protection laws and privacy rights within the EU. These regulations have had a profound impact on privacy rights and data-handling practices across member states, setting a global standard for data protection.

EU Data Protection Directive

The EU Data Protection Directive aimed to harmonize data protection laws across EU member states, providing a framework for the protection of individuals' privacy rights in the processing of personal data. The directive laid the foundation for the modern data protection regime in the EU and served as the precursor to GDPR, which replaced it in 2018. For more details, you can visit this link: `https://commission.europa.eu/law/law-topic/data-protection/data-protection-eu_en`

Key features of the **EU Data Protection Directive (95/46/EC)** include the following:

- **Scope and purpose**: The directive applied to the processing of personal data, regardless of the means used, by automated or non-automated means. Its purpose was to protect the fundamental rights and freedom of individuals, particularly their right to privacy, in relation to the processing of personal data.

- **Data protection principles**: The directive established key principles governing the processing of personal data, including the requirement for data to be processed fairly and lawfully, collected for specified and legitimate purposes, adequate, relevant, non-excessive, accurate, and kept up to date.

- **Rights of data subjects**: Individuals, referred to as "data subjects," were granted certain rights, including the right to know whether data about them was being processed, the right to access that data, and the right to rectify inaccuracies.

- **Notification requirements**: Data controllers were required to notify relevant authorities of their data-processing activities. Additionally, data controllers were required to obtain the consent of data subjects for the processing of their personal data, with certain exceptions.

- **Cross-border data transfers**: The directive included provisions regulating the transfer of personal data to countries outside the **European Economic Area** (**EEA**). Transfers to non-EEA countries were allowed only if the receiving country ensured an adequate level of data protection.

- **Data protection authorities**: The directive required the establishment of independent national data protection authorities in each EU member state. These authorities were responsible for ensuring compliance with the directive and handling data protection issues.

- **Sensitive data**: The directive recognized the importance of protecting sensitive personal data, such as information about a person's racial or ethnic origin, political opinions, religious beliefs, and health. Processing sensitive data was subject to stricter conditions.

The EU Data Protection Directive laid the groundwork for a more consistent and comprehensive approach to data protection within the EU. Over time, it became clear that advancements in technology and changes in the digital landscape necessitated an updated and more robust legal framework. This led to the development and implementation of GDPR, which became applicable on May 25, 2018, replacing the EU Data Protection Directive and introducing more stringent data protection standards.

GDPR

GDPR is a comprehensive data protection and privacy regulation enacted by the EU. It aims to provide individuals with greater control over their personal data and establish a consistent framework for data protection across EU member states.

The key components of GDPR are as follows:

- **Data controller**: A data controller is an entity or organization that determines the purposes, conditions, and means of processing personal data. The controller is responsible for ensuring that the processing complies with GDPR's principles.

- **Data processor**: A data processor is an entity that processes personal data on behalf of the data controller. Processors act under the authority of the controller and are obligated to implement appropriate security measures to protect the data.

- **Data subject**: A data subject is an individual whose personal data is processed. GDPR grants data subjects various rights to control and manage their personal information.

The core elements of GDPR are as follows:

- **Lawfulness, fairness, and transparency**: Personal data must be processed lawfully, fairly, and transparently. Data controllers must provide clear and accessible information to data subjects regarding the processing of their data.

- **Purpose limitation**: Personal data should be collected for specified, explicit, and legitimate purposes. Any further processing should be compatible with the initial purposes.

- **Data minimization**: Data controllers should only collect and process the minimum amount of personal data necessary for the intended purposes.

- **Accuracy**: Personal data must be accurate and kept up to date. Controllers are responsible for taking reasonable steps to ensure the accuracy of the data.

- **Storage limitation**: Personal data should be kept in a form that permits identification for no longer than necessary for the purposes for which it is processed.

- **Integrity and confidentiality**: Controllers and processors must implement appropriate security measures to protect personal data from unauthorized access, disclosure, alteration, and destruction.

- **Accountability**: Data controllers are responsible for demonstrating compliance with the principles of GDPR. This involves maintaining documentation, conducting impact assessments, and appointing a data protection officer in certain cases.

- **Data subject rights**: GDPR grants data subjects various rights, including the right to access their data, the right to rectify inaccuracies, the right to erasure (right to be forgotten), the right to restrict processing, the right to data portability, and the right to object to processing.

- **Data breach notification**: Organizations must report data breaches to supervisory authorities and, in certain cases, notify affected data subjects without undue delay.

- **Cross-border data transfers**: GDPR provides mechanisms for the transfer of personal data to countries outside the EEA while ensuring an adequate level of data protection.

- **Consent**: Data controllers must obtain explicit and freely given consent from data subjects for processing their personal data. Consent can be withdrawn at any time.

- **Data Protection Impact Assessments (DPIAs)**: DPIAs are required for high-risk processing activities to assess potential impacts on data subjects and implement measures to mitigate risks.

GDPR grants various privacy rights to data subjects, ensuring greater control over their personal data, as follows:

- **Right to information**: Data subjects have the right to be informed about the processing of their personal data, including the purpose of processing, legal basis, recipients of the data, and data retention period.

- **Right of access**: Data subjects have the right to obtain confirmation from the data controller of whether their personal data is being processed and access to that data.

- **Right to rectification**: Data subjects can request the correction of inaccurate or incomplete personal data held by the data controller.

- **Right to erasure (right to be forgotten)**: Data subjects have the right to request the deletion of their personal data under specific circumstances, such as when the data is no longer necessary for the purposes for which it was collected.

- **Right to restriction of processing**: Data subjects can request the limitation of processing in certain situations, such as when they contest the accuracy of the data or when processing is unlawful.

- **Right to data portability**: Data subjects have the right to receive their personal data in a structured, commonly used, and machine-readable format and can request the transfer of this data to another data controller.

- **Right to object**: Data subjects can object to the processing of their personal data, including profiling, for reasons relating to their particular situation.

- **Rights related to automated decision-making, including profiling**: Data subjects have the right not to be subject to decisions based solely on automated processing, including profiling, that produce legal effects concerning them or similarly significantly affect them.

- **Right to withdraw consent**: When processing is based on consent, data subjects have the right to withdraw their consent at any time. Withdrawal does not affect the lawfulness of processing based on consent before its withdrawal.

GDPR aims to establish a robust and uniform data protection framework, emphasizing transparency, accountability, and individuals' rights in the processing of personal data. It has global implications as organizations worldwide that handle EU residents' data are subject to its provisions.

Asia-Pacific

The **Asia-Pacific Economic Cooperation** (**APEC**) Privacy Framework is a set of privacy principles developed by the APEC economies to facilitate the flow of personal information across borders while ensuring the protection of individuals' privacy rights. APEC is a regional economic forum comprised of 21 member economies from the Asia-Pacific region.

The APEC Privacy Framework was endorsed by APEC in 2005 and is designed to serve as a reference for member economies in developing and implementing their domestic privacy laws and policies. It is not a binding treaty but provides a set of principles and guidelines for the protection of personal information. The framework is intended to promote interoperability among APEC economies by fostering a common understanding of privacy principles.

Key elements of the APEC Privacy Framework include the following:

- **Preventing harm**: The framework emphasizes the importance of preventing harm to individuals resulting from the misuse of or unauthorized access to their personal information.

- **Notice**: Individuals should be informed about the collection, use, and disclosure of their personal information, including the purposes for which it is processed.

- **Collection limitation**: The collection of personal information should be limited to what is necessary for the purposes identified, and it should be obtained by lawful and fair means.

- **Use of personal information**: Organizations should only use personal information for the purposes disclosed to individuals, and they should not use it for purposes incompatible with the original ones.

- **Choice**: Individuals should have the opportunity to exercise choices regarding the collection, use, and disclosure of their personal information, and their choices should be respected.

- **Integrity of personal information**: Organizations should take reasonable steps to ensure the accuracy and completeness of personal information and to keep it up to date.

- **Security safeguards**: Organizations should implement reasonable security safeguards to protect personal information against risks, such as unauthorized access, disclosure, alteration, and destruction.

- **Access and correction**: Individuals should have the right to access their personal information held by organizations and request the correction of inaccuracies.

- **Accountability**: Organizations are accountable for complying with the privacy principles, and they should have mechanisms in place to demonstrate their adherence to these principles.

- **Cross-border data flows**: The APEC Privacy Framework encourages member economies to facilitate the cross-border flow of personal information while ensuring effective privacy protection.

The APEC Privacy Framework reflects a cooperative and flexible approach to privacy protection, recognizing the diverse legal and cultural contexts of the APEC economies. It serves as a model for privacy frameworks and legislation in the region, influencing the development of privacy laws in APEC member economies.

United States

The **United States (U.S.)** does not have a comprehensive federal privacy law that provides a comprehensive framework for the protection of personal data similar to the GDPR in the EU. Privacy regulations in the U.S. are often sector-specific and vary across industries and states. Several states have enacted their own privacy statutes to address various aspects of consumer data protection. Here are a few examples of state-level privacy laws in the U.S.:

- **California Consumer Privacy Act (CCPA)**: Enacted in 2018, the CCPA is one of the most comprehensive state privacy laws. It grants California residents certain rights over their personal information held by businesses, including the right to know what information is collected, the right to opt out of the sale of their data, and the right to request the deletion of their information.

- **Virginia Consumer Data Protection Act (VCDPA)**: The VCDPA, effective from January 2023, is another comprehensive state privacy law. It grants Virginia residents certain rights similar to the CCPA, including the right to access, correct, delete, and opt out of the processing of their personal data.

- **New York Stop Hacks and Improve Electronic Data Security (SHIELD) Act**: The SHIELD Act, enacted in 2019, enhances data breach notification requirements for businesses that handle the private information of New York residents. It requires businesses to implement reasonable data security safeguards.

HIPAA

HIPAA is a U.S. federal law that was enacted in 1996. HIPAA is primarily known for its provisions aimed at safeguarding the privacy and security of individuals' health information. The law has several key components:

- **Privacy Rule**: The HIPAA Privacy Rule establishes national standards to protect individuals' medical records and other personal health information. It sets limits on the use and disclosure of such information by covered entities, which include healthcare providers, health plans, and healthcare clearing houses.

- **Security Rule**: The HIPAA Security Rule complements the Privacy Rule by establishing national standards for the security of **electronic Protected Health Information (ePHI)**. It outlines specific safeguards that covered entities and their business associates must implement to ensure the confidentiality, integrity, and availability of electronic health information.

- **Breach Notification Rule**: Covered entities are required to notify affected individuals, the Secretary of the Department of **Health and Human Services (HHS)**, and, in certain circumstances, the media, following the discovery of a breach. The notification must be made without unreasonable delay and no later than 60 days after the discovery of the breach.

- **Enforcement Rule**: The Enforcement Rule outlines procedures for investigating complaints and imposing civil and criminal penalties for HIPAA violations. The **Office for Civil Rights** (**OCR**) within the Department of HHS is responsible for enforcing HIPAA.

- **Health Information Technology for Economic and Clinical Health (HITECH)**: The HITECH Act was enacted as part of the American Recovery and Reinvestment Act of 2009 and strengthened HIPAA by expanding its privacy and security provisions. HITECH introduced significant changes and enhancements to HIPAA, aiming to promote the adoption and meaningful use of **Electronic Health Records** (**EHRs**) and strengthen the privacy and security protections for electronic health information.

HIPAA's primary goals are to ensure the confidentiality and security of individuals' health information, promote the use of EHRs, and protect the rights of patients. Covered entities and their business associates are required to comply with HIPAA regulations, and failure to do so can result in significant penalties and fines.

Gramm–Leach–Bliley Act (GLBA)

GLBA, also known as the Financial Services Modernization Act of 1999, is a U.S. federal law that addresses the privacy and security of consumers' personal financial information. The primary purpose of GLBA is to enhance consumer privacy and promote the security of sensitive financial information held by financial institutions. The core rules that constitute GLBA are as follows:

- Financial privacy rule: This requires financial institutions to provide consumers with privacy notices that explain the institution's information-sharing practices. Consumers have the right to opt out of having their non-public personal information shared with non-affiliated third parties.

- Safeguards rule: This mandates that financial institutions develop, implement, and maintain a comprehensive information security program to protect the confidentiality and integrity of customer information. This includes assessing risks, designing safeguards, and regularly monitoring and adjusting the program.

- Pretexting provisions: GLBA includes provisions to combat "pretexting," which is the practice of obtaining consumers' personal financial information under false pretenses. It prohibits the use of false, fictitious, or fraudulent statements or documents to obtain customers' non-public personal information.

- Exceptions for certain institutions: GLBA provides certain exceptions for certain institutions, such as those engaged solely in activities such as processing transactions, servicing accounts, or providing advisory services.

- Regulatory oversight: The **Federal Trade Commission** (**FTC**) has regulatory oversight of GLBA and is responsible for enforcing its provisions. Other federal regulatory agencies also have authority over specific types of financial institutions.

- Penalties for non-compliance: Financial institutions that fail to comply with GLBA's privacy and security requirements may face penalties, including fines and regulatory actions. The severity of penalties depends on the nature and extent of the violation.

Stored Communications Act (SCA)

SCA is a U.S. federal law that is part of the **Electronic Communications Privacy Act** (**ECPA**) enacted in 1986. SCA specifically addresses the privacy of electronic communications held in electronic storage by third-party service providers. The primary objective of SCA is to protect the privacy of individuals' communications and the content of their electronic messages.

Key provisions and components of SCA include the following:

- **Definition of Electronic Communication Service (ECS) and Remote Computing Service (RCS)**: SCA distinguishes between two types of services: ECS and RCS. ECS providers offer services that enable the creation, transmission, or reception of wire or electronic communications, while RCS providers offer computer storage or processing services.

- **Protection of stored communications**: SCA prohibits unauthorized access to or disclosure of stored wire and electronic communications and the contents of those communications. It establishes privacy protections for emails and other electronic messages stored on servers operated by service providers.

- **Government access and warrant requirements**: SCA outlines the procedures for the government to obtain access to stored communications. In general, government entities, including law enforcement, need a warrant to access the content of stored communications.

- **Voluntary disclosure by service providers**: SCA allows service providers to voluntarily disclose customer communications to specified entities, such as law enforcement, under certain circumstances. Providers may also disclose non-content information (e.g., subscriber information) to other entities in specific situations.

- **User consent and lawful access**: SCA permits service providers to disclose the content of communications to the intended recipient or with the lawful consent of the originator. Additionally, providers may disclose communications to their own employees during the ordinary course of business.

- **Penalties for violations**: The SCA imposes penalties, including fines and imprisonment, for unauthorized access to stored communications and the contents thereof. It provides legal recourse for individuals whose privacy rights have been violated.

Clarifying Lawful Overseas Use of Data (CLOUD) Act

The CLOUD Act is a U.S. federal law that was enacted in 2018. It was designed to address the challenges associated with cross-border data access and the requirements of law enforcement agencies to access electronic data held by service providers, even when that data is stored in foreign countries.

Key features of the CLOUD Act include the following:

- **Cross-border data access**: The CLOUD Act facilitates cross-border data access by providing mechanisms for U.S. law enforcement to access electronic data held by U.S.-based service providers, regardless of where the data is stored globally. It allows for the retrieval of data even if it is located in countries that have data protection laws.

- **Executive agreements**: The CLOUD Act authorizes the U.S. government to enter into executive agreements with foreign governments to streamline the process of obtaining electronic data for the purpose of law enforcement investigations. These agreements are intended to establish reciprocal arrangements for data access.

- **Probable cause standard**: The CLOUD Act imposes a probable cause standard for the issuance of warrants to access electronic data. This aligns with Fourth Amendment protections against unreasonable searches and seizures.

- **Notification requirements**: The Act includes provisions requiring service providers to notify foreign governments when they receive legal processes seeking the data of foreign nationals. However, there are exceptions to this requirement, including situations where notice may be delayed.

- **Human rights considerations**: The CLOUD Act includes considerations for protecting human rights. Before entering into executive agreements, the U.S. government is required to assess whether a foreign government has a record of respecting privacy and human rights.

- **Amendments to SCA**: The Act includes amendments to SCA to align it with the new provisions related to cross-border data access.

The CLOUD Act was enacted to modernize and clarify the legal framework for obtaining electronic data in the context of today's globalized digital environment. It aims to balance the needs of law enforcement with privacy and human rights considerations. However, it has also sparked debates and concerns about the potential impact on privacy and the extraterritorial reach of U.S. law.

EU-U.S. Privacy Shield

The EU-U.S. Privacy Shield was a framework for regulating the transfer of personal data from the EU to the U.S. It was designed to provide a mechanism for U.S. companies to meet EU data protection requirements when processing the personal data of EU residents.

Key features of the EU-U.S. Privacy Shield included the following:

- **Privacy principles**: The framework outlined a set of privacy principles that participating U.S. organizations had to adhere to. These principles included Notice, Choice, Accountability for Onward Transfer, Security, Data Integrity and Purpose Limitation, Access and Recourse, Enforcement, and Liability.

- **Self-certification**: U.S. companies that wished to receive and process personal data from the EU had to self-certify their compliance with the Privacy Shield principles. This involved publicly declaring their commitment to the framework and submitting to the oversight of the U.S. Department of Commerce.

- **Oversight and enforcement**: The U.S. Department of Commerce, in collaboration with the FTC, administered and enforced the Privacy Shield. Companies found to be non-compliant could face corrective actions and sanctions.

- **Redress mechanism**: Individuals in the EU had the right to access a no-cost, independent dispute resolution mechanism for addressing complaints related to the processing of their personal data. This included the option of arbitration.

- **Annual review**: The framework underwent an annual review by the European Commission and U.S. authorities to assess its functionality, effectiveness, and any necessary improvements.

It's important to note that the **Court of Justice of the European Union (CJEU)** invalidated the EU-U.S. Privacy Shield on July 16, 2020, in the "Schrems II" case. The court expressed concerns about the privacy safeguards related to U.S. surveillance practices. Following this decision, alternative mechanisms, such as **Standard Contractual Clauses (SCCs)** and **Binding Corporate Rules (BCRs)**, continue to be used for legal data transfers from the EU to the U.S. Organizations engaged in transatlantic data transfers should carefully assess and ensure compliance with the current legal landscape.

The Safe Harbor framework was a predecessor to the Privacy Shield and also facilitated data transfers between the EU and the U.S. It was established in 2000, allowing U.S. companies to self-certify compliance with certain privacy principles. However, the Safe Harbor framework faced criticism and legal challenges, leading to its invalidation by the CJEU in October 2015 in the "Schrems I" case. The court raised concerns about the level of protection afforded to personal data under the framework, particularly in the context of U.S. surveillance programs.

SCCs and BCRs are legal mechanisms used to facilitate the transfer of personal data from the EEA to countries outside the EEA, where the level of data protection may not be deemed adequate by European data protection authorities. These mechanisms are essential for ensuring compliance with GDPR and other data protection laws.

SCCs are standard sets of contractual terms and conditions that have been approved by the European Commission. They are designed to ensure an adequate level of data protection when personal data is transferred from a data controller or processor in the EEA to a data controller or processor outside the EEA. SCCs can be used for data transfers between two data controllers, or between a data controller and a data processor. The European Commission has adopted several sets of SCCs for different types of data transfer arrangements. Organizations can use the SCCs relevant to their specific situation. While SCCs provide a standardized framework, organizations can include additional clauses to address specific aspects of their data processing activities.

BCRs are internal rules and policies for the protection of personal data that are adopted by multinational companies and approved by relevant data protection authorities. BCRs are particularly suitable for organizations with a global presence, as they provide a way to ensure consistent data protection standards across the entire corporate group. BCRs are typically used for intra-group data transfers, allowing multinational companies to transfer personal data between their various entities in a way that complies with data protection regulations. BCRs need to be approved by the relevant data protection authority in each EU member state where the data controller operates. BCRs must include specific elements outlined in GDPR, and they require a commitment to enforceable rights and obligations regarding the protection of personal data.

Australia

Australian privacy law primarily revolves around the Privacy Act 1988, which regulates the handling of personal information by Australian government agencies and private sector organizations. The Act establishes the **Australian Privacy Principles** (**APPs**), which outline the standards, rights, and obligations concerning the collection, use, disclosure, and storage of personal information.

Key features of the Australian Privacy Act include the following:

- **APPs**: The APPs are a set of 13 principles that govern the handling of personal information. They cover various aspects, including the open and transparent management of personal information, the purposes for which information can be collected, consent, data quality, security, access, and correction.

- **Notifiable Data Breaches scheme**: The Privacy Act includes mandatory data breach notification requirements. If an eligible data breach occurs, organizations are required to notify affected individuals and the **Office of the Australian Information Commissioner** (**OAIC**).

- **Privacy Commissioner**: The OAIC is responsible for overseeing privacy-related matters and ensuring compliance with the Privacy Act. The Privacy Commissioner has powers to investigate privacy complaints, conduct assessments, and promote awareness of privacy issues.

- **Extraterritorial application**: The Privacy Act has extraterritorial application, meaning that it can apply to overseas entities that collect or hold personal information about individuals in Australia. This includes businesses operating outside of Australia but offering goods or services to Australian residents.

- **Credit reporting**: Part IIIA of the Privacy Act regulates the handling of credit-related personal information. It includes provisions related to credit-reporting bodies, credit providers, and individuals' rights regarding their credit information.

Canada

Canada has privacy laws at both the federal and provincial levels. The key federal privacy law in Canada is the **Personal Information Protection and Electronic Documents Act (PIPEDA)**. Additionally, some provinces have their own privacy laws, such as Alberta: Personal Information Protection Act and British Columbia: Personal Information Protection Act, that apply to private sector organizations operating within their jurisdictions.

PIPEDA is the primary federal privacy law governing the collection, use, and disclosure of personal information by private sector organizations engaged in commercial activities across Canada. It applies to organizations that operate in federally regulated industries, such as telecommunications, banking, and transportation, as well as organizations that engage in interprovincial or international trade. PIPEDA sets out principles for the fair and transparent handling of personal information. These principles include obtaining consent for the collection and use of personal information, limiting the collection to what is necessary for the identified purposes, safeguarding the information, and providing individuals with access to their personal information. PIPEDA also addresses the transfer of personal information across borders and requires organizations to inform individuals about the purposes for which their information is being transferred.

Jurisdictional Differences in Data Privacy

As a cloud security professional, recognizing that different laws and regulations apply based on the location of the data (stored, traveled, and processed), user, data owner, data processor, **Cloud Service Provider** (CSP), and other contractors handling data is essential. Understanding the legal and regulatory landscape is crucial when dealing with data in the cloud. It is imperative to be aware of where the company's data is stored and processed, considering that cloud services may distribute data across servers in different regions or countries. A thorough understanding of how data flows across borders is necessary, as different jurisdictions have varying rules on the transfer of personal data.

Compliance with regulations governing cross-border data transfers, such as the EU's GDPR requirements, is a key responsibility. Ensuring that CSPs adhere to the applicable regulations is essential for maintaining a secure and compliant data environment. Staying informed about the privacy and data protection laws in the jurisdictions where the company operates or where its customers are located is crucial for ongoing compliance.

Additionally, being aware of any data residency requirements imposed by specific jurisdictions is important, as some countries may mandate that certain types of data remain within their borders. Collaborating closely with legal counsel is highly recommended to interpret and apply privacy laws correctly. Legal professionals can provide valuable guidance on compliance strategies and assist in addressing any legal challenges that may arise in the ever-evolving landscape of data protection and privacy laws.

Standard Privacy Requirements

Organizations prioritize adherence to essential frameworks such as ISO/IEC 27018, GAPP, and GDPR as standard privacy requirements. These frameworks are crucial for safeguarding individuals' privacy and ensuring the responsible management of personal information.

International Organization for Standardization/International Electrotechnical Commission (ISO/IEC) 27018

ISO/IEC 27018 is an international standard that provides guidelines for protecting PII in the context of cloud computing. Specifically, it addresses the privacy concerns related to the processing of PII by CSPs. Here are the key aspects of ISO/IEC 27018:

- ISO/IEC 27018 aims to **establish a set of controls and practices for CSPs** to protect the privacy of individuals whose PII is processed in the cloud. It emphasizes transparency, consent, and security in handling PII.

- The standard is applicable to organizations acting as PII processors in a cloud computing environment. It covers various **aspects of PII processing, from collection and storage to disclosure and disposal**.

- ISO/IEC 27018 places importance on obtaining clear and informed **consent** from individuals for the processing of their PII. It also requires CSPs to provide transparent and easily accessible notices regarding the purpose and methods of PII collection and processing.

- The standard outlines specific **security measures** that CSPs should implement to protect PII. This includes encryption, access controls, and measures to prevent unauthorized access, disclosure, alteration, and destruction of PII.

- ISO/IEC 27018 encourages **data portability** by requiring CSPs to support the easy transfer of PII to other providers upon the data subject's request. Additionally, it mandates the secure deletion of PII when it is no longer needed for the specified purposes.

- The standard addresses the engagement of **third-party subprocessors** by CSPs. It requires CSPs to assess and manage the privacy practices of these third parties, ensuring they adhere to the same privacy and security controls.

- ISO/IEC 27018 recommends regular **audits and assessments** to ensure compliance with the standard. This includes internal audits and assessments by independent third parties to verify that the controls are effectively implemented.

ISO/IEC 27018 serves as a valuable tool for CSPs to demonstrate their commitment to privacy and build trust with customers. It aligns with broader information security standards and provides a framework for addressing the unique challenges of processing PII in cloud environments.

Generally Accepted Privacy Principles (GAPP)

GAPP is a framework established by the **American Institute of Certified Public Accountants (AICPA)** to guide organizations in developing and maintaining effective privacy policies. GAPP addresses various aspects of privacy management, and it often aligns with other industry standards, including **Service Organization Control (SOC)** reports. Here is GAPP's comprehensive set of principles:

- **Management**: Organizations are required to designate responsibility for privacy management and establish policies and procedures to achieve privacy objectives.

- **Notice**: Clear and concise notices must be provided to individuals about the organization's privacy practices, including the purposes of data collection and how the information will be used.

- **Choice and consent**: Individuals should have the opportunity to opt in or opt out of the collection and use of their personal information, allowing them to control how their data is handled.

- **Collection**: The collection of personal information should be limited to what is necessary for the stated purposes. Organizations must obtain consent before collecting sensitive information.

- **Use, retention, and disposal**: Organizations should use personal information only for the purposes disclosed in the notice. Data retention periods should be defined, and disposal procedures must be secure.

- **Access**: Individuals have the right to access their personal information and request corrections if inaccuracies are identified.

- **Disclosure to third parties**: Organizations should disclose personal information to third parties only for specified purposes and with the consent of the individuals involved.

- **Security**: Adequate safeguards must be implemented to protect personal information from unauthorized access, disclosure, alteration, and destruction.

- **Quality**: Organizations should strive to maintain accurate and complete records of personal information, taking steps to ensure data quality.

- **Monitoring and enforcement**: Regular monitoring and enforcement mechanisms should be in place to ensure compliance with privacy policies and applicable laws.

Privacy Impact Assessments (PIAs)

PIAs are systematic assessments conducted to identify and evaluate the potential privacy risks and implications of a particular project, program, or system. These assessments are integral to privacy management and compliance with privacy regulations. The **International Association of Privacy Professionals (IAPP)** provides guidance and best practices for conducting PIAs. Here is an overview of PIAs based on IAPP principles:

- **Purpose and scope**: Clearly define the purpose and scope of the PIA, outlining the specific project, program, or system under assessment. Identify the goals, objectives, and intended outcomes.

- **Data mapping**: Conduct a thorough data mapping exercise to understand what personal information is collected, processed, stored, and shared. Identify the sources of data, its flow, and any third parties involved.

- **Stakeholder involvement**: Involve key stakeholders throughout the PIA process. This includes representatives from legal, IT, security, compliance, and any other relevant departments. Input from data subjects and privacy advocates may also be considered.

- **Legal and regulatory compliance**: Assess the project's compliance with applicable privacy laws and regulations. Ensure that the PIA aligns with legal requirements and identify any necessary adjustments to ensure compliance.

- **Risk assessment**: Conduct a comprehensive risk assessment to identify potential privacy risks and threats. Evaluate the likelihood and impact of these risks on individuals and the organization. Prioritize risks for mitigation.

- **Privacy controls and safeguards**: Propose and implement privacy controls and safeguards to mitigate identified risks. This may include technical measures, policy adjustments, or procedural changes to enhance privacy protection.

- **Data subject rights**: Consider and address the rights of data subjects. Ensure that individuals are informed about their rights and how they can exercise them. Implement mechanisms for individuals to access, correct, or delete their personal information.

- **Documentation and reporting**: Maintain thorough documentation of the PIA process, findings, and mitigation strategies. Prepare a report that summarizes the assessment, including recommendations and any changes made to address privacy concerns.

- **Training and awareness**: Provide training and awareness programs for employees involved in the project to ensure they understand and adhere to privacy principles. This includes awareness of data protection obligations and the importance of privacy by design.

- **Review and monitoring**: Regularly review and monitor the project to ensure ongoing compliance with privacy requirements. Periodically update the PIA as needed, especially when there are significant changes to the project or regulatory landscape.

The CCSP exam emphasizes the importance of understanding the legal and regulatory landscape surrounding cloud security. This includes:

- **Laws**: Laws are rules and regulations created and enforced by a governing body, typically a government, to maintain order and protect the rights and well-being of its citizens. An example is GDPR in the EU, which regulates the processing of personal data.

- **Regulations**: Regulations are detailed rules derived from laws, issued by government agencies or bodies. They provide specific guidelines on how to comply with the overarching laws. An example is the **Federal Risk and Authorization Management Program (FedRAMP)**. In the U.S., FedRAMP provides a standardized approach to security assessment, authorization, and continuous monitoring for cloud products and services.

- **Standards**: Standards are established criteria or guidelines for products, services, or systems to ensure quality, safety, performance, interoperability, and so on. Compliance with standards often demonstrates adherence to best practices and industry benchmarks. Standard-setting organizations can be independent bodies, government agencies, industry groups, or international entities that include experts, industry professionals, government representatives, and other stakeholders to develop and maintain standards. An example is ISO/IEC 27001, an international standard for **Information Security Management Systems** (**ISMSs**), providing a systematic approach to managing sensitive company information.

- **Frameworks**: Frameworks are structured sets of guidelines, best practices, and processes designed to help organizations achieve specific objectives or develop and implement policies. An example is the **National Institute of Standards and Technology** (**NIST**) Cybersecurity Framework for managing and improving cybersecurity risk.

The laws can be categorized into the following types:

- **Criminal law**: Criminal laws deal with offenses against the state or society. They define actions that are considered crimes and prescribe punishments for individuals convicted of committing those crimes. Examples include theft, assault, murder, and drug trafficking.

- **Civil law**: Civil laws regulate disputes between private parties, such as individuals or organizations. They address issues such as contracts, property disputes, family matters, and personal injury cases. Examples include contract law, family law, and property law.

- **Constitutional law**: Constitutional laws establish the framework and principles of government. They define the powers and limits of different branches of government and protect individual rights. An example is the U.S. Constitution. The U.S. Constitution is the supreme law of the land, providing the framework for the federal government and guaranteeing fundamental rights to its citizens. The Constitution begins with the Preamble, which outlines the purposes and goals of the document, including the establishment of justice, the promotion of domestic tranquility, the provision for the common defense, the promotion of general welfare, and the securing of the blessings of liberty to posterity. The Constitution is divided into seven articles, each addressing different aspects of the government's structure and powers. The first three articles establish the three branches of government: the Legislative Branch (Article I), the Executive Branch (Article II), and the Judicial Branch (Article III).

- **Administrative law**: Administrative laws govern the actions of government agencies. They regulate the implementation and enforcement of laws, ensuring that administrative decisions are fair and legal. Examples include environmental regulations and HIPPA.

- **International law**: International laws govern the relationships between nations. They address issues such as treaties, diplomatic relations, human rights, and the conduct of states in the international community. For example, the Wassenaar Arrangement on Export Controls for Conventional Arms and Dual-Use Goods and Technologies aims to promote transparency and responsibility in the export of conventional arms and dual-use technologies, including those related to information security.

- **Contract law**: Contract laws govern agreements between parties. They specify the terms and conditions of a contract and establish the legal obligations of the parties involved. A breach of contract occurs when one party fails to fulfill its contractual obligations without a legal excuse. Examples include business contracts, employment agreements, and vendor contracts.

- **Tort law**: Tort laws deal with civil wrongs that cause harm or loss to individuals. They provide remedies for those who have been wronged by the actions of others. Duty of care is a legal obligation requiring individuals or entities to act reasonably and responsibly to prevent foreseeable harm to others. In the context of IT, a duty of care might apply when handling sensitive data, and ensuring the security of digital systems. A breach of duty occurs when there is a failure to fulfill the legal obligation to exercise reasonable care, resulting in a violation of the duty of care. Damages refer to the harm or losses suffered by the party that has been wronged. Causation establishes the link between the breach of duty and the damages suffered. It demonstrates that the harm or loss was a direct result of the defendant's failure to meet the duty of care. Examples include negligence, defamation, and product liability.

- **Family law**: Family laws regulate matters related to family relationships and domestic issues. They cover areas such as marriage, divorce, child custody, and adoption. Examples include divorce laws and child custody laws.

- **Intellectual property law**: Intellectual property laws protect the rights of individuals or entities in their creations or inventions. They cover patents, trademarks, copyrights, and trade secrets. Examples include copyright laws and patent laws.

There are multiple types of liabilities that you need to be familiar with:

- **Legal liability** refers to the legal obligation one has to perform certain acts or responsibilities and can arise when a person or entity is found to be in violation of laws, regulations, or contractual agreements.

- **Criminal liability** involves the legal responsibility for committing a crime. Crimes are offenses against the state or society, and those found guilty may face penalties such as fines, imprisonment, or other sanctions. For example, if an individual is charged and convicted of theft, they may be criminally liable and subject to criminal penalties.

- **Civil liability** pertains to legal responsibility for actions that cause harm to others, resulting in a civil lawsuit. The goal is often to compensate the injured party for losses rather than impose criminal penalties. For example, a person injured in a car accident may file a civil lawsuit seeking damages for medical expenses, pain, and suffering, alleging negligence on the part of the other driver.

- **Contractual liability** arises from a breach of contract, where one party fails to fulfill the terms and conditions specified in a legally binding agreement. The non-breaching party may seek remedies such as compensation or specific performance. For example, if a company fails to deliver goods as agreed in a contract, it may be held contractually liable and required to compensate the other party for losses incurred.

Statutory, regulatory, and contractual requirements are distinct categories of legal obligations that organizations must adhere to in various aspects of their operations:

- **Statutory requirements** are laws enacted by legislative bodies, such as federal, state, or local governments. These laws are binding and mandatory for individuals and organizations within the jurisdiction. Statutory requirements may include laws such as HIPAA in the U.S. or GDPR in the EU.

- **Regulatory requirements** are rules and standards set forth by regulatory bodies or agencies authorized by the government to oversee specific industries or sectors. These regulations are designed to ensure compliance with laws and standards to protect public interests. Regulatory requirements include compliance with FedRAMP. FedRAMP establishes security standards for cloud services used by U.S. government agencies to ensure the confidentiality, integrity, and availability of federal data.

- **Contractual requirements** are obligations stipulated in agreements or contracts between parties. These requirements are specific to the terms and conditions agreed upon by the parties involved and may cover a wide range of topics, including performance standards, confidentiality, and data protection. For example, compliance with PCI-DSS requirements is mandated as a contractual requirement between entities engaging in transactions involving payment cards.

HITRUST, or the **Health Information Trust Alliance**, is an organization that developed and maintains the **Common Security Framework (CSF)**, a comprehensive and certifiable framework for safeguarding sensitive healthcare information. The HITRUST CSF is designed to address the specific security, privacy, and regulatory challenges within the healthcare industry. Here are some key aspects of HITRUST:

- **CSF**: The CSF is a set of controls, frameworks, and standards that integrates various existing frameworks, standards, and regulations, such as ISO, NIST, and HIPAA, into a comprehensive security framework tailored for the healthcare sector.

- **Risk management and compliance**: HITRUST provides a risk management and compliance methodology that organizations in the healthcare industry can use to assess and manage their information security and privacy risks. It offers a systematic approach to identifying, assessing, and mitigating risks associated with handling sensitive health information.

- **Certification program**: HITRUST offers a certification program that allows organizations to undergo a thorough assessment of their security controls against the HITRUST CSF. Achieving HITRUST certification demonstrates an organization's commitment to meeting high standards for information security and privacy.

- **Industry collaboration**: HITRUST promotes collaboration among healthcare organizations, government agencies, and other stakeholders to establish and improve security and privacy standards. It works to develop and update the CSF based on evolving threats, technologies, and regulatory requirements.

- **Privacy and security assurance**: HITRUST aims to provide assurance to both healthcare organizations and their customers that sensitive health information is handled securely and in compliance with industry regulations.

- **Third-party risk management**: The HITRUST CSF includes controls and processes to manage third-party risks effectively. This is particularly important in the healthcare industry, where organizations often share information with various partners, including vendors and service providers.

Summary

In this chapter, you learned about privacy issues and understood the differences between contractual and regulated private data. You examined specific country legislations, such as HIPAA, the CLOUD Act, GDPR, and PIPEDA, to understand their approaches to protecting private information. You discovered jurisdictional disparities in data privacy, along with standard privacy requirements and internationally recognized privacy frameworks such as ISO/IEC 27018 and GAPP. Additionally, the chapter discussed the importance of PIAs in managing privacy concerns across diverse regulatory frameworks. In the next chapter, you will learn how audits function in cloud computing.

Exam Readiness Drill – Chapter Review Questions

Apart from a solid understanding of key concepts, being able to think quickly under time pressure is a skill that will help you ace your certification exam. That is why working on these skills early on in your learning journey is key.

Chapter review questions are designed to improve your test-taking skills progressively with each chapter you learn and review your understanding of key concepts in the chapter at the same time. You'll find these at the end of each chapter.

> **How to Access These Materials**
>
> To learn how to access these resources, head over to the chapter titled *Chapter 25, Accessing the Online Resources*.

To open the Chapter Review Questions for this chapter, perform the following steps:

1. Click the link – `https://packt.link/CCSPE1_CH23`.

   Alternatively, you can scan the following **QR code** (*Figure 23.1*):

Figure 23.1 – QR code that opens Chapter Review Questions for logged-in users

2. Once you log in, you'll see a page similar to the one shown in *Figure 23.2*:

Figure 23.2 – Chapter Review Questions for Chapter 23

3. Once ready, start the following practice drills, re-attempting the quiz multiple times.

Exam Readiness Drill

For the first three attempts, don't worry about the time limit.

ATTEMPT 1

The first time, aim for at least **40%**. Look at the answers you got wrong and read the relevant sections in the chapter again to fix your learning gaps.

ATTEMPT 2

The second time, aim for at least **60%**. Look at the answers you got wrong and read the relevant sections in the chapter again to fix any remaining learning gaps.

ATTEMPT 3

The third time, aim for at least **75%**. Once you score 75% or more, you start working on your timing.

> **Tip**
>
> You may take more than **three** attempts to reach 75%. That's okay. Just review the relevant sections in the chapter till you get there.

Working On Timing

Target: Your aim is to keep the score the same while trying to answer these questions as quickly as possible. Here's an example of how your next attempts should look like:

Attempt	Score	Time Taken
Attempt 5	77%	21 mins 30 seconds
Attempt 6	78%	18 mins 34 seconds
Attempt 7	76%	14 mins 44 seconds

Table 23.1 – Sample timing practice drills on the online platform

> **Note**
>
> The time limits shown in the above table are just examples. Set your own time limits with each attempt based on the time limit of the quiz on the website.

With each new attempt, your score should stay above **75%** while your "time taken" to complete should "decrease". Repeat as many attempts as you want till you feel confident dealing with the time pressure.

24

Cloud Audit Processes and Methodologies

In this chapter, you'll learn about **audit controls**. You will see how virtualization and cloud technologies are transforming audits, presenting unique assurance challenges. You'll also learn about different types of audit reports and the limitations set by audit scope statements. The chapter will guide you through audit planning, including detailed gap analysis and insights into internal information security management. Finally, you'll learn the compliance requirements for highly regulated industries, highlighting how distributed **Information Technology (IT)** is reshaping the audit landscape. These topics aren't just crucial for the CCSP exam; they also help you prepare for cloud audits.

By the end of this chapter, you will be able to confidently answer questions on the following:

- Internal and external audit controls
- Impact of audit requirements
- Assurance challenges of virtualization and the cloud
- Types of audit reports
- Restrictions of audit scope statements
- Gap analysis
- Audit planning
- Internal **Information Security Management System (ISMS)**
- Policies
- Identification and involvement of relevant stakeholders
- Specialized compliance requirements for highly regulated industries
- Impact of distributed IT model

Understanding the Audit Process, Methodologies, and Required Adaptations for a Cloud Environment

Auditing is a systematic examination or inspection of financial, operational, or information system processes to ensure compliance with established standards, policies, regulations, or best practices. The primary purpose of auditing is to provide an independent and objective assessment of an organization's activities, controls, and processes. Auditing activities involve examining historical records, transactions, and processes to identify discrepancies, errors, unauthorized activities, crimes, or potential areas of improvement, thereby demonstrating a commitment to fulfilling due care obligations. Auditing serves as a detective control designed to identify and detect issues or deviations after they have occurred. Auditing is often mandated and required by various regulations and standards to ensure transparency, accountability, and compliance with established rules.

Audits can be categorized based on what they audit and the areas they focus on. Typically, they are conducted to ensure compliance with relevant laws, standards, and regulations in their respective areas of focus. Here are some common types of audits based on their scope:

- A **financial audit** examines an organization's financial statements and accounting records to ensure accuracy and compliance with accounting standards.

- An **operational audit** evaluates the efficiency and effectiveness of an organization's internal processes and operational activities.

- A **compliance audit** ensures that an organization complies with relevant laws, regulations, and internal policies.

- An **IT audit** examines an organization's information systems, IT infrastructure, and data management practices.

- A **quality audit** reviews organizational processes and procedures to ensure adherence to quality standards.

- A **supply chain audit** examines the processes within the supply chain, including procurement, production, and distribution.

- An **information security audit** evaluates an organization's information security policies, procedures, and controls. This includes assessing the security of IT systems, networks, and data to identify vulnerabilities and potential risks.

Audits can also be categorized into internal and external audits based on who conducts them.

Internal and External Audit Controls

Both internal and external audit perspectives play vital roles in ensuring organizational compliance and efficiency. For instance, a company may conduct regular internal audits to assess its financial health and operational efficiency, while external audits are often mandated by regulatory bodies to provide independent verification of financial statements.

Internal Audit Controls

Internal audits are conducted by the internal audit team that is part of the organization. One of the objectives is to assess and evaluate the organization's adherence to its internal policies, procedures, and standards. Internal audits delve into various areas, including data protection policies, employee training, and internal controls, aiming to identify potential risks, gaps, or deficiencies in compliance efforts. These audits are conducted regularly as part of an ongoing monitoring process, contributing valuable insights for the continuous improvement of internal compliance programs.

Internal audits play a crucial role in preparing organizations for external audits by proactively assessing internal processes, controls, and adherence to policies. This preparation enables the organization to catch and fix any issues before they show up on a more formal external audit report. It not only enhances overall compliance but also streamlines the external audit process for a more efficient and effective examination by external auditors. Internal audits can be more cost effective compared to external audits since internal auditors are already familiar with the organization's operations. However, external audits serve specific purposes as well, which you'll see in the next section.

Internal auditors may face challenges in maintaining complete objectivity, particularly if their relationships within the organization influence assessments or if they were involved in the design or implementation of certain functions, indicating limited independence in their evaluations.

External Audit Controls

External audits offer an independent and unbiased assessment of an organization's compliance with external regulations, industry standards, and contractual obligations. Conducted by external auditors, this independent assessment ensures objectivity, free from internal biases, enhancing the overall credibility of the audit findings. These audits are typically conducted by third-party audit firms or external compliance professionals who are not affiliated with the organization. External audits are sometimes mandatory as per certain regulations, requiring organizations to undergo an independent assessment conducted by external auditors. The primary objective of external compliance audits is to validate that the organization meets the requirements set forth by regulatory bodies and external stakeholders. Focusing on specific regulatory frameworks, industry standards, and contractual agreements relevant to the organization's operations, external compliance audits offer an objective evaluation of compliance efforts. Conducted periodically, often annually, or as required by regulatory standards, external compliance audits enhance transparency and accountability, providing assurance to external stakeholders and demonstrating the organization's commitment to meeting external compliance requirements.

External audits can be expensive, especially for organizations with complex operations or those requiring specialized expertise. External audits may take a considerable amount of time, leading to delays in addressing identified issues or implementing improvements.

Together, internal and external compliance audit controls form a comprehensive approach to managing and ensuring compliance within the dynamic regulatory landscape.

Impact of Audit Requirements

Audit requirements refer to the criteria, standards, and procedures that organizations must adhere to when conducting audits. These requirements are typically defined by various factors, including regulatory bodies, industry standards, internal policies, and contractual obligations. Audit requirements serve as guidelines for assessing and validating the effectiveness of an organization's controls, processes, and systems.

Audit requirements in cloud environments have a profound impact on organizations, especially considering the unique features and challenges inherent in cloud service offerings. The extent and objectives of audits are pivotal in confirming the efficacy of controls in these environments.

The global dispersion of **Cloud Service Providers** (**CSPs**) poses unique challenges for audits. Traditional audit methods, which may involve physical visits to data centers, become impractical for globally dispersed CSPs. Even if the customers could visit the CSP's data center, it's tough to find the actual machines hosting their data. This necessitates innovative approaches to effectively manage and conduct audits in diverse geographical locations. Additionally, audit requirements become complex due to conflicting and numerous regulations. Navigating through varied regulatory landscapes adds intricacy to audits, requiring a nuanced understanding of different compliance obligations.

Multitenancy, which involves multiple users sharing the same hardware infrastructure, adds complexity to cloud environments. In such shared environments, robust controls are crucial to prevent unintended data exposure and ensure the security and privacy of each tenant's information.

The distinct nature of cloud environments demands innovation in auditing practices. Cloud auditors need to develop new methodologies to effectively audit cloud environments, considering factors such as virtualization, multitenancy, and dynamic configurations.

Identify Assurance Challenges of Virtualization and the Cloud

Conducting audits in the cloud poses challenges because organizations depend heavily on CSPs. This reliance creates a situation where organizations may find it challenging to conduct independent audits as they do not have direct control over these aspects. The shared responsibility model in cloud services further complicates the audit process, as organizations must navigate the complexities of auditing within a collaborative framework with CSPs.

The dynamic nature of cloud environments often incorporates emerging technologies, making it challenging for traditional auditing practices to keep pace. Technologies such as serverless computing, containerization, and advanced automation introduce new layers of complexity, requiring auditors to continually adapt their methodologies to effectively assess the security, compliance, and performance of these innovative features.

The hypervisor, a critical component in virtualized environments, presents unique challenges for auditing in the cloud. As the layer that enables multiple VMs to run on a single physical server, the hypervisor introduces complexities in assessing the security and isolation of these virtualized instances. Traditional audit practices designed for physical environments may struggle to adapt to the intricacies of hypervisor-based infrastructures. Ensuring that the hypervisor is configured securely, VMs are properly segregated, and access controls are effectively enforced become paramount. Additionally, understanding and validating the hypervisor's role in managing resource allocation and performance optimization adds layers of complexity to the audit process. The CSP is generally responsible for the security and integrity of the hypervisor itself and the customer is responsible for auditing the VMs hosted on the hypervisor, including ensuring proper configuration, access controls, and isolation between VMs.

Sampling is a strategic auditing approach that involves selecting a representative subset from a larger population to draw conclusions about the entire group. This method enables auditors to assess the characteristics, controls, and compliance of a system or environment efficiently, without the need to examine every individual element. Cloud computing, with its dynamic and scalable nature, introduces unique challenges to traditional sampling practices. Unlike traditional on-premises environments, cloud infrastructures are characterized by rapid changes, elastic resource scaling, and a diverse array of services. These factors complicate the process of selecting representative samples that accurately reflect the entire cloud environment.

In traditional settings, sampling methods are often designed for relatively static environments, where configurations and resources change less frequently. However, the dynamic nature of cloud operations means that a snapshot taken at a specific moment may quickly become outdated. The ability to scale resources up or down based on demand adds an extra layer of complexity, making it challenging to capture the variability in resource configurations. As a result, traditional sampling approaches may struggle to keep pace with the continuously evolving and elastic nature of cloud computing. The shared responsibility model in cloud services also adds complexity, as auditors need to ensure that their samples effectively cover the responsibilities of both the CSP and the customer.

Types of Audit Reports

An **audit report** is a formal document that summarizes the findings, conclusions, and recommendations of an audit. It is prepared by auditors who have examined and assessed the financial statements, internal controls, or other aspects of an organization's operations. The report provides an objective and independent opinion on the accuracy, completeness, and reliability of the information under review. There are various types of audit reports that organizations use to communicate the findings of their audits.

Statement on Standards for Attestation Engagements (SSAE) and Service Organization Control (SOC)

SSAE is a set of standards and guidelines issued by the **American Institute of Certified Public Accountants (AICPA)**. SSAE provides a framework for auditors when performing attestation engagements, which involve providing assurance on assertions made by an entity. One of the key components of SSAE is SSAE 18, which replaced the previous standard, SSAE 16. SSAE 18 outlines the requirements for reporting on controls at service organizations. Service organizations are entities that provide services to other organizations, and these services may impact the financial reporting of those client organizations. For example, CSPs, data centers, and managed service providers often undergo SSAE 18 audits to assure their clients about the effectiveness of their controls.

SSAE 18 brought enhancements and clarifications to the attestation engagement process, with a focus on service organizations and their controls. It outlines the requirements for reporting on controls, emphasizing the importance of the system and organization controls framework. SSAE 18 introduced distinctions between Type I and Type II reports. Under SSAE and SOC, both Type I and Type II reports are common. A Type I report assesses the suitability of the design of controls at a specific point in time, while a Type II report evaluates the operational effectiveness of these controls over a specified period. This distinction aligns with the broader SSAE requirements and allows organizations to choose the type of report that best suits their needs and the needs of their clients.

SSAE and SOC are intricately connected, with SOC reports being a specific application of the broader SSAE framework. SSAE 18 provides the guidelines and requirements for reporting on controls at service organizations, while SOC reports operationalize these standards, providing detailed insights into the effectiveness of controls within specific domains. These reports are essential for service providers to demonstrate the effectiveness of their internal controls and the security of the services they offer to their clients.

There are different types of SOC reports, each tailored to specific needs:

- **SOC 1**: This focuses on controls relevant to financial reporting. It is commonly used when a service organization's services are likely to be relevant to their clients' financial statements. SOC 1 reports are associated with the SSAE 18 standard.

- **SOC 2**: This addresses controls related to security, availability, processing integrity, confidentiality, and privacy. This type of report is particularly relevant for technology and cloud computing organizations. It ensures that these organizations have effective controls in place to protect client data. Due to the sensitive nature of the information disclosed in SOC 2 reports, organizations may require parties requesting access to such reports to sign a **Non-Disclosure Agreement (NDA)**.

- **SOC 3**: This is similar to SOC 2, but it provides a more concise report that can be shared publicly. It includes a seal that demonstrates adherence to AICPA trust services criteria. SOC 3 reports are often used for marketing and public communication purposes.

While these standards and reports originated in the United States, they are recognized and adopted by organizations globally.

The SOC 2 reports can further be bifurcated into two types, as follows:

- **SOC 2 Type I**: A SOC 2 Type I report assesses the suitability of the design of controls at a specific point in time. It provides an understanding of the service organization's system and the suitability of the design of controls to meet the specified criteria.

- **SOC 2 Type II**: Building on Type I, a SOC 2 Type II report goes further to evaluate the operational effectiveness of these controls over a specified period. This type of report provides a more comprehensive view by assessing how well the controls were implemented and operated during the review period, typically over a minimum of six months.

AICPA has established the **trust services criteria**, which are a set of principles designed to guide organizations in building trust and confidence in their service delivery and system reliability. The five **trust services principles** are as follows:

- **Security**: The system is protected against unauthorized access (both physical and logical). The objective is to ensure that the system and the information it processes are protected from unauthorized access, damage, or interference, whether intentional or accidental.

- **Availability**: The system is available for operation and use as committed or agreed. The objective is to ensure that the system is available for operation and use when needed, and that disruptions to availability are minimized or promptly addressed.

- **Processing integrity**: System processing is complete, valid, accurate, timely, and authorized. The objective is to ensure that system processing is accurate, valid, and authorized, and that data is processed in a manner consistent with organizational expectations.

- **Confidentiality**: Information designated as confidential is protected as committed or agreed. The objective is to ensure that information designated as confidential is protected from unauthorized disclosure, both during processing and at rest.

- **Privacy**: Personal information is collected, used, retained, disclosed, and disposed of in conformity with the commitments in the entity's privacy notice. The objective is to ensure that personal information is handled in accordance with the organization's privacy policies and commitments, and in compliance with applicable privacy laws and regulations.

International Standard on Assurance Engagements (ISAE)

ISAE is a set of international standards issued by the **International Auditing and Assurance Standards Board (IAASB)**. ISAE covers a wide range of assurance engagements, including those related to internal controls, non-financial information, and other areas where stakeholders seek assurance beyond traditional financial audits. The primary objective of ISAE is to establish a framework for practitioners to provide assurance to users regarding the reliability of information or the effectiveness of controls, processes, or systems.

ISAE is designed for global applicability, and it serves as a reference for assurance practitioners worldwide. It is particularly relevant to providing consistency in assurance practices across different jurisdictions.

There are various ISAE standards, each addressing specific types of assurance engagements. For example, ISAE 3000 deals with assurance engagements related to non-financial information, while ISAE 3402 specifically addresses assurance engagements for controls at service organizations. ISAE 3402 is closely aligned with SOC 2.

Cloud Security Alliance (CSA) Security, Trust, Assurance, and Risk (STAR)

The CSA STAR audit reports provide stakeholders with valuable insights into the security controls implemented by the CSP. These reports are essential for customers, regulators, and other interested parties to assess the security posture and practices of a CSP. It refers to the **Cloud Controls Matrix (CCM)**, which serves as the basis for evaluating security controls. The criteria align with industry-recognized standards and best practices. The CSA STAR certification program includes various levels, each offering a different degree of assurance:

- **CSA STAR Level 1 – Self-Assessment**: In this initial level, CSPs conduct a self-assessment against the CCM. This level relies on self-reported information, and the provider documents its security controls, policies, and procedures.

- **CSA STAR Level 2 – Third-Party Certification**: Level 2 involves a more rigorous third-party assessment conducted by an accredited certification body. The certification body evaluates the implementation and effectiveness of security controls based on the criteria defined in the CCM. The audit is designed to provide a higher level of assurance compared to self-assessment.

- **CSA STAR Continuous Monitoring (CM)**: The CM level emphasizes ongoing monitoring of security controls and practices. It ensures that the CSP maintains a consistent and effective security posture over time. CM enhances the dynamic nature of security and provides stakeholders with up-to-date information.

- **CSA STAR Attestation**: The STAR Attestation level represents an advanced certification that goes beyond the baseline CCM requirements. It involves an examination of the CSP's compliance with specific industry standards and regulations, demonstrating a higher commitment to security and compliance.

It's important to note that these levels are designed to provide flexibility for organizations at different stages of maturity in their security practices. Achieving a higher level of CSA STAR certification indicates a more comprehensive and robust implementation of security controls.

Restrictions of Audit Scope Statements

The audit scope defines the boundaries and objectives of an audit, outlining what will be examined, the extent of the examination, and the specific criteria or standards against which the audit will be conducted. The process of defining the audit scope is a collaborative effort between the organization being audited and the auditors. This collaboration ensures that the audit scope aligns with the organizational goals, regulatory requirements, and areas of concern.

The key elements to include in an audit scope are as follows:

- Clearly state the overall objective of the audit and its specific purpose. Define what the audit aims to achieve and what questions it seeks to answer.

- Specify the criteria or standards against which the audit will be conducted. This could include regulatory requirements, industry standards, internal policies, or best practices.

- Define the period covered by the audit. Specify the start and end dates to establish the temporal boundaries within which the examination will take place.

- Identify the specific entities, departments, or processes that will be subject to the audit. Provide a detailed list of the areas that fall within the scope of the examination.

- Clearly state any areas, processes, or entities that are explicitly excluded from the audit. This helps manage expectations and prevents misunderstandings about the audit's coverage.

- Specify the level of detail or depth at which the audit will be conducted. This could include a high-level overview or a more detailed examination of specific transactions, processes, or controls.

- Provide an overview of the audit methodology that will be employed. This includes the approach to be taken, the techniques used, and any specific audit procedures.

- Clearly indicate the responsible parties who will sign off and approve the audit scope. This ensures that all stakeholders are in agreement with the defined boundaries and objectives.

- Outline the specific criteria that must be met for acceptance or successful completion of the audit. This ensures a clear understanding of the expectations and standards that entities or processes should meet.

- Clearly articulate the expected deliverables from the audit process. This may include reports, documentation, or recommendations that stakeholders can anticipate receiving upon completion of the audit.

- If applicable, specify the classification levels associated with the information or data involved in the audit. This is particularly important for audits dealing with sensitive or classified information.

- If the audit involves a security assessment, specify the particular security requirements that will be assessed. This could include evaluating access controls and encryption measures.

Audit scope restrictions refer to the limitations and boundaries placed on the extent and depth of an audit. Limiting the audit scope is a strategic and practical decision made for several reasons, each contributing to the efficiency and effectiveness of the audit process. Here are some key reasons why organizations choose to limit the audit scope:

- **Resource constraints**: Audits can be resource-intensive and are frequently associated with significant costs. Limited resources, including time, budget, and personnel, often necessitate a focused audit scope. By narrowing the scope, organizations can optimize the use of available resources and ensure a thorough examination of practical limitations.

- **Risk management**: Focusing on high-risk areas allows organizations to prioritize their efforts in addressing potential vulnerabilities and threats. Limiting the scope to critical processes or sensitive information helps manage and mitigate risks more effectively.

- **Regulatory compliance**: Compliance audits are often bound by specific regulations and standards. Limiting the scope to the relevant regulatory requirements ensures that the audit remains aligned with legal obligations and industry standards.

- **Strategic objectives**: Organizations may strategically choose to audit specific areas that align with their immediate goals and objectives. This targeted approach ensures that the audit outcomes directly contribute to the organization's strategic priorities.

- **Operational efficiency**: A limited scope allows for a more focused and efficient audit process. Auditors can dedicate their efforts to in-depth examinations, reducing the risk of oversight or superficial analysis that could occur with a broader scope.

- **Timely completion**: A focused scope enhances the likelihood of completing the audit within the established timeframe. Broader scopes may lead to extended timelines and potential delays, while a targeted approach promotes timely and effective audit delivery.

- **Performance impact**: Auditing of systems can affect system performance. Limiting the scope minimizes the impact on operational systems, reducing the risk of disruptions or slowdowns during the audit.

- **System readiness**: Certain systems may not be fully prepared for extensive audits as they may not be fully built or operational in live environments.

- **Audit time restrictions**: During business-critical hours, time restrictions on audits become imperative to prevent substantial disruptions to essential operational periods. Recognizing the sensitivity of key business hours, organizations strategically limit audit activities during these times.

Gap Analysis

Gap analysis is a strategic assessment process that involves evaluating the differences or "gaps" between the current state of a system, process, or organization and its desired future state.

Control analysis within gap analysis involves assessing the effectiveness and implementation of controls in place. It identifies gaps in control measures and helps evaluate vulnerabilities, ensuring that the organization's security and compliance requirements are met. The analysis compares existing control mechanisms with established standards or regulatory requirements, highlighting areas where controls may be insufficient or non-compliant.

Establishing baselines involves defining a reference point or standard against which future progress or changes can be measured. Baselines provide a stable foundation for performance evaluation and improvement initiatives. They serve as benchmarks for identifying deviations or gaps in ongoing processes. Organizations create baselines by documenting current processes, performance metrics, and control measures.

Gap analysis can be conducted prior to undergoing a formal external audit by regulatory bodies, certification authorities, or external stakeholders to proactively identify and address any potential gaps in compliance, control measures, or documentation that may be scrutinized during the external audit. Additionally, gap analysis is beneficial during strategic planning to align capabilities with objectives, during change management to evaluate impacts, in compliance assessments for regulatory adherence, in continuous improvement for process effectiveness, in project implementation to measure impact, in technology adoption for compatibility, in risk management for vulnerability identification, in training for skills gap analysis, in customer satisfaction for quality assurance, and in budgeting to prioritize investments and resource allocation.

Some commonly employed frameworks for gap analysis include ISO 27001, which focuses on ISMS, ISO 27002, which provides a code of practice for information security controls, and the NIST **Cybersecurity Framework (CSF)**.

Audit Planning

Audit planning is a systematic process that involves defining the scope, objectives, and methodologies of an audit to ensure its effectiveness and efficiency. The key components and objectives of audit planning include the following:

- **Scope**: Clearly outline the boundaries and extent of the audit, specifying the areas, processes, or departments to be examined. Consider geographical locations, regulatory requirements, and any other relevant factors that may impact the scope.

- **Objective setting**: Establish specific and measurable audit objectives aligned with organizational goals, regulatory compliance, and stakeholder expectations. Ensure that the objectives contribute to the overall improvement of processes, controls, and performance.

- **Resource allocation**: Allocate human and technological resources in alignment with the audit scope and objectives. Ensure the audit team possesses the necessary skills and expertise to address the identified risks and objectives.

- **Documentation of processes**: Document audit methodologies, procedures, and decision criteria for consistent recording of audit activities. Establish standardized documentation protocols to maintain clarity and transparency.

- **Training and development**: Provide training to the audit team on specific audit methodologies, tools, and organizational nuances. Foster continuous learning and development within the audit function to enhance skills and expertise.

- **Audit plan development**: Formulate a detailed audit plan outlining the approach, methodologies, timelines, and resources required. Allocate roles and responsibilities within the audit team, specifying tasks for each team member.

- **Technology integration**: Leverage audit technology and data analytics tools to enhance efficiency and effectiveness. Ensure that audit software is compatible with organizational systems and supports data-driven analysis.

- **Readiness assessment**: A readiness assessment report highlights systems that are prepared for auditing, recognizing that some systems may not yet be audit ready due to ongoing construction or not being fully operational in the production environment.

- **Expected outcomes**: Clearly document the anticipated outcomes of the audit, including improved controls, enhanced compliance, and actionable recommendations for organizational improvement.

Here are the audit phases after planning:

- **Fieldwork**: Following the planning phase, auditors move into the fieldwork stage, where they execute the audit plan. This involves actively collecting data, conducting interviews, and rigorously testing controls. The focus is on verifying compliance with established policies, procedures, and regulatory requirements.

- **Data analysis**: Along with fieldwork, auditors can leverage data analytics tools to process and analyze large datasets efficiently. This approach enables them to identify patterns, trends, anomalies, and potential areas of concern within the data.

- **Issue identification**: Auditors systematically identify and document issues discovered during the audit process. This includes discrepancies, non-compliance instances, or weaknesses in controls. Issues are categorized and prioritized based on their impact and significance.

- **Audit reporting**: The reporting phase in an audit involves the synthesis of audit findings into a comprehensive report. Auditors draft a document that outlines the scope of the audit, details audit procedures, and presents key findings and conclusions. Emphasis is placed on clarity and accuracy, supported by evidence obtained during the audit. The report incorporates management's responses, undergoes rigorous internal review, and is communicated to both internal stakeholders and, if applicable, external entities. The final version is then distributed to relevant parties, contributing to transparency and informed decision-making. The reporting phase is essential for conveying audit outcomes, fostering accountability, and facilitating a clear understanding of the organization's financial health and control environment.

- **Follow-up and monitoring**: Post-audit, there is a focus on monitoring the implementation of corrective actions and recommendations. Follow-up audits or reviews are conducted to assess the effectiveness of remediation efforts. Timely and thorough addressing of identified issues is crucial to maintaining a robust control environment.

- **Lessons learned and continuous improvement**: A crucial aspect of the post-audit phase involves conducting a lessons learned session. This retrospective analysis gathers insights from the audit process, identifying areas for improvement in planning, execution, and reporting. Implementing changes or enhancements based on these lessons ensures continuous improvement in future audits.

Internal ISMS

ISMS is a systematic approach to managing sensitive company information, ensuring its confidentiality, integrity, and availability. It involves a set of policies, procedures, processes, people, and controls designed to protect and manage an organization's information assets. The primary goal of an ISMS is to establish a framework that enables the organization to identify, assess, and mitigate information security risks.

The functions of an ISMS typically include the following:

- **Risk assessment and management**: Identifying and evaluating potential risks to information security, and implementing measures to mitigate or manage these risks effectively

- **Security policies and procedures**: Establishing a set of guidelines, rules, and procedures that govern how information assets should be handled, stored, and protected

- **Access controls**: Implementing mechanisms to control and monitor access to sensitive information, ensuring that only authorized individuals have appropriate levels of access

- **Incident response and management**: Developing a plan to respond to and recover from security incidents, minimizing the impact of potential breaches or disruptions

- **Security awareness and training**: Educating employees and stakeholders about security policies and best practices to promote a security-conscious culture within the organization

- **Security monitoring and auditing**: Continuously monitoring and auditing information systems to detect and respond to security incidents, as well as ensuring compliance with security policies

- **Business continuity and disaster recovery planning**: Developing strategies and plans to ensure the availability of critical information and systems in the event of a disaster or disruptive incident

- **Compliance management**: Ensuring that the organization complies with relevant laws, regulations, and industry standards related to information security

- **Security controls implementation**: Implementing technical and procedural controls to safeguard information assets, such as encryption, firewalls, and secure coding practices

- **Continuous improvement**: Regularly reviewing and updating the ISMS to adapt to changes in the organization's environment, technology landscape, and emerging security threats

International standards such as ISO/IEC 27001 provide a framework for establishing, implementing, maintaining, and continually improving an ISMS. ISO/IEC 27001 provides a systematic and risk-based approach to managing information security within an organization. Organizations seeking ISO/IEC 27001 certification undergo a formal assessment process by an accredited certification body. Certification provides external validation that the organization's ISMS complies with the requirements of the ISO/IEC 27001 standard, instilling confidence in customers, partners, and other stakeholders regarding the organization's commitment to information security.

An internal ISMS refers to the set of processes, policies, procedures, people, and controls established within an organization to safeguard its information assets. Unlike the ISO/IEC 27001 standard, which provides a globally recognized framework for implementing an ISMS, an internal ISMS is developed and tailored by an organization to meet its specific needs and requirements. It is a dynamic and evolving system that adapts to the organization's needs and the changing landscape of information security.

Internal Information Security Controls System

An internal information security controls system is a subset of an ISMS that specifically focuses on the development, implementation, and monitoring of security controls within an organization. It is concerned with ensuring the effectiveness of security controls to protect sensitive information and mitigate information security risks.

Scoping controls involves reviewing the controls within an established framework to determine which controls are relevant and applicable to the organization. The objective is to identify and select controls that align with the organization's operational context, risk profile, and regulatory environment. It helps in defining the boundaries of ISMS and streamlining efforts toward implementing the most relevant controls. For example, if an organization operates in an industry where certain controls are not applicable due to the nature of its business, scoping helps exclude those controls from the implementation scope.

Tailoring controls refers to the process of customizing and adapting the applicable controls to match the organization's specific circumstances, needs, and risk appetite. The goal is to ensure that the controls are not only relevant but also practical and feasible within the organization's unique operational environment. Tailoring allows organizations to implement controls in a manner that aligns with their business processes and objectives. For example, if a framework prescribes a generic control related to access management, tailoring involves customizing the specific access control measures to suit the organization's internal structure, roles, and responsibilities.

ISO/IEC 27002 is an international standard that provides a set of guidelines and best practices for establishing, implementing, maintaining, and improving an ISMS. It offers a comprehensive set of information security controls and objectives that organizations can use to address various aspects of information security.

Policies

Policies are foundational elements within an organization, providing a structured framework that dictates the conduct of business operations, employee behavior, and organizational processes. They embody the character of an organization, reflecting its values, priorities, and commitment to ethical practices, safety, and quality. By setting clear, actionable guidelines, policies help to mitigate risks, enhance efficiency, and promote a unified direction for all stakeholders. The content and scope of policies may vary based on the organization's industry, size, culture, and specific operational needs.

Policies serve as a reference point for acceptable and unacceptable actions within the organization. They help employees understand their roles and responsibilities, ensuring that individual actions align with the organization's values and objectives. Clear policies communicate the expectations the organization has from its employees, partners, and stakeholders. This clarity helps in setting standards for performance, conduct, and decision-making.

Consistency in operations and decision-making is crucial for maintaining quality and fairness. Policies provide a standardized approach to recurring situations, ensuring that actions and decisions are not arbitrary but based on predefined standards. Many policies are developed in response to legal and regulatory requirements. They ensure that the organization and its employees comply with laws, regulations, and industry standards, thereby avoiding legal issues and penalties.

Policies are a reflection of an organization's culture and values. They play a significant role in shaping the organizational environment and influencing how employees interact with each other and with external stakeholders. By outlining procedures for handling various situations, policies help in identifying, assessing, and managing risks. This proactive approach to risk management protects the organization's assets and reputation.

Organizational Policies

Organizational policies are overarching guidelines that set the tone for the entire organization, shaping its identity and culture.

Here are some examples of organizational policies:

- Code of conduct defines the ethical standards and expected behavior for employees, ensuring integrity, honesty, and professionalism in all organizational activities

- Diversity and inclusion policy promotes a diverse and inclusive workplace by outlining the organization's commitment to equality, non-discrimination, and fostering an inclusive culture

- Ethics policy establishes the ethical principles and values that govern decision-making and behavior, guiding employees to uphold the highest standards of integrity

- Anti-harassment policy provides guidelines to prevent and address harassment in the workplace, fostering a safe and respectful environment for all employees

- Remote work policy outlines the guidelines, expectations, and security measures for employees working remotely, ensuring continuity and productivity outside the traditional office setting

Functional Policies

Functional policies are specific guidelines and rules tailored to individual departments or functions within an organization. These policies address the unique requirements, processes, and challenges of each functional area, providing a framework for efficient and effective operations within that department.

Here are some examples of functional policies:

- Data classification policy establishes criteria for categorizing data based on sensitivity, guiding security measures to protect confidential information

- Email use policy governs proper and secure email communication, addressing etiquette, data sharing, and protection against security threats

- Incident response policy defines procedures for identifying, reporting, and responding to security incidents, facilitating a coordinated and effective resolution

- Password policy sets guidelines for creating and managing secure passwords, enhancing access controls and overall cybersecurity

- Vulnerability management policy guides the systematic identification, assessment, and mitigation of vulnerabilities, enhancing overall system resilience

- Endpoint management policy focuses on securing end-user devices, providing guidelines for configuration, monitoring, and protection

- Encryption policy establishes criteria for encrypting data in transit and at rest, enhancing confidentiality, integrity, and compliance

Cloud Computing Policies

The introduction of cloud computing has transformed the landscape of organizational policies, impacting both CSPs and customers. As businesses increasingly rely on cloud services for their operations, it becomes imperative for customers to thoroughly review the policies established by CSPs to ensure alignment with their own organizational requirements and regulatory obligations, particularly in areas such as data security, privacy, access controls, and incident response. By ensuring alignment between CSP policies and their own organizational policies, customers can mitigate risks, maintain compliance, and safeguard their data effectively.

On the other hand, CSPs bear the responsibility of setting robust policies that meet the needs and expectations of their customers. CSPs are responsible for ensuring that their policies incorporate regulatory requirements and industry best practices to provide customers with a secure and compliant cloud environment. Moreover, CSPs should adopt a proactive approach by continuously updating and enhancing their policies to address emerging threats, technological advancements, and evolving regulatory frameworks. By staying ahead of the curve and maintaining stringent policies, CSPs can instill confidence in their customers, build trust, and foster long-term relationships based on reliability and security.

There are key terms around cloud computing policies that are important with respect to the CCSP exam. The definitions of these key terms are listed here:

- **Policies**: High-level statements that outline an organization's objectives, principles, and expectations regarding a specific area of operation or the entire organization

- **Standards**: Specific criteria or requirements that must be met to achieve compliance with policies, often established as benchmarks for performance or quality

- **Procedures**: Detailed, step-by-step instructions for carrying out specific tasks or activities in accordance with policies and standards

- **Guidelines**: General recommendations, suggestions, or best practices that provide flexibility in achieving objectives while aligning with policies and standards

- **Baselines**: Reference points or starting positions used to measure deviations, often serving as the foundation for developing policies, standards, and procedures

Identification and Involvement of Relevant Stakeholders

Various stakeholders, each with distinct roles and responsibilities, contribute to the success of the audit.

Senior management plays a pivotal role by providing oversight, direction, and approval at critical junctures. Senior management is responsible for approving the initiation of audits and ensuring alignment with organizational objectives. They allocate necessary resources, including personnel and budget, to support the audit process, emphasizing the importance of the audit within the broader strategic framework. Their involvement extends to the final approval of audit findings and recommendations, ensuring that the organization's leadership is actively engaged in the decision-making process.

Security professionals play a crucial role in ensuring the organization's information security measures align with industry standards and best practices. Their role involves collaborating with auditors to assess the effectiveness of security controls. Security professionals implement and manage security controls, assess vulnerabilities, and provide expertise on the organization's information security posture. In collaboration with auditors, they work to address findings related to security practices.

The legal team provides essential guidance to guarantee the organization's operations and policies comply with relevant laws and regulations throughout the audit process. Their role includes addressing legal implications stemming from audit findings. Responsibilities encompass a comprehensive review of policies and procedures for legal compliance, addressing potential legal risks identified during the audit, and offering guidance on corrective actions. The legal team's involvement ensures that the organization maintains legal integrity and alignment with regulatory requirements.

Employees' role involves providing accurate information, cooperating with auditors, and implementing recommended improvements. To fulfill these responsibilities, employees adhere to established procedures, share relevant documentation, and offer valuable insights into daily operations.

Specialized Compliance Requirements for Highly Regulated Industries

In highly regulated industries, organizations face a complex landscape of compliance standards designed to safeguard critical infrastructure, protect sensitive data, and ensure the secure handling of transactions. Notably, the **North American Electric Reliability Corporation Critical Infrastructure Protection (NERC CIP)** standards are tailored to fortify the security and reliability of the power grid. These standards encompass rigorous cybersecurity measures, covering facets such as access controls and incident response.

Healthcare, a sector handling vast amounts of sensitive information, adheres to the **Health Insurance Portability and Accountability Act (HIPAA)** and the **Health Information Technology for Economic and Clinical Health (HITECH)** Act. HIPAA sets stringent controls on the protection of patient information, requiring robust measures in risk assessment, encryption, and privacy practices. HITECH builds upon these provisions, emphasizing the adoption of **Electronic Health Records (EHRs)** and introducing obligations for breach notifications.

For entities involved in payment card transactions, the **Payment Card Industry Data Security Standard** (**PCI DSS**) establishes a comprehensive framework to safeguard cardholder data and secure payment processes. PCI DSS mandates adherence to strict security controls, including encryption, access management, and routine security assessments.

Navigating these specialized compliance requirements demands a strategic and proactive approach. Organizations must continuously assess and enhance their cybersecurity measures, conduct regular risk assessments, and stay vigilant to changes in the regulatory landscape. Non-compliance not only exposes organizations to financial penalties but also poses reputational risks and legal consequences. By prioritizing compliance and adopting robust security practices, entities in highly regulated industries can effectively mitigate risks and foster a secure operational environment.

Impact of Distributed IT Model

The distributed IT model in cloud computing has a profound impact on auditing processes, introducing challenges and opportunities that arise from the dynamic and geographically dispersed nature of cloud environments. Key factors influencing auditing in the cloud include diverse geographical locations and the crossing over of legal jurisdictions.

Cloud services often leverage a global network of data centers, making it challenging for auditors to physically inspect and assess the security measures in each location. Data stored and processed in the cloud may traverse multiple legal jurisdictions, each with its own set of data protection and privacy laws. Ensuring compliance with diverse regulations becomes complex. Organizations often rely on third-party services within the cloud ecosystem. Auditing becomes more complex as it involves assessing the security and compliance measures of multiple interconnected entities. Establishing and maintaining comprehensive audit trails across diverse geographical locations can be challenging, impacting the ability to trace and investigate security incidents. Ensuring consistent compliance with regulatory standards becomes complex when operating in a distributed cloud environment. Auditors need to verify that organizations adhere to the same level of compliance across all geographical locations.

The automatic replication of data is a common practice in cloud environments for redundancy and disaster recovery. While customers can manage automatic data replication themselves, especially in most IaaS and PaaS models, it may impact compliance efforts, especially if data crosses legal jurisdictions. Auditors need to verify that replicated data complies with relevant regulations and data protection laws.

While CSPs play a crucial role in implementing and maintaining security measures, the customer retains the final accountability for safeguarding their data.

Summary

In this chapter, you explored how audits function in cloud computing. You started by examining internal and external audit controls, outlining their specific roles and applications. You gained an understanding of the impact of audit requirements and the unique challenges presented by virtualization and cloud technologies.

You also learned about various types of audit reports and their limitations in the cloud. The chapter discussed gap analysis. Furthermore, you delved into policies such as organization, functional, and cloud computing policies.

You also learned about specialized compliance requirements for highly regulated industries such as NERC CIP and HIPAA. Finally, you examined the transformative effects of distributed IT in cloud computing, uncovering both challenges and opportunities within the audit landscape.

Exam Readiness Drill – Chapter Review Questions

Apart from a solid understanding of key concepts, being able to think quickly under time pressure is a skill that will help you ace your certification exam. That is why working on these skills early on in your learning journey is key.

Chapter review questions are designed to improve your test-taking skills progressively with each chapter you learn and review your understanding of key concepts in the chapter at the same time. You'll find these at the end of each chapter.

> **How to Access These Materials**
>
> To learn how to access these resources, head over to the chapter titled *Chapter 25, Accessing the Online Resources*.

To open the Chapter Review Questions for this chapter, perform the following steps:

1. Click the link – `https://packt.link/CCSPE1_CH24`.

   Alternatively, you can scan the following **QR code** (*Figure 24.1*):

Figure 24.1 – QR code that opens Chapter Review Questions for logged-in users

2. Once you log in, you'll see a page similar to the one shown in *Figure 24.2*:

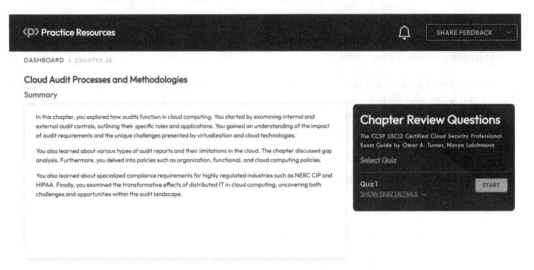

Figure 24.2 – Chapter Review Questions for Chapter 24

3. Once ready, start the following practice drills, re-attempting the quiz multiple times.

Exam Readiness Drill

For the first three attempts, don't worry about the time limit.

ATTEMPT 1

The first time, aim for at least **40%**. Look at the answers you got wrong and read the relevant sections in the chapter again to fix your learning gaps.

ATTEMPT 2

The second time, aim for at least **60%**. Look at the answers you got wrong and read the relevant sections in the chapter again to fix any remaining learning gaps.

ATTEMPT 3

The third time, aim for at least **75%**. Once you score 75% or more, you start working on your timing.

> **Tip**
>
> You may take more than **three** attempts to reach 75%. That's okay. Just review the relevant sections in the chapter till you get there.

Working On Timing

Target: Your aim is to keep the score the same while trying to answer these questions as quickly as possible. Here's an example of how your next attempts should look like:

Attempt	Score	Time Taken
Attempt 5	77%	21 mins 30 seconds
Attempt 6	78%	18 mins 34 seconds
Attempt 7	76%	14 mins 44 seconds

Table 24.1 – Sample timing practice drills on the online platform

> **Note**
>
> The time limits shown in the above table are just examples. Set your own time limits with each attempt based on the time limit of the quiz on the website.

With each new attempt, your score should stay above **75%** while your "time taken" to complete should "decrease". Repeat as many attempts as you want till you feel confident dealing with the time pressure.

25

Accessing the Online Practice Resources

Your copy of *CCSP (ISC)2 Certified Cloud Security Professional: Exam Guide* comes with free online practice resources. Use these to hone your exam readiness even further by attempting practice questions on the companion website. The website is user-friendly and can be accessed from mobile, desktop, and tablet devices. It also includes interactive timers for an exam-like experience.

How to Access These Materials

Here's how you can start accessing these resources depending on your source of purchase.

Purchased from Packt Store (packtpub.com)

If you've bought the book from the Packt store (`packtpub.com`) eBook or Print, head to `https://packt.link/ccspunlock`. There, log in using the same Packt account you created or used to purchase the book.

Packt+ Subscription

If you're a *Packt+ subscriber*, you can head over to the same link (`https://packt.link/ccsppractice`), log in with your `Packt ID`, and start using the resources. You will have access to them as long as your subscription is active.

If you face any issues accessing your free resources, contact us at `customercare@packt.com`.

Purchased from Amazon and Other Sources

If you've purchased from sources other than the ones mentioned above (like *Amazon*), you'll need to unlock the resources first by entering your unique sign-up code provided in this section. **Unlocking takes less than 10 minutes, can be done from any device, and needs to be done only once**. Follow these five easy steps to complete the process:

STEP 1

Open the link `https://packt.link/ccspunlock` OR scan the following **QR code** (*Figure 25.1*):

Figure 25.1 – QR code for the page that lets you unlock this book's free online content.

Either of those links will lead to the following page as shown in *Figure 25.2*:

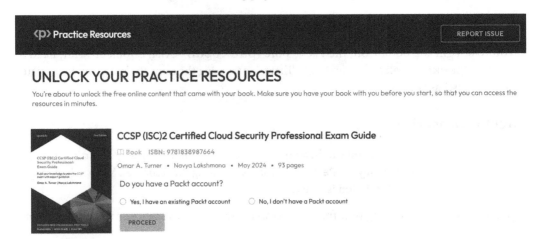

Figure 25.2 – Unlock page for the online practice resources

STEP 2

If you already have a Packt account, select the option `Yes, I have an existing Packt account`. If not, select the option `No, I don't have a Packt account`.

If you don't have a Packt account, you'll be prompted to create a new account on the next page. It's free and only takes a minute to create.

Click `Proceed` after selecting one of those options.

STEP 3

After you've created your account or logged in to an existing one, you'll be directed to the following page as shown in *Figure 25.3*.

Make a note of your unique unlock code:

```
IGM4585
```

Type in or copy this code into the text box labeled 'Enter Unique Code':

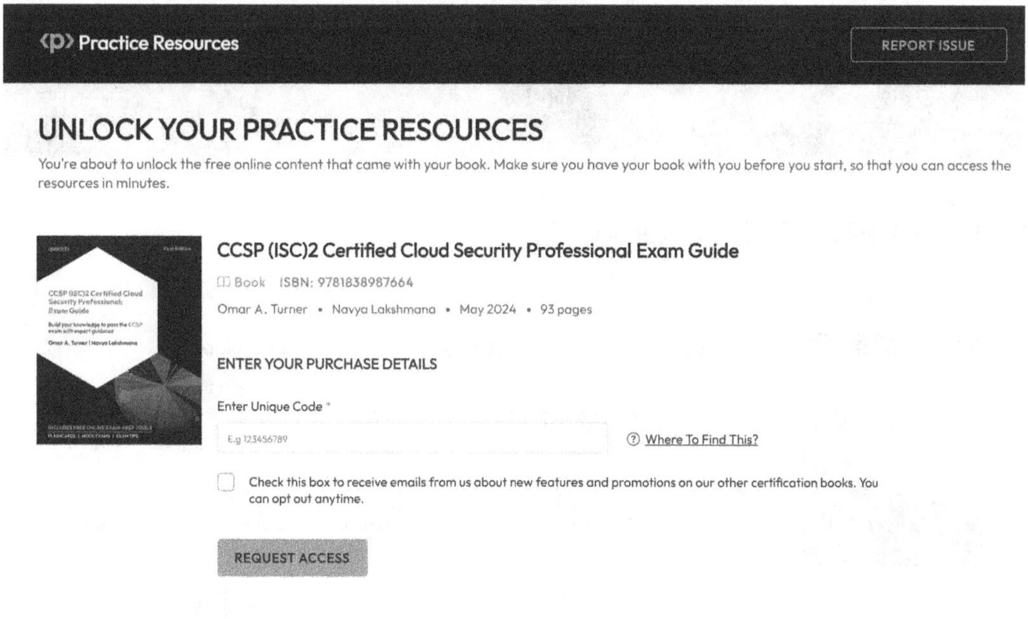

‹p› Practice Resources   REPORT ISSUE

UNLOCK YOUR PRACTICE RESOURCES

You're about to unlock the free online content that came with your book. Make sure you have your book with you before you start, so that you can access the resources in minutes.

CCSP (ISC)2 Certified Cloud Security Professional Exam Guide

Book   ISBN: 9781838987664

Omar A. Turner • Navya Lakshmana • May 2024 • 93 pages

ENTER YOUR PURCHASE DETAILS

Enter Unique Code *

E.g 123456789   ⑦ Where To Find This?

☐ Check this box to receive emails from us about new features and promotions on our other certification books. You can opt out anytime.

REQUEST ACCESS

Figure 25.3 – Enter your unique sign-up code to unlock the resources

> **Troubleshooting tip**
>
> After creating an account, if your connection drops off or you accidentally close the page, you can reopen the page shown in *Figure 25.2* and select `Yes, I have an existing account`. Then, sign in with the account you had created before you closed the page. You'll be redirected to the screen shown in *Figure 25.3*.

STEP 4

> **Note**
>
> You may choose to opt into emails regarding feature updates and offers on our other certification books. We don't spam, and it's easy to opt out at any time.

Click `Request Access`.

STEP 5

If the code you entered is correct, you'll see a button that says, `OPEN PRACTICE RESOURCES`, as shown in *Figure 25.4*:

PACKT PRACTICE RESOURCES

You've just unlocked the free online content that came with your book.

CCSP (ISC)2 Certified Cloud Security Professional Exam Guide

Book   ISBN: 9781838987664

Omar A. Turner • Navya Lakshmana • May 2024 • 93 pages

⊘ Unlock Successful
Click the following link to access your practice resources at any time.

Pro Tip: You can switch seamlessly between the ebook version of the book and the practice resources. You'll find the ebook version of this title in your <u>Owned Content</u>

OPEN PRACTICE RESOURCES ↗

Figure 25.4 – Page that shows up after a successful unlock

Click the OPEN PRACTICE RESOURCES link to start using your free online content. You'll be redirected to the Dashboard shown in *Figure 25.5*:

Figure 25.5 – Dashboard page for CCSP practice resources

Bookmark this link

Now that you've unlocked the resources, you can come back to them anytime by visiting `https://packt.link/ccsppractice` or scanning the following QR code provided in *Figure 25.6*:

Figure 25.6 – QR code to bookmark practice resources website

Troubleshooting Tips

If you're facing issues unlocking, here are three things you can do:

- Double-check your unique code. All unique codes in our books are case-sensitive and your code needs to match exactly as it is shown in *STEP 3*.

- If that doesn't work, use the `Report Issue` button located at the top-right corner of the page.

- If you're not able to open the unlock page at all, write to `customercare@packt.com` and mention the name of the book.

Share Feedback

If you find any issues with the platform, the book, or any of the practice materials, you can click the `Share Feedback` button from any page and reach out to us. If you have any suggestions for improvement, you can share those as well.

Back to the Book

To make switching between the book and practice resources easy, we've added a link that takes you back to the book (*Figure 25.7*). Click it to open your book in Packt's online reader. Your reading position is synced so you can jump right back to where you left off when you last opened the book.

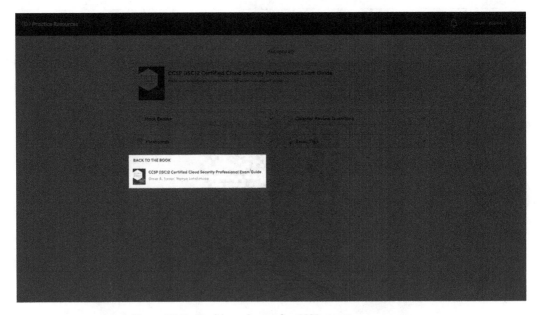

Figure 25.7 – Dashboard page for CCSP practice resources

> **Note**
>
> Certain elements of the website might change over time and thus may end up looking different from how they are represented in the screenshots of this book.

Index

D

F

N

O

S

W

X

Z

www.packtpub.com

Subscribe to our online digital library for full access to over 7,000 books and videos, as well as industry leading tools to help you plan your personal development and advance your career. For more information, please visit our website.

Why subscribe?

- Spend less time learning and more time coding with practical eBooks and Videos from over 4,000 industry professionals

- Improve your learning with Skill Plans built especially for you

- Get a free eBook or video every month

- Fully searchable for easy access to vital information

- Copy and paste, print, and bookmark content

At www.packtpub.com, you can also read a collection of free technical articles, sign up for a range of free newsletters, and receive exclusive discounts and offers on Packt books and eBooks.

Other Books You May Enjoy

If you enjoyed this book, you may be interested in these other books by Packt:

CompTIA Security+ SY0-701 Certification Guide, Third Edition

Ian Neil

ISBN: 978-1-83546-153-2

- Differentiate between various security control types
- Apply mitigation techniques for enterprise security
- Evaluate security implications of architecture models
- Protect data by leveraging strategies and concepts
- Implement resilience and recovery in security
- Automate and orchestrate for running secure operations
- Execute processes for third-party risk assessment and management
- Conduct various audits and assessments with specific purposes

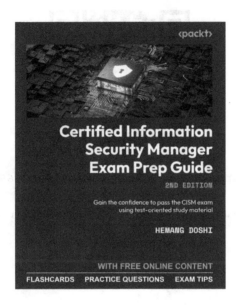

Certified Information Security Manager Exam Prep Guide

Hemang Doshi

ISBN: 978-1-80461-063-3

- Understand core exam objectives to prepare for the CISM exam with confidence
- Get to grips with detailed procedural guidelines for effective information security incident management
- Execute information security governance in an efficient manner
- Strengthen your preparation for the CISM exam using interactive flashcards and practice questions
- Conceptualize complex topics through diagrams and examples
- Find out how to integrate governance, risk management, and compliance functions

Share Your Thoughts

Now you've finished *CCSP (ISC)2 Certified Cloud Security Professional Exam Guide*, we'd love to hear your thoughts! Scan the QR code below to go straight to the Amazon review page for this book and share your feedback or leave a review on the site that you purchased it from.

https://packt.link/r/1838987665

Your review is important to us and the tech community and will help us make sure we're delivering excellent quality content.

Download a Free PDF Copy of This Book

Thanks for purchasing this book!

Do you like to read on the go but are unable to carry your print books everywhere?

Is your eBook purchase not compatible with the device of your choice?

Don't worry, now with every Packt book you get a DRM-free PDF version of that book at no cost.

Read anywhere, any place, on any device. Search, copy, and paste code from your favorite technical books directly into your application.

The perks don't stop there, you can get exclusive access to discounts, newsletters, and great free content in your inbox daily.

Follow these simple steps to get the benefits:

1. Scan the QR code or visit the link below:

https://packt.link/free-ebook/9781838987664

2. Submit your proof of purchase.
3. That's it! We'll send your free PDF and other benefits to your email directly.

www.ingramcontent.com/pod-product-compliance
Lightning Source LLC
Chambersburg PA
CBHW060637060326
40690CB00020B/4431